Lecture Notes
in Control and Information Sciences 348

Editors: M. Thoma · M. Morari

Heinz Teutsch

Modal Array Signal Processing: Principles and Applications of Acoustic Wavefield Decomposition

With 94 Figures

Springer

Author

Dr. Heinz Teutsch

Avaya Labs Research
233 Mount Airy Road
Basking Ridge, NJ 07920
USA

teutsch@research.avayalabs.com

ISSN 0170-8643

ISBN-10 3-540-40893-2 **Springer Berlin Heidelberg New York**
ISBN-13 978-3-540-40893-2 **Springer Berlin Heidelberg New York**

Library of Congress Control Number: 2006932727

Springer is a part of Springer Science+Business Media

springer.com

Typesetting: Data conversion by author.
Final processing by PTP-Berlin Protago-TEX-Production GmbH, Germany (www.ptp-berlin.com)
Cover-Design: WMXDesign GmbH, Heidelberg
Printed on acid-free paper 89/3141/Yu - 5 4 3 2 1 0

Preface

Many applications of acoustic signal processing, such as teleconferencing and surveillance systems, require an estimate of several parameters present in the observed acoustic scene. The most important parameters are the location of the acoustic sources as well as the number of currently active sources. These parameters have been traditionally estimated by means of classical array signal processing using microphone arrays. The term 'classical' is used here to classify algorithms that are derived from the microphone signals directly.

In contrast, the algorithms presented in this book are not solely based on classical array signal processing but also on the principles of wave propagation and wave scattering offered by the science of classical acoustics. Novel so-called modal array signal processing algorithms can now be derived from the fundamental solutions of the acoustic wave equation.

Algorithms for parameter estimation solely based on the paradigm of classical array signal processing, often suffer from the problem that they rely on a narrowband assumption underlying the signal model, which limit their usability when broadband signals, such as speech, are present in the wavefield under observation. It will be shown that by additionally considering the paradigm of classical acoustics, algorithms can be derived that are inherently broadband, hence allowing for ubiquitous application.

Moreover, it will be shown that parameter estimation, based on these novel methods, has the potential to unambiguously detect and localize multiple simultaneously active wideband sources in the array's full field-of-view.

In order to take advantage of the paradigms that can be found in the field of acoustics, special array geometries are considered here, i.e. circular and spherical geometries for the analysis of wave propagation in two and three spatial dimensions, respectively.

A rigorous derivation of modal array signal processing algorithms for parameter estimation as well as performance evaluations by means of simulations and measurements using an actual real-time capable implementation are discussed.

Acknowledgments

I would like to thank Prof. Walter Kellermann of the Friedrich-Alexander-University in Erlangen, Germany who, with his inspirations and great patience has supported me throughout my career in his research group. He showed faith in my work even during times where research funding was critical. Thanks go to Dr. Rudolf Rabenstein for funding my first two and a half years throughout the CARROUSO project.

I am very indebted to Dr. James L. Flanagan of Rutgers University, New Jersey, USA, who was willing to dedicate a big chunk of his busy schedule to the review of this work. Thanks also go to Prof. Wolfgang Koch, Prof. em. Adolf W. Lohmann, Prof. Reinhard Lerch, and Dr. Manfred Kaltenbacher, all from the University of Erlangen, Germany, for showing interest in my work and for finding the time to participate in the defense of my PhD thesis.

I am grateful to Dr. Gary Elko for many fruitful discussions and for making my memorable stay at Bell Laboratories, New Jersey, USA, possible.

I would also like to thank all my former colleagues at the Telecommunications Laboratory for creating a very enjoyable and inspiring atmosphere. I am thankful for the unresting dedication of the supportive staff. In particular, I would like to thank Mrs. Ursula Arnold for making the impossible possible, as well as Rüdiger Nägel and Manfred Lindner for helping me with the design of numerous hardware components.

Thanks also go to my new colleagues and friends at Avaya Labs for welcoming me at the Multimedia Technologies Research Department.

I am very grateful to my friends and family for their unconditional support and for sharing unforgettable moments with me.

And finally, my deepest thanks go to Debbie for giving me her heart and for accepting the countless days that we had to spend so far apart.

Basking Ridge
July 2006

Heinz Teutsch

Acknowledgements

[illegible]

Contents

Abbreviations and Acronyms

ACC	All curtains closed
ACO	All curtains open
AED	Adaptive eigenvalue decomposition
AIC	Akaike information criteria
ASA	Acoustic scene analysis
BM	Blocking matrix
CC	Cross-correlation
CDB	Constant directivity beamformer
CRLB	Cramér-Rao lower bound
DMA	Differential microphone array
DOA	Direction-of-arrival
DSB	Delay-and-sum beamformer
EBF	Eigen-beamformer
EB-GSC	Eigenbeam generalized side-lobe canceler
ESPRIT	Estimation of signal parameters via rotational invariance techniques
FFT	Fast Fourier transform
FIR	Finite impulse response
FBF	Fixed beamformer
FBSS	Forward-backward spatial smoothing
FSB	Filter-and-sum beamformer
GCC	Generalized cross-correlation
GSC	Generalized side-lobe canceler
IC	Interference canceler
LCMP	Linearly constrained minimum power
LCMV	Linearly constrained minimum variance
LMS	Least-mean square
LS	Least-squares
MDL	Minimum description length
MIMO	Multiple-input-multiple-output
MLS	Maximum-length sequence
MUSIC	Multiple signal classification
NOS	Number of sources
PE	Parameter estimation
PHAT	Phase transform
SCOT	Smoothed coherence transform
SDB	Superdirective beamformer
SNR	Signal-to-noise ratio
SVD	Singular value decomposition
TDOA	Time-difference-of-arrival
TLS	Total least-squares
ULA	Uniform linear array

WE	Waveform estimation
WFD	Wavefield decomposition
WFS	Wavefield synthesis
WNG	White noise gain
WSS	Wide-sense stationary

Notations and Conventions

Conventions

The following conventions are used throughout this book.

- Time-domain scalar quantities are denoted by lowercase characters, e.g. $a(t)$.
- Frequency-domain scalar quantities are denoted by uppercase characters, e.g. $A(\omega)$.
- Time-domain vector quantities are denoted by boldface lowercase characters, e.g. $\boldsymbol{a}(t)$.
- Frequency-domain vector quantities are denoted by boldface uppercase characters, e.g. $\boldsymbol{A}(\omega)$.
- Time-domain matrix quantities are denoted by underlined, boldface lowercase characters, e.g. $\underline{\boldsymbol{a}}(t)$.
- Frequency-domain matrix quantities are denoted by underlined, boldface uppercase characters, e.g. $\underline{\boldsymbol{A}}(\omega)$.
- All vectors are assumed to be column vectors.
- The denotation $0(1)M$ means $0, 1, \ldots, M$, where M is an integer.

Mathematical operations

$(\cdot)^T$	Vector or matrix transposition
$(\cdot)^*$	Conjugate complex of a vector or matrix
$(\cdot)^H$	Hermitian operation, i.e. conjugate complex transposed, of a vector or matrix
$(\cdot)^{(-1)}$	Matrix inverse
$(\cdot)^\dagger$	Pseudo-inverse of a matrix
$\mathrm{diag}\{\cdot\}$	Diagonal matrix where the entries are defined by the argument (vector)
$\mathrm{span}\{\cdot\}$	Space which is spanned by the columns of the argument (matrix)
$\mathrm{tr}\{\cdot\}$	Trace of a matrix, i.e. sum of elements on main diagonal
$\mathrm{rank}\{\cdot\}$	Rank of the argument (matrix)
$\odot$	Hadamard product, i.e. element-wise multiplication of matrix elements

$\lvert\cdot\rvert$	Absolute value of the argument (scalar)
$\lVert\cdot\rVert \triangleq \lVert\cdot\rVert_2$	2-norm of the argument (vector or matrix)
$\lVert\cdot\rVert_F$	Frobenius norm of the argument (matrix)
$E\{\cdot\}$	Expectation operator
$\hat{(\cdot)}$	Estimated quantity
$\mathcal{F}_t, \mathcal{F}_\tau$	Fourier transform with respect to t or τ
∇	Nabla operator (gradient)
∇^2	Laplace operator
$\partial/\partial x$	Partial derivative with respect to x
$*$	Convolution operator
$f'(x)$	Derivative with respect to x
$\dot{f}(t)$	Temporal derivative of $f(t)$

List of symbols

Variables and constants

$A(\omega)$	Array gain
$A_w(\omega)$	White noise gain of an array
β	Room boundary reflection coefficient
$B(\omega, \vartheta, \varphi)$	Beampattern
c	Speed of sound
$c_{x_1 x_2}(t)$	Correlation between signal $x_1(t)$ and $x_2(t)$
$\underline{\boldsymbol{c}}_{\boldsymbol{xx}}(\mathrm{t})$	Spatio-temporal correlation matrix
$\underline{\boldsymbol{C}}_{\mathrm{CRLB}}(\boldsymbol{\Theta})$	Cramér-Rao lower bound with respect to parameter set $\boldsymbol{\Theta}$
d	Sensor spacing
$D(\omega, \vartheta, \varphi)$	Directivity
$D_I(\omega, \vartheta, \varphi)$	Directivity index
$\delta(\cdot)$	Dirac delta function
$\boldsymbol{e}_\nu$	Unit vector in direction ν
f	Frequency
$F_n(kR, \varphi)$	Decomposed circular harmonics, order n
$\overset{\circ}{F}_n\ (kR, \varphi)$	Decomposed circular harmonics, order n, unbaffled aperture
$\overset{\bullet}{F}_n\ (kR, \varphi)$	Decomposed circular harmonics, order n, baffled aperture
$\overset{\circ}{\mathcal{F}}_n\ (kR, \varphi)$	Decomposed circular harmonics, order n, unbaffled directional aperture
$g_k(\boldsymbol{r}\vert\boldsymbol{r}_0)$	Green's function (response of source/sink positioned at $\boldsymbol{r}_0$ with respect to point of observation $\boldsymbol{r}$)
$\underline{\boldsymbol{\Gamma}}_{\boldsymbol{ZZ}}(\omega)$	Coherence matrix of signal vector $\boldsymbol{Z}(\omega)$

$\overset{\circ}{G}{}_n^m(kR,\vartheta,\varphi)$	Decomposed spherical harmonics, order n, degree m, unbaffled aperture
$\overset{\bullet}{G}{}_n^m(kR,\vartheta,\varphi)$	Decomposed spherical harmonics, order n, baffled aperture
$\overset{\circ}{\mathcal{G}}{}_n^m(kR,\vartheta,\varphi)$	Decomposed spherical harmonics, order n, unbaffled directional aperture
i	Complex number, $i^2=-1$
I	Number of sources
$\underline{\boldsymbol{I}}_\mu$	Quadratic identity matrix of dimension μ
$\underline{\bar{\boldsymbol{I}}}_\mu$	Quadratic exchange matrix of dimension μ
$\boldsymbol{k}$	Wavenumber vector
k	Wavenumber, $k=\lVert\boldsymbol{k}\rVert=2\pi f/c$
K	Number of data snapshots
$\underline{\boldsymbol{K}}_s, \underline{\boldsymbol{K}}^a, \underline{\boldsymbol{K}}^b$	Selection matrices
λ	Wavelength
L	Length of a cylinder
M	Number of sensors
n, m	Order and mode
$\mathcal{N}, \mathcal{M}$	Maximum order and maximum mode
ω	Temporal-radian frequency
$p(t), P(\omega)$	Pressure
$\underline{\boldsymbol{Q}}_M$	Unitary matrix of dimension M
$Q(\psi), \tilde{Q}(z)$	MUSIC (pseudo) spectrum
$\boldsymbol{r}$	three-dimensional position vector
t	Continuous time
τ	Time delay
T	Time interval of observation
(r,θ,ϕ)	Spherical coordinates
(ϱ,ϕ,z)	Cylindrical coordinates
R	Radius of cylinder or sphere
$\underline{\boldsymbol{S}}_{\boldsymbol{ZZ}}(\omega)$	Spatio-spectral correlation matrix of signal vector $\boldsymbol{Z}(\omega)$
σ_s^2, σ_w^2	Signal variance, noise variance
(ϑ,φ)	Direction-of-arrival in three-dimensional space
$\underline{\boldsymbol{U}}_N(\omega)$	Noise subspace
$\underline{\boldsymbol{U}}_S(\omega)$	Signal subspace
$\boldsymbol{V}(\omega,\Theta)$	Array manifold vector
$\underline{\boldsymbol{V}}(\omega,\Theta)$	Array manifold matrix
w, W	Particle displacement
$\boldsymbol{W}(\omega)$	Sensor weighting
ξ	Eigenvector
(x,y,z)	Cartesian coordinates
ζ	Eigenvalue
Z, Z_0	Acoustic impedance, specific acoustic impedance

Special Functions

All definitions of special functions are given in Appendix B.

$J_n(x)$	Bessel function of order n with respect to argument x
$H_n(x)$	Hankel function of order n with respect to argument x
$Y_n(x)$	Neumann function of order n with respect to argument x
$j_n(x)$	Spherical Bessel function of order n with respect to argument x
$h_n(x)$	Spherical Hankel function of order n with respect to argument x
$y_n(x)$	Spherical Neumann function of order n with respect to argument x
$P_n(x)$	Legendre Polynomial of order n with respect to argument x
$P_n^m(x)$	Associated Legendre function of order n and degree m with respect to argument x
$Y_n^m(x)$	Spherical harmonics of order n and degree m with respect to argument x

1

Introduction

The term 'acoustic scene analysis' (ASA) describes the task of extracting information contained in an acoustic wavefield, such as the waveform itself or parameter describing the source of the wavefield. Since acoustic wavefields are processes spread-out in space and time it follows quite naturally that ASA is predominantly performed by evaluating the signals captured by a number of spatially distinct microphones, i.e. microphone arrays. A standard and widely applied vehicle for evaluating the microphone array signals is built upon classical array signal processing techniques [JD93, Tre02]. In this context, the term 'classical' is used to denote signal processing algorithms, to be introduced below, that are applied directly to the individual microphones comprising the array. In contrast, the algorithms to be derived in this book are applied to signals that are obtained by transforming the microphone signals into a domain defined by the eigen-solutions of the acoustic wave equation in two- and three spatial dimensions.

ASA, as considered in this book, can be grouped into three distinct classes.

1. *Supervised ASA* or *waveform estimation.* Here, ASA is concerned with the extraction of the desired source signal from the observed wavefield by means of beamforming and beamsteering techniques. Note that although the signal processing steps necessary for the extraction of the desired source involves an estimation of filter coefficients, the primary task is to extract the signal itself. Therefore, the term 'waveform estimation' is chosen in this context. Beamforming is an attempt to add the desired signal coherently at the output of the beamformer while the noise is added incoherently. Beamforming methods include, delay-and-sum beamforming [Dol46, Fla85, FJZE85], filter-and-sum beamforming [DB88, VSd96, WKW01], as well as beamforming using differential [Elk04, TE01a] and superdirective arrays [CZK86, CZO87, BS01]. Both signal-independent as well as signal-dependent implementations are used in practice, see e.g. [Her05]. All of the above mentioned beamformers will be discussed in Chapter 4. For performing the signal alignment, i.e. the

beamsteering step, a prerequisite for this type of ASA is the knowledge of the array's geometry and the location of the desired acoustic source, hence the term 'supervised'. Typical scenarios set for supervised ASA are teleconferencing applications and front-ends to distant-talker speech recognition systems. Note that supervised ASA techniques typically assume a single desired source signal to be present in a wavefield.

2. *Semi-supervised ASA* or *parameter estimation.* In this context, ASA deals with the number and location of possibly multiple simultaneously active acoustic sources using microphone arrays. While the localization of a single desired acoustic source has been the subject of research for the last several decades – mainly by applying the notion of the time-difference-of-arrival between microphone pairs [MS69, KC76, Ben00] – the problem of estimating of the number and positions of multiple acoustic sources using non-heuristic methods has not been under investigation only until very recently [BAS+05, TK05a, TK05b, TK05c]. The discussion and an attempt to build a framework for this problem is the main focus of this book. A typical application of semi-supervised ASA is acoustic surveillance. Parameter estimation also serves as a preprocessing step for the techniques based on supervised ASA.
3. *Unsupervised ASA* or *blind system identification / blind source separation.* This rather new and emerging area of research does, in principle, not require any knowledge of the parameters present in the wavefield, nor does it rely on the geometry of the microphone array. The desired information is extracted by adaptively estimating the system comprising the acoustic source(s) and the acoustic environment observed by the microphones. This class of ASA is not treated further in this book. Blind system identification and its application to parameter estimation, using classical array signal processing techniques, is described in [BAK05b, BAS+05]. Blind source separation and its application to waveform estimation by extracting the desired acoustic signals from a mixture of signals is described in detail, e.g., in [BAK04, BAK05a].

Note that the ASA considered here is *not* to be confused with auditory scene analysis which is a technique for encoding Structured Audio bitstreams from acoustic data, as described within the ISO/IEC 14496 MPEG-4 context [MPE].

The work presented in this book is, in contrast to the algorithms as considered in the above mentioned references, not solely based on classical array signal processing but also on the principles of wave propagation and wave scattering offered by the science of classical acoustics. Algorithms for parameter estimation solely based on the pillar 'classical array signal processing', often suffer from the problem that they rely on a narrowband assumption underlying the signal model, which limit their usability when broadband signals, such as speech, are present in the wavefield under observation. It will be subsequently shown that by erecting a second pillar 'classical acoustics', algorithms can be derived that are inherently broadband, hence allowing for ubiquitous application. The possibility to formulate parameter estimation algorithms that inherently allow for wideband source models is a previously unknown result of the fusion between classical array signal processing and methods of classical acoustics. This discovery is one of the main contributions of this book.

Moreover, it will be shown that acoustic source localization, based on these novel methods, has the potential to unambiguously localize multiple simultaneously active wideband sources in the array's full field-of-view.

In order to take advantage of the paradigms that can be found in the field of acoustics, special array geometries are considered here, i.e. circular and spherical geometries for wave propagation in two and three spatial dimensions, respectively. The more generic scenario of three-dimensional wave propagation can be cast into two-dimensional wave propagation problems whenever two-dimensional wavefields provide reasonable models for propagating wavefields, such as in closed rooms, where ceiling and floor reflections are sufficiently attenuated. The main advantage offered by the reduction in dimensionality is the significant decrease in array complexity.

The work presented in this book is organized as follows. The first two chapters are concerned with several concepts of classical acoustics. Chapter 2 introduces the notion of acoustic wave propagation and wave scattering in cylindrical and spherical coordinates, respectively. Subject of discussion are infinite-length and finite-length cylindrical radiators and scatterers as well as spherical radiators and scatterers. These physical foundations are required for the introduction of wavefield decomposition utilizing circular and spherical apertures, see Chapter 3. A natural way of analyzing a two- or three-dimensional wavefield is to decompose it into an orthogonal set of eigenfunctions corresponding to the acoustic wave equation in circular or spherical coordinates. In subsequent chapters, it will be shown that the decomposed components, the *eigenbeams*, can be used directly to yield novel array signal processing algorithms, denoted as eigenbeam – or *modal* – array signal processing. The decomposition, which is in essence equivalent to a Fourier analysis, is performed by either circular or spherical apertures. Both continuous and, more relevant in practice, sampled continuous apertures are considered. It will be shown that the properties of acoustic scattering can be advantageously utilized in the decomposition process. Note that the Fourier analysis of acoustic wavefields is sometimes referred to as Fourier acoustics [Wil99].

Chapter 4 summarizes the main techniques, properties, and problems of acoustic scene analysis using classical array signal processing. Following the introduction of fundamental signal models and assumptions, various waveform and parameter estimation techniques are presented. Parameter estimation focuses on two distinct signal processing tasks, the first one being the localization of possibly multiple simultaneously active sources. The second task is the estimation of the number of active sources in a wavefield.

In Chapter 5, the fundamental techniques shown in Chapter 3 are combined with the algorithms derived in Chapter 4 to yield novel algorithms for ASA using modal array signal processing, most notably source localization of multiple wideband acoustic sources as well as the estimation of the number of active sources in a wavefield. Derivations and performance evaluations in terms of simulations for the new waveform estimation techniques as well as for the new parameter estimation methods are detailed.

Chapter 6 presents a practical and real-time capable system for performing ASA tasks utilizing a circularly symmetric microphone array mounted into a finite-length rigid cylindrical baffle. This system is used for evaluating the waveform and parameter estimation algorithms outlined in Chapter 5 by considering measurements performed in real acoustic environments. Additional results of the performance evaluations are reproduced in Appendix F. During the course of the performance evaluation it will become clear that unprecedented potential lies in the paradigm of modal array signal processing applied to compact systems in real acoustic environments.

2

Acoustic Wavefields

This chapter is concerned with acoustic radiators and scatterers of cylindrical and spherical shape which forms the basis for the discussion of wavefield decomposition to be detailed in Chapter 3. Acoustic wavefields in cylindrical and spherical coordinate systems are considered that travel through a fluid, most notably air, and that interact with objects and structures resulting in scattering phenomena. In order to arrive at tractable mathematical descriptions of problems related to acoustic radiation and scattering a few basic assumptions are introduced here [JF93]:

1. All fluids and materials considered are assumed to obey linear equations. This restriction limits the following discussions to small-signal disturbances of the transmitting medium.
2. The media the acoustic wavefields interact with are assumed to be homogeneous.
3. Steady-state conditions are assumed which means that initial transient effects can be neglected.
4. For scattering problems, only local interactions between a wavefield propagating in a homogeneous medium and bodies within this medium are assumed.

It should be noted that all of the above assumptions are reasonably standard in many textbooks on linear acoustics, such as [MI68], [JF93], [Wil99], and [Bla00]. This book further assumes lossless transport of acoustic energy, thereby neglecting the effects of attenuation and dispersion the wave may be affected by as it propagates through a medium.

This chapter is organized as follows. Section 2.1 lays the mathematical and physical foundations by introducing Euler's equation and the acoustic wave equation in an arbitrary orthogonal coordinate system. Section 2.2 introduces the notion of point sources and point sinks in two and three dimensions. Section 2.3 considers acoustic wavefields in cylindrical coordinates. Cylindrical coordinates are applicable to scenarios where the region under observation, e.g. the radiator itself, is extended in one dimension and compact in the other two. An example is a vibrating cylindrical object where its radius is much smaller than its length. Section 2.4 details acoustic wavefields in spherical coordinates. Spherical coordinates are appropriate when the region of interest is finite in all three dimensions, such as spherical radiators.

2.1 Mathematical Foundations

A periodic time-dependent driving force resulting in a harmonic particle displacement of a field point $Q(q_1, q_2, q_3)$ is assumed that can be described in an arbitrary orthogonal coordinate system (q_1, q_2, q_3). This means that the field point Q can be interpreted in the Cartesian coordinate system, (x, y, z), as well as in all other orthogonal coordinate systems (q_1, q_2, q_3), for example cylindrical (ϱ, ϕ, z) or spherical (r, θ, ϕ), as [Arf85],

$$x = x(q_1, q_2, q_3), \tag{2.1a}$$
$$y = y(q_1, q_2, q_3), \tag{2.1b}$$
$$z = z(q_1, q_2, q_3). \tag{2.1c}$$

For cylindrical coordinate systems,

$$x = \varrho \cos \phi, \tag{2.2a}$$
$$y = \varrho \sin \phi, \tag{2.2b}$$
$$z = z, \tag{2.2c}$$

and,

$$x = r \sin \theta \cos \phi, \tag{2.3a}$$
$$y = r \sin \theta \sin \phi, \tag{2.3b}$$
$$z = r \cos \theta, \tag{2.3c}$$

for spherical coordinate systems, cf. Fig. 2.2 and Fig. 2.1, respectively. Note that the z-coordinate has been omitted in Fig. 2.2.

In the coordinate system of interest, the time-varying position of the displacement vector, $\boldsymbol{w}(\boldsymbol{r}_Q, t)$, representing the field point $Q(q_1, q_2, q_3)$ due to the driving force can be written as [JF93],

$$\boldsymbol{w}(\boldsymbol{r}_Q, t) = [w_1(\boldsymbol{r}_Q)\boldsymbol{e}_{q_1} + w_2(\boldsymbol{r}_Q)\boldsymbol{e}_{q_2} + w_3(\boldsymbol{r}_Q)\boldsymbol{e}_{q_3}]\, e^{-i\omega t}, \tag{2.4}$$

where $\boldsymbol{e}_{q_1}, \boldsymbol{e}_{q_2}$, and $\boldsymbol{e}_{q_3}$ are the unit vectors in the directions of the respective coordinate axis, and $\boldsymbol{r}_Q$ is the position vector associated with $Q(q_1, q_2, q_3)$ in that coordinate system. The time variable is denoted as t, ω is the temporal radian frequency, and $i^2 = -1$. In the following, only the first component of the displacement, i.e.,

$$w(q_1, q_2, q_3, t) \triangleq w_1(\boldsymbol{r}_Q)\, e^{-i\omega t}, \tag{2.5}$$

is taken into consideration in all further developments. This component corresponds to the radial, or normal, component of the displacement vector in cylindrical and spherical coordinate systems. Since mainly longitudinal waves are of interest in this book, only the normal component of the displacement (radially oriented driving force) is considered.

Most analyses of acoustic wavefields are based on two fundamental equations, Euler's equation of motion and the acoustic wave equation. Both equations are introduced in the following.

2.1.1 Euler's Equation

Assuming linearity of the medium and negligible body forces such as gravity, *Euler's equation*, also known as the momentum equation, can be written in the time-domain as [Wil99],

$$-\varrho_0 \frac{\partial^2}{\partial t^2} w(q_1, q_2, q_3, t) = \boldsymbol{\nabla}\, p(q_1, q_2, q_3, t), \tag{2.6}$$

where ϱ_0 is the fluid density and $p(\cdot)$ denotes the pressure. [1] In general, the gradient or Nabla operator, $\boldsymbol{\nabla}$, is defined as [Arf85],

$$\boldsymbol{\nabla} = \frac{1}{a_1}\frac{\partial}{\partial q_1}\boldsymbol{e}_{q_1} + \frac{1}{a_2}\frac{\partial}{\partial q_2}\boldsymbol{e}_{q_2} + \frac{1}{a_3}\frac{\partial}{\partial q_3}\boldsymbol{e}_{q_3}, \tag{2.7}$$

where,

$$a_l^2 = \left(\frac{\partial x}{\partial q_l}\right)^2 + \left(\frac{\partial y}{\partial q_l}\right)^2 + \left(\frac{\partial z}{\partial q_l}\right)^2, \quad l = 1, 2, 3. \tag{2.8}$$

The following discussions do not exploit the relationship between the particle displacement and pressure but considers the relationship between particle velocity,

$$\dot{w}(q_1, q_2, q_3, t) \triangleq \frac{\partial}{\partial t} w(q_1, q_2, q_3, t), \tag{2.9}$$

and pressure $p(q_1, q_2, q_3, t)$. Consequently, Eq. (2.6) can be written as,

$$-\varrho_0 \frac{\partial}{\partial t} \dot{w}(q_1, q_2, q_3, t) = \boldsymbol{\nabla}\, p(q_1, q_2, q_3, t). \tag{2.10}$$

Considering the steady-state assumption, the analysis of Euler's equation can be carried over to the frequency-domain by utilizing the temporal Fourier transform, as defined in Section A.1. By applying the inverse temporal Fourier transform, Eq. (A.2), to the particle velocity and calculating the time-derivative one obtains,

$$\frac{\partial}{\partial t}\dot{w}(q_1, q_2, q_3, t) = \frac{1}{2\pi}\int_{-\infty}^{\infty} -i\omega \dot{W}(q_1, q_2, q_3, \omega)\, e^{-i\omega t}\, d\omega. \tag{2.11}$$

[1] Note the fundamentally different meanings of the symbols ϱ_0 and ϱ.

It therefore follows that,

$$\mathcal{F}_t\left\{\frac{\partial}{\partial t}\dot{w}(q_1,q_2,q_3,t)\right\} = -i\omega\dot{W}(q_1,q_2,q_3,\omega), \tag{2.12}$$

where $\mathcal{F}_t\{\xi(t)\} = \Xi(\omega)$ denotes the temporal Fourier transform of the argument with respect to the time variable (subscript t). As a result, in the frequency-domain Eq. (2.10) becomes,

$$\dot{W}(q_1,q_2,q_3,\omega) = \frac{1}{i\varrho_0 ck}\boldsymbol{\nabla}P(q_1,q_2,q_3,\omega), \tag{2.13}$$

where $k = \omega/c$ is the wavenumber, c is the speed of sound, and $P(q_1,q_2,q_3,\omega) = \mathcal{F}_t\{p(q_1,q_2,q_3,t)\}$.

2.1.2 The Acoustic Wave Equation

The second important equation in acoustic wavefield decomposition is the *acoustic wave equation*. Considering the above mentioned assumptions (1-4) in the introductory paragraph of Chapter 2, the homogeneous acoustic wave equation in the time-domain reads [Wil99],

$$\left(\boldsymbol{\nabla}^2 - \frac{1}{c^2}\frac{\partial^2}{\partial t^2}\right)p(q_1,q_2,q_3,t) = 0, \tag{2.14}$$

where, in general, [Arf85],

$$\boldsymbol{\nabla}^2 = \frac{1}{a_1a_2a_3}\left[\frac{\partial}{\partial q_1}\left(\frac{a_2a_3}{a_1}\frac{\partial}{\partial q_1}\right) + \frac{\partial}{\partial q_2}\left(\frac{a_1a_3}{a_2}\frac{\partial}{\partial q_2}\right) + \frac{\partial}{\partial q_3}\left(\frac{a_1a_2}{a_3}\frac{\partial}{\partial q_3}\right)\right]. \tag{2.15}$$

$\boldsymbol{\nabla}^2$ is also known as the Laplace operator. A detailed derivation of the acoustic wave equation, in which Euler's equation, Eq. (2.6) or Eq. (2.13), is of major importance, can be found in [Bla00]. Here and throughout this work, steady-state conditions are assumed and the explicit time dependence is dropped for notational convenience. Transforming Eq. (2.14) into the frequency-domain using the temporal Fourier transform one obtains the so-called homogeneous Helmholtz equation [MF53],

$$\boldsymbol{\nabla}^2P(q_1,q_2,q_3,\omega) + k^2P(q_1,q_2,q_3,\omega) = 0. \tag{2.16}$$

Specific solutions of the acoustic wave equation in cylindrical and spherical coordinate systems are given in Sections 2.3 and 2.4, respectively.

2.2 Point Sources and Plane-Waves

The homogeneous Helmholtz equation, Eq. (2.16), assumes a source-free medium, i.e. the right-hand side of Eq. (2.16) is zero. In most acoustical problems there is a source, or a superposition of multiple sources, that generate the wavefield under observation. Therefore, some non-zero term needs to be added to the right-hand side of the homogeneous Helmholtz equation which then becomes an inhomogeneous Helmholtz equation. An acoustic source is defined as a region of space that is in contact with the medium, in which the wavefield is to be evaluated, where *new* acoustic energy is generated. This energy is radiated outward, into the medium, with respect to the source [MI68]. Two models for acoustic sources are widely used, i.e. point sources and plane-waves. Point sources can be regarded as infinitesimally small objects in two- or three-dimensional space that radiate acoustic energy, which spreads cylindrically or spherically, respectively. If the point sources are located infinitely far from the point(s) of observation, the cylindrically/spherically spreading wave-front becomes planar, i.e. the points of identical phase travel on a plane. These two models are studied in the following sections.

2.2.1 Point Sources in Three Dimensions

A point source/sink [2] with position vector $\boldsymbol{r}_0$, cf. Fig. 2.1, is defined as the solution of the inhomogeneous Helmholtz equation [MF53],

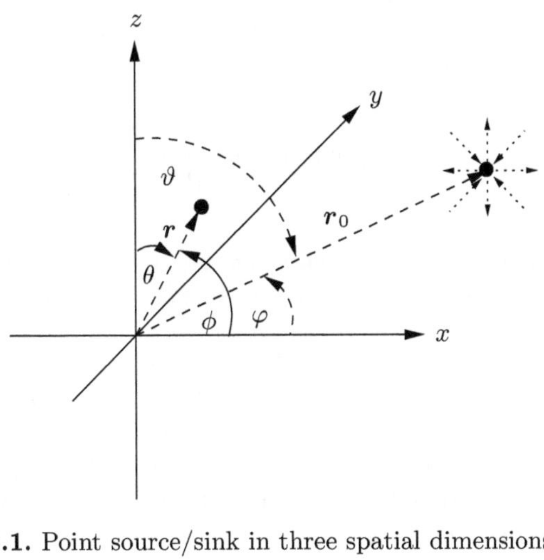

Fig. 2.1. Point source/sink in three spatial dimensions

[2] Note that the term *sink* in this context is merely used to describe that acoustic energy is traveling *toward* a particular point in space, i.e. in the direction of $\boldsymbol{r}_0$.

$$\nabla^2 g_k(\boldsymbol{r}|\boldsymbol{r}_0) + k^2 g_k(\boldsymbol{r}|\boldsymbol{r}_0) = -\delta(\boldsymbol{r} - \boldsymbol{r}_0), \tag{2.17}$$

where $\boldsymbol{r}$ is the point of observation and $g_k(\boldsymbol{r}|\boldsymbol{r}_0)$ is called a Green's function, which also satisfies the homogeneous Helmholtz equation, Eq. (2.16), everywhere except for $\boldsymbol{r} = \boldsymbol{r}_0$. The subscript k is used to stress the dependency of the Green's function on the wavenumber. $\delta(\boldsymbol{r} - \boldsymbol{r}_0)$ denotes a three dimensional Dirac delta function, representing a point source located at $\boldsymbol{r}_0$. This *free-field* Green's function that represent point sources must also satisfy the so-called Sommerfeld radiation condition [MF53],

$$\lim_{\|\boldsymbol{r}-\boldsymbol{r}_0\|\to\infty} \|\boldsymbol{r} - \boldsymbol{r}_0\|^{\frac{D-1}{2}} \left(\frac{\partial}{\partial \|\boldsymbol{r} - \boldsymbol{r}_0\|} g_k(\boldsymbol{r}|\boldsymbol{r}_0) - ik g_k(\boldsymbol{r}|\boldsymbol{r}_0) \right) = 0, \tag{2.18}$$

where $D = 2, 3$ denotes the dimension of the problem. Equation (2.18) provides a *boundary condition* at infinity and therefore ensures that sources radiate sound waves instead of absorbing them. These waves are also called outgoing, or outward traveling, waves. It is shown in [MF53] that a solution of Eq. (2.17) that also satisfies Eq. (2.18) is given by,

$$g_k(\boldsymbol{r}|\boldsymbol{r}_0) = \frac{e^{ik\|\boldsymbol{r}-\boldsymbol{r}_0\|}}{4\pi\|\boldsymbol{r} - \boldsymbol{r}_0\|}. \tag{2.19}$$

Equation (2.19) is an expression for the free-field (i.e. boundary at infinity) response of a point source located at $\boldsymbol{r}_0$ evaluated at the observation point $\boldsymbol{r}$ in three dimensions. Incoming, or inward traveling, waves also satisfy Eq. (2.17) and can be written as,

$$g_k(\boldsymbol{r}|\boldsymbol{r}_0) = \frac{e^{-ik\|\boldsymbol{r}-\boldsymbol{r}_0\|}}{4\pi\|\boldsymbol{r} - \boldsymbol{r}_0\|}. \tag{2.20}$$

More solutions of the inhomogeneous Helmholtz equation, Eq. (2.17), can be formulated by considering any linear combination of Eq. (2.19) and Eq. (2.20).

Note that Eq. (2.19) and Eq. (2.20) are expressions for point sources and sinks in arbitrary orthogonal but separable coordinate systems.

For the special case of spherical coordinate systems, it is shown in [MF53] that for $r_0 = \|\boldsymbol{r}_0\| > r = \|\boldsymbol{r}\|$, a point source, cf. Eq. (2.19), positioned at $(r_0, \vartheta, \varphi)$ can be expanded into eigenfunctions in spherical coordinates at the observation point (r, θ, ϕ) as,

$$\frac{e^{ik\|\boldsymbol{r}-\boldsymbol{r}_0\|}}{\|\boldsymbol{r} - \boldsymbol{r}_0\|} = 4\pi ik \sum_{n=0}^{\infty} j_n(kr) h_n^{(1)}(kr_0) \sum_{m=-n}^{n} Y_n^m(\theta, \phi) Y_n^m(\vartheta, \varphi)^*, \tag{2.21}$$

where * denotes the conjugate complex operator, and where $j_n(\cdot)$ is the spherical Bessel function of order n, see Section B.2. $h_n^{(1)}(\cdot)$ denotes the spherical Hankel function of the first kind of order n and is defined as [AS72],

$$h_n^{(1)}(\cdot) = j_n(\cdot) + i\, y_n(\cdot), \tag{2.22}$$

where $y_n(\cdot)$ is the spherical Neumann function of order n, cf. Section B.2. $Y_n^m(\theta, \phi)$ denotes the so-called spherical harmonic of order n and degree m satisfying $n \geq |m|$, and is defined as, cf. Section B.3,

$$Y_n^m(\theta, \phi) \triangleq \sqrt{\frac{(2n+1)}{4\pi} \frac{(n-m)!}{(n+m)!}}\, P_n^m(\cos\theta)\, e^{im\phi}, \tag{2.23}$$

where $P_n^m(\cos\theta)$ is the associated Legendre function of order n and degree m, see Section B.3. Refer to Section 2.4 for an in-depth discussion of solutions of the acoustic wave equation in spherical coordinates. For now it is sufficient to note that by identifying the coefficients in Eq. (2.104) – the most general solution of the homogeneous Helmholtz equation in spherical coordinates – it can be easily seen that Eq. (2.21) satisfies the wave equation. Note also that Eqs. (2.19) and (2.20) are solutions of the inhomogeneous Helmholtz equation under free-field conditions. Note that the term 'free-field' means that there are no other boundary conditions than implied by the Sommerfeld radiation condition. Free-field propagation can generally not be assumed, e.g., when the sources that generate a wavefield are positioned within an enclosure.

Analogously, an expansion of Eq. (2.20) for $r_0 < r$ yields,

$$\frac{e^{-ik\|\boldsymbol{r}-\boldsymbol{r}_0\|}}{\|\boldsymbol{r}-\boldsymbol{r}_0\|} = 4\pi(-i)k \sum_{n=0}^{\infty} j_n(kr)h_n^{(2)}(kr_0) \sum_{m=-n}^{n} Y_n^m(\theta,\phi)^* Y_n^m(\vartheta,\varphi), \tag{2.24}$$

where,

$$h_n^{(2)}(\cdot) = j_n(\cdot) - i\, y_n(\cdot), \tag{2.25}$$

is the n-th-order spherical Hankel function of the second kind, n being an integer. If the medium is bounded by surfaces that introduce certain boundary conditions, the most general solution of Eq. (2.17) is [MI68],

$$G_k(\boldsymbol{r}|\boldsymbol{r}_0) = g_k(\boldsymbol{r}|\boldsymbol{r}_0) + \xi(\boldsymbol{r}), \tag{2.26}$$

where $\xi(\boldsymbol{r})$ is a solution of the homogeneous Helmholtz equation $\boldsymbol{\nabla}^2\xi + k^2\,\xi = 0$.

An interesting property of the Green's function is the principle of reciprocity [MF53], i.e.,

$$G_k(\boldsymbol{r}|\boldsymbol{r}_0) = G_k(\boldsymbol{r}_0|\boldsymbol{r}). \tag{2.27}$$

This principle means that the position of the source and receiver (point of observation) are interchangeable.

2.2.2 Plane-Waves in Three Dimensions

Plane-waves can be interpreted as point sources, cf. Eq. (2.21), where $r_0 = \|\boldsymbol{r}_0\| \rightarrow \infty$. While the denominator of the left-hand side of Eq. (2.21) can then be approximated by $\|\boldsymbol{r} - \boldsymbol{r}_0\| \approx r_0$, the phase term of the exponential needs a separate discussion since it is an oscillating function with respect to $\|\boldsymbol{r} - \boldsymbol{r}_0\|$. According to [CK98] it holds that,

$$\|\boldsymbol{r} - \boldsymbol{r}_0\| = \sqrt{\|\boldsymbol{r}\|^2 - 2\boldsymbol{r}^T\boldsymbol{r}_0 + \|\boldsymbol{r}_0\|^2}, \tag{2.28}$$

which can be approximated by,

$$\|\boldsymbol{r} - \boldsymbol{r}_0\| \approx \|\boldsymbol{r}_0\| - \boldsymbol{e}_{\boldsymbol{r}_0}^T \boldsymbol{r}, \tag{2.29}$$

where $\boldsymbol{e}_{\boldsymbol{r}_0}$ denotes a unit vector in the direction of $\boldsymbol{r}_0$ and where $()^T$ denotes transposition. Performing the farfield transition for the left-hand side of Eq. (2.21), therefore yields,

$$\frac{e^{ik\|\boldsymbol{r}-\boldsymbol{r}_0\|}}{\|\boldsymbol{r} - \boldsymbol{r}_0\|} \approx \frac{e^{ikr_0}}{r_0} e^{-ik\boldsymbol{e}_{\boldsymbol{r}_0}^T \boldsymbol{r}} = \frac{e^{ikr_0}}{r_0} e^{-i\boldsymbol{k}^T\boldsymbol{r}}, \tag{2.30}$$

where $\boldsymbol{k}$ is the three-dimensional wavenumber vector and $k = \|\boldsymbol{k}\|$. Applying the large argument limit of the spherical Hankel function, Eq. (B.43), for the right-hand side of Eq. (2.21), it follows that,

$$\frac{e^{ikr_0}}{r_0} e^{-i\boldsymbol{k}^T\boldsymbol{r}} = 4\pi ik \frac{e^{ikr_0}}{kr_0} \sum_{n=0}^{\infty} (-i)^{n+1} j_n(kr) \sum_{m=-n}^{n} Y_n^m(\theta,\phi) Y_n^m(\vartheta,\varphi)^*, \tag{2.31}$$

which can be written as,

$$e^{-i\boldsymbol{k}^T\boldsymbol{r}} = 4\pi \sum_{n=0}^{\infty} (-i)^n j_n(kr) \sum_{m=-n}^{n} Y_n^m(\theta,\phi) Y_n^m(\vartheta,\varphi)^*. \tag{2.32}$$

Therefore, a plane-wave with unity amplitude due to a point source at infinity can be expressed as,

$$P_{\text{out}_3}(\boldsymbol{k}^T\boldsymbol{r}) = e^{-i\boldsymbol{k}^T\boldsymbol{r}}, \tag{2.33}$$

where the subscript denotes an outgoing plane-wave in three spatial dimensions. Since the Green's function, $g_k(\boldsymbol{r}|\boldsymbol{r}_0)$, can be identified with a free-field transfer function describing the channel between source and receiver in the sense of system theory, Eq. (2.33) can be considered as a response, i.e. pressure, observed at $\boldsymbol{r}$ due to a point source at infinity.

Applying the preceding discussion to sinks at infinity results in an analogous expression for incoming plane-waves, i.e.,

$$P_{\text{in}_3}(\boldsymbol{k}^T\boldsymbol{r}) = e^{i\boldsymbol{k}^T\boldsymbol{r}}. \tag{2.34}$$

The plane-wave expansion in analogy to Eq. (2.32) then yields,

$$\begin{aligned} e^{i\boldsymbol{k}^T\boldsymbol{r}} &= 4\pi \sum_{n=0}^{\infty} i^n j_n(kr) \sum_{m=-n}^{n} Y_n^m(\theta,\phi)^* Y_n^m(\vartheta,\varphi) \\ &= 4\pi \sum_{n=0}^{\infty} i^n j_n(kr) \sum_{m=-n}^{n} Y_n^m(\theta,\phi) Y_n^m(\vartheta,\varphi)^*. \end{aligned} \tag{2.35}$$

Note that the conjugate complex operator can be applied to either spherical harmonic. This statement applies also to Eq. (2.21) and Eq. (2.24). Again, it can be easily seen that Eqs. (2.32) and (2.35) are solutions of the wave equation in spherical coordinates, see Eq. (2.104), evaluated at the observation point (r,θ,ϕ).

2.2.3 Point Sources and Plane-Waves in Two Dimensions

Figure 2.2 illustrates a point source/sink in two spatial dimensions. Then, a general solution of Eq. (2.17) can be expressed as [MF53],

$$g_k(\boldsymbol{\varrho}|\boldsymbol{\varrho}_0) = \frac{i}{4} H_0^{(1,2)}(k\|\boldsymbol{\varrho} - \boldsymbol{\varrho}_0\|), \tag{2.36}$$

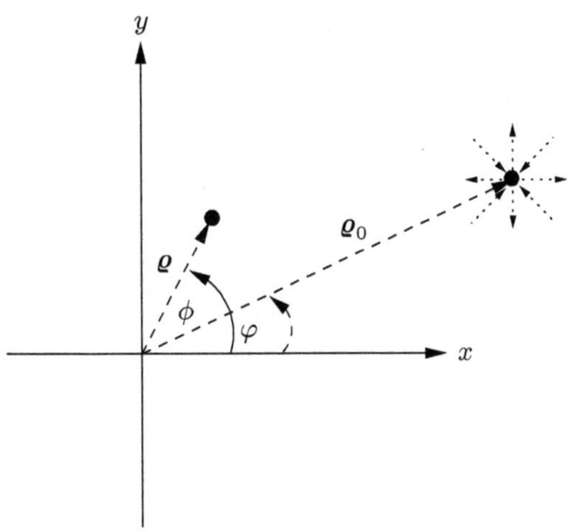

Fig. 2.2. Point source/sink in two spatial dimensions

where the superscript $^{(1)}$ is used for outward traveling waves – resulting from a point source – and the superscript $^{(2)}$ is used for inward traveling waves – corresponding to a sink. For $\varrho_0 = \|\boldsymbol{\varrho}_0\| > \varrho = \|\boldsymbol{\varrho}\|$ an expansion of this solution can be written as [CK98],

$$H_0^{(1)}(k\|\boldsymbol{\varrho} - \boldsymbol{\varrho}_0\|) = \sum_{n=-\infty}^{\infty} e^{in(\phi-\varphi)} J_n(k\varrho) H_n^{(1)}(k\varrho_0), \tag{2.37}$$

where, cf. Appendix B.1,

$$H_n^{(1)}(\cdot) = J_n(\cdot) + i\,Y_n(\cdot), \tag{2.38}$$
$$H_n^{(2)}(\cdot) = J_n(\cdot) - i\,Y_n(\cdot), \tag{2.39}$$

are the n-th order Hankel functions of the first and second kind, respectively. $J_n(\cdot)$ and $Y_n(\cdot)$ are the n-th order Bessel and Neumann functions, respectively. The angle φ is the direction-of-propagation of the source with respect to the origin.

As in the three-dimensional case, this result can be identified with the solution of the homogeneous Helmholtz equation in cylindrical coordinates, Eq. (2.54), with $z = 0$. Refer to Section 2.3 for an in-depth discussion of solutions of the acoustic wave equation in cylindrical coordinates.

Following the same line of argumentation as in the previous section from Eq. (2.28) to Eq. (2.30), and by inserting Eq. (B.24) into Eq. (2.37) it follows that,

$$\sqrt{\frac{2}{\pi k \varrho_0}}\, e^{i(k\varrho_0 - \boldsymbol{k}^T\boldsymbol{\varrho})}\, e^{-i\pi/4} = \sqrt{\frac{2}{\pi k \varrho_0}}\, e^{ik\varrho_0}\, e^{-i\pi/4} \sum_{n=-\infty}^{\infty} (-i)^n e^{in(\phi-\varphi)} J_n(k\varrho), \tag{2.40}$$

and, as a result for the plane-wave expansion,

$$e^{-i\boldsymbol{k}^T\boldsymbol{\varrho}} = \sum_{n=-\infty}^{\infty} (-i)^n J_n(k\varrho)\, e^{in(\phi-\varphi)}. \tag{2.41}$$

Thus, in two dimensions, a plane-wave with unity amplitude due to a point *source* at infinity can be approximated by,

$$P_{\mathrm{out}_2}(\boldsymbol{k}^T\boldsymbol{\varrho}) = e^{-i\boldsymbol{k}^T\boldsymbol{\varrho}}. \tag{2.42}$$

Equivalently, a plane-wave due to a sink at infinity can be approximated by,

$$P_{\mathrm{in}_2}(\boldsymbol{k}^T\boldsymbol{\varrho}) = e^{i\boldsymbol{k}^T\boldsymbol{\varrho}}, \tag{2.43}$$

and a plane-wave expansion of Eq. (2.43), in analogy to Eq. (2.41) that will be used extensively in the following discussions can be written as [CK98],

$$e^{i\boldsymbol{k}^T\boldsymbol{\varrho}} = \sum_{n=-\infty}^{\infty} i^n J_n(k\varrho)\, e^{-in(\phi-\varphi)} = \sum_{n=-\infty}^{\infty} i^n J_n(k\varrho)\, e^{in(\phi-\varphi)}, \tag{2.44}$$

where in the last step Eq. (B.6) and $(-1)^n i^{-n} = i^n$ has been used. See also Section 3.1.1 for a derivation of this result.

2.3 Acoustic Wave Equation in Cylindrical Coordinates

This section examines wavefields in cylindrical coordinates. Both wavefields produced by a cylindrically shaped object and modifications of wavefields due to the presence of a cylindrical object in a wavefield are discussed. The results obtained here will be of considerable importance in subsequent chapters.

Section 2.3.1 presents a general solution of the acoustic wave equation. Section 2.3.2 quantifies the wavefield of an infinite-length and finite-length cylindrical radiator. Scattering phenomena of infinite-length and finite-length rigid cylinders are outlined in Section 2.3.3.

Although not further pursued in this book, the total wavefield due to a so-called 'pressure-release' cylinder is derived at the end of Section 2.1.2 for completeness of discussion. Pressure-release boundaries force the acoustic pressure on its surface to vanish.

2.3.1 General Solution of the Acoustic Wave Equation in Cylindrical Coordinates

The cylindrical coordinate system and the geometric model of a cylindrical radiator or scatterer with radius R and length $2L$ are shown in Fig. 2.3. The relation of the cylindrical coordinate system with respect to the Cartesian coordinate system was given in Eq. (2.2). The field point $Q(\varrho, \phi, z)$, located on the cylinder surface, is also depicted. Transferring the results from Section 2.1 to the particular case of a cylindrical coordinate system gives for Euler's equation, Eq. (2.13),

$$\dot{W}(\varrho, \phi, z, \omega) = \frac{1}{i\varrho_0 ck} \boldsymbol{\nabla} P(\varrho, \phi, z, \omega), \tag{2.45}$$

where with Eqs. (2.7), (2.2), and (2.8),

$$\boldsymbol{\nabla} = \frac{\partial}{\partial \varrho}\boldsymbol{e}_\varrho + \frac{1}{\varrho}\frac{\partial}{\partial \phi}\boldsymbol{e}_\phi + \frac{\partial}{\partial z}\boldsymbol{e}_z. \tag{2.46}$$

Note again the fundamentally different meanings of the symbols ϱ and ϱ_0.

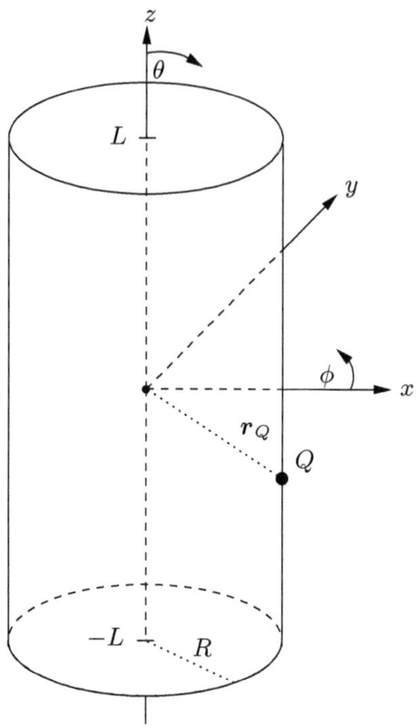

Fig. 2.3. Geometric model of a cylindrical radiator/scatterer

For the homogeneous Helmholtz equation, cf. Eq. (2.16), it follows that,

$$\nabla^2 P(\varrho,\phi,z,\omega) + k^2 P(\varrho,\phi,z,\omega) = 0, \tag{2.47}$$

where, with Eq. (2.15) and Eq. (2.8),

$$\nabla^2 = \frac{\partial^2}{\partial\varrho^2} + \frac{1}{\varrho}\frac{\partial}{\partial\varrho} + \frac{1}{\varrho^2}\frac{\partial^2}{\partial\phi^2} + \frac{\partial^2}{\partial z^2}. \tag{2.48}$$

It can be shown that the general steady-state solution of the acoustic wave equation, Eq. (2.14) or Eq. (2.16), in cylindrical coordinates in the frequency-domain can be expressed as [Wil99],

$$\begin{aligned} P(\varrho,\phi,z) &= \frac{1}{2\pi}\sum_{n=-\infty}^{\infty} e^{in\phi} \\ &\times \int_{-\infty}^{\infty}\left[A_n(k_z)\,e^{ik_z z}H_n^{(1)}(k_\varrho\varrho) + B_n(k_z)\,e^{ik_z z}H_n^{(2)}(k_\varrho\varrho)\right]dk_z, \end{aligned} \tag{2.49}$$

where,

$$k_\varrho = \sqrt{k^2 - k_z^2}, \tag{2.50}$$

are the radial components of the wave-vector $\boldsymbol{k}$. A comment on the form of Eq. (2.49) is in order. Since $e^{in\phi}$ is 2π-periodic for any ϕ, n must be an integer.

Thus, a sum takes all contributions of the wavenumbers in ϕ-direction into account. Since no such restriction applies to $e^{ik_z z}$ an integral is necessary to sum up all contributions of the wavenumbers in z-direction, k_z. Note that in Eq. (2.49) the dependency of the pressure on ω or, equivalently, on k has been omitted. Whenever possible, this notational convenience will be adapted throughout this section.

By defining the time-dependence to be $e^{-i\omega t}$ rather than $e^{+i\omega t}$ in the time-domain formulation of the displacement vector, cf. Section 2.1, the Hankel function of the first kind represents outgoing cylindrical waves and the Hankel function of the second kind needs to be considered for incoming cylindrical waves. This convention will be of importance in later developments. By expressing the Hankel functions using Bessel and Neumann functions as defined by Eqs. (B.4) and (B.5), Eq. (2.49) can be written as,

$$P(\varrho, \phi, z) = \frac{1}{2\pi} \sum_{n=-\infty}^{\infty} e^{in\phi} \times \int_{-\infty}^{\infty} \left[C_n(k_z)\, e^{ik_z z} J_n(k_\varrho \varrho) + D_n(k_z)\, e^{ik_z z} Y_n(k_\varrho \varrho) \right] dk_z, \tag{2.51}$$

where,

$$C_n(k_z) = A_n(k_z) + B_n(k_z) \tag{2.52}$$

$$D_n(k_z) = i[A_n(k_z) - B_n(k_z)]. \tag{2.53}$$

$A_n(k_z)$ and $B_n(k_z)$ in Eq. (2.49), and implicitly in Eq. (2.51) are specified by the boundary conditions of the problem, i.e. the behavior of the acoustic wavefield at a boundary between two media.

Figure 2.4 depicts the boundary value problem, where the annular disk region specifies the area of validity of the homogeneous wave equation. All sources denoted by S_I are located within the cylindrical surface bounded by R_I and all sources denoted by S_E are located outside the cylindrical surface bounded by R_E. The resulting annular disk specifies the area of validity of the homogeneous acoustic wave equation. This general scenario can be simplified for many boundary value problems of interest.

Interior boundary value problems are considered when all sources are located outside the cylindrical surface bounded by R_E, i.e. all sources S_I are absent. In this case, the area of validity includes the origin. Since all systems under consideration are physical systems and both Hankel functions in Eq. (2.49), $H_n^{(1,2)}(\cdot)$, are infinite at the origin (see Section B.1), Eq. (2.51) is used. Since the Neumann function, $Y_n(\cdot)$, is also infinite at the origin, it is necessary that $D_n(k_\varrho \varrho) = 0$ and therefore Eq. (2.51) simplifies to,

$$P(\varrho, \phi, z) = \frac{1}{2\pi} \sum_{n=-\infty}^{\infty} e^{in\phi} \int_{-\infty}^{\infty} C_n(k_z)\, e^{ik_z z} J_n(k_\varrho \varrho)\, dk_z. \tag{2.54}$$

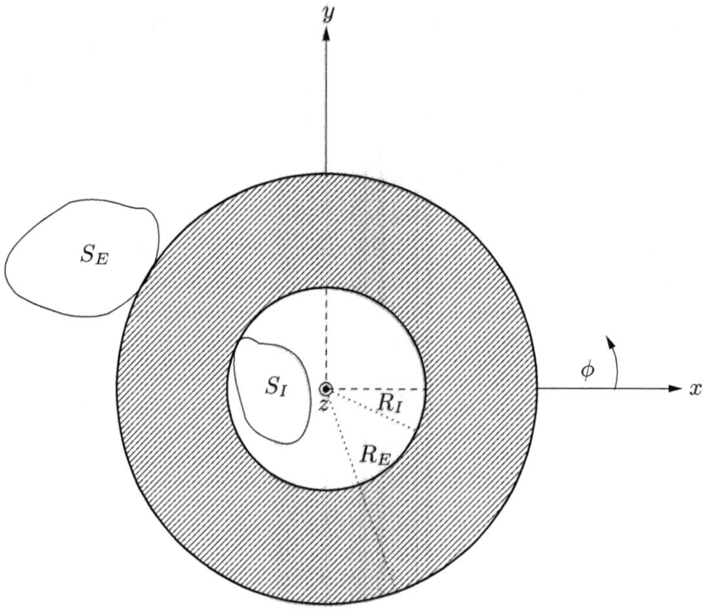

Fig. 2.4. Boundary value problem. The annular disk region specifies the area of validity of the homogeneous wave equation (after [Wil99])

Exterior boundary problems exist when the cylindrical surface completely encloses all sources, cf. S_I in Fig. 2.4, and the sources S_E are absent. Now, the solution of the homogeneous acoustic wave equation is determined by Eq. (2.49), where $B_n(k_z) = 0$, since incoming waves cannot exist in this case. Exterior boundary problems reduce the general solution of the wave equation to,

$$\begin{aligned} P(\varrho, \phi, z) &= \frac{1}{2\pi} \sum_{n=-\infty}^{\infty} e^{in\phi} \int_{-\infty}^{\infty} A_n(k_z)\, e^{ik_z z} H_n^{(1)}(k_\varrho \varrho)\, dk_z \\ &\triangleq \frac{1}{2\pi} \sum_{n=-\infty}^{\infty} e^{in\phi} \int_{-\infty}^{\infty} A_n(k_z)\, e^{ik_z z} H_n(k_\varrho \varrho)\, dk_z . \end{aligned} \tag{2.55}$$

Since most of the following discussions will not involve Hankel functions of the second kind, the superscript $^{(1)}$ will be omitted in further derivations, except when indicated otherwise.

Note that it can be seen from Eq. (2.50) that the radial component of the wavenumber, k_ϱ, can be both real ($k > k_z$) or purely imaginary ($k < k_z$). These two cases correspond to cylindrical wave radiation and the so-called *evanescent* wave radiation [Wil99], respectively. Further explanation is given in Section 2.3.2.

2.3.2 The Cylindrical Radiator

This section examines infinite-length and finite-length cylindrical radiators following the discussion presented in [JF93]. Since steady-state conditions and a periodic time-dependent radially acting driving force resulting in a harmonic particle displacement of the field point on the cylinder surface at $\varrho = R$, $Q(R, \phi, z)$, see Fig. 2.3, are assumed, the resulting particle velocity is spatially periodic. It is therefore assumed that the spatial periodicity of the particle velocity is confined to the circumferential coordinate, ϕ. Then, the particle velocity can be expanded into a Fourier series in ϕ as,

$$\dot{W}(\varrho, \phi, z) = \sum_{n=-\infty}^{\infty} \dot{W}_n f(z) e^{in\phi}, \tag{2.56}$$

where the z-dependence is accounted for by the term $f(z)$, and $\dot{W}_n$ denotes the Fourier coefficients. The pressure field for an infinite-length and finite-length cylindrical radiator is derived in Section 2.3.2. In the case of an infinite-length cylindrical radiator, the z-dependence is assumed to be periodic, whereas in the case of a finite-length cylindrical radiator this assumption, in general, cannot be made. This is due to the fact that the particle velocity distribution on the cylinder surface is limited to the length of the cylinder.

The Cylindrical Radiator of Infinite Length

In this section, the special case of a periodic $f(z)$ in z of Eq. (2.56) is examined. Spatial periodicity in both ϕ- and z-coordinates can be accounted for by a double Fourier series as,

$$\dot{W}(\varrho, \phi, z) = \sum_{n=-\infty}^{\infty} \sum_{m=-\infty}^{\infty} \dot{W}_{mn} e^{ik_m z} e^{in\phi}, \tag{2.57}$$

with Fourier coefficients $\dot{W}_{mn}$. Specifically, the continuous wavenumber, k_z, is thereby replaced by the discrete wavenumber k_m.

The pressure of a cylindrical radiator of infinite length can be obtained by applying the appropriate boundary condition. Because of the spatial periodicity of the problem, the pressure, Eq. (2.55), becomes,

$$P(\varrho, \phi, z) = \sum_{n=-\infty}^{\infty} \sum_{m=-\infty}^{\infty} A_n(k_m) H_n(k_\varrho \varrho) e^{in\phi} e^{ik_m z}. \tag{2.58}$$

The spatial derivative with respect to the radial component on the cylinder surface is,

$$\begin{aligned} \nabla P(\varrho, \phi, z)|_{\varrho=R} &= \frac{\partial}{\partial \varrho} P(\varrho, \phi, z)|_{\varrho=R} \\ &= k_\varrho \sum_{n=-\infty}^{\infty} \sum_{m=-\infty}^{\infty} A_n(k_m) H_n^{'}(k_\varrho R) e^{in\phi} e^{ik_m z}, \end{aligned} \tag{2.59}$$

where the prime denotes the derivative with respect to the argument. Applying Euler's equation, Eq. (2.45), to Eq. (2.57) and Eq. (2.59) with $k_\varrho = \sqrt{k^2 - k_m^2}$ results in,

$$A_n(k_m) = \frac{i\varrho_0 ck\, \dot{W}_{mn}}{\sqrt{k^2 - k_m^2} H_n'(\sqrt{k^2 - k_m^2} R)} . \tag{2.60}$$

Plugging this result into Eq. (2.58) finally yields the pressure field of an infinite-length cylindrical radiator with spatially periodic configuration, i.e.,

$$P(\varrho, \phi, z) = i\varrho_0 ck \sum_{n=-\infty}^{\infty} \sum_{m=-\infty}^{\infty} \frac{\dot{W}_{mn} H_n(\sqrt{k^2 - k_m^2}\, \varrho)}{\sqrt{k^2 - k_m^2} H_n'(\sqrt{k^2 - k_m^2}\, R)} e^{ik_m z} e^{in\phi} . \tag{2.61}$$

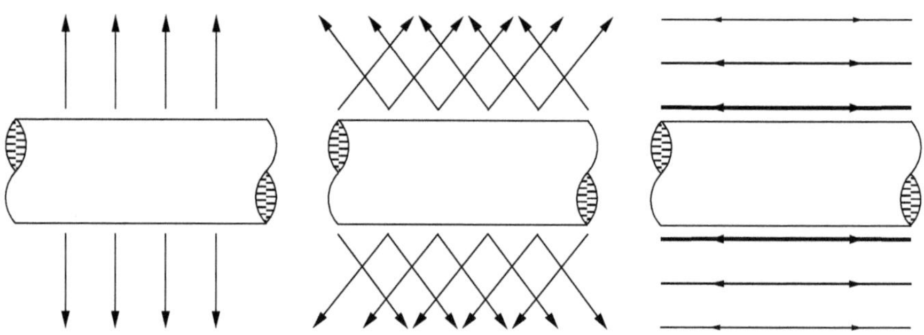

(a) cylindrical wave ($k \gg k_m$) (b) conical wave ($k > k_m$) (c) evanescent wave ($k < k_m$)

Fig. 2.5. Wave-fronts emitted by cylindrical radiators with spatially periodic particle velocity (after [JF93])

Two forms of wavefields can be deduced from Eq. (2.61), i.e. cylindrically spreading wave-fronts ($k > k_m$) as well as evanescent waves ($k < k_m$) [Wil99], where pressure decays exponentially with respect to distance ϱ, see Fig. 2.5. Cylindrical radially spreading wave-fronts can be further subdivided into purely radial ($k \gg k_m$), Fig. 2.5(a), and conically ($k > k_m$), spreading wave-fronts, Fig. 2.5(b), see [JF93]. Evanescent waves do not radiate any power into the farfield since the associated pressure field decays exponentially with growing distance from the radiator, Fig. 2.5(c). [Wil99] derives a condition for the presence of evanescent waves which is,

$$k\varrho = \frac{2\pi\varrho}{\lambda} < n. \tag{2.62}$$

The left-hand side of this inequality denotes the number of wavelengths which fit around the circumference of the wavefront at radius ϱ [Wil99].

Pressure Field from an Omnidirectionally Radiating Line Source

Here, a special case of an infinite-length cylindrical radiator with finite radius, i.e. a line source, is discussed. For simplicity, it is assumed that the surface velocity is independent of z and ϕ, i.e. the surface velocity is said to be *omnidirectional.* More complicated cases with surface velocity distributions other than of omnidirectional nature are discussed in [JF93] and [Wil99]. In this simple case, it follows that,

$$\dot{W}_{mn} = \begin{cases} \dot{W}_{00}, & m = n = 0 \\ 0, & \text{else} \end{cases} . \tag{2.63}$$

Equation (2.61) then becomes with $k_\varrho = k$,

$$P(\varrho, \phi, z) = P(\varrho) = i\varrho_0 c \frac{\dot{W}_{00} H_0(k\varrho)}{H_0'(kR)}. \tag{2.64}$$

The transition to line sources can be performed by considering the limiting case of small R. Applying Eq. (B.20) with $n = 0$ yields,

$$H_0'(kR) \approx \frac{2i}{\pi k R}, \tag{2.65}$$

and therefore the pressure of a line source can be approximated by,

$$P_L(\varrho) \triangleq P(\varrho, \phi, z) = \frac{\pi}{2} \varrho_0 c \dot{W}_{00} k R \, H_0(k\varrho) \sim H_0(k\varrho). \tag{2.66}$$

Finally, an expression for the pressure of a line source in the farfield, $\varrho \to \infty$, is given. A large argument approximation for the zeroth-order Hankel function of the first kind is with Eq. (B.24),

$$H_0(k\varrho) \approx \sqrt{\frac{2}{\pi k \varrho}} \, e^{i(k\varrho - \pi/4)}, \tag{2.67}$$

and therefore,

$$P_{L\infty}(\varrho) \approx \sqrt{\frac{k\pi}{2}} \varrho_0 c R \dot{W}_{00} e^{-i\pi/4} \cdot \sqrt{\frac{1}{\varrho}} \, e^{ik\varrho} \sim \sqrt{\frac{1}{\varrho}} \, e^{ik\varrho}. \tag{2.68}$$

Equation (2.68) is an expression for the outgoing wavefield generated by an omnidirectionally radiating, i.e. pulsating, line source in cylindrical coordinates. The factor $1/\sqrt{\varrho}$ is responsible for a 3 dB attenuation of the wavefield when the point of observation is subsequently doubled in distance from the source. This attenuation is typical for cylindrically spreading wavefields.

Figure 2.6 shows the position of a bandlimited impulse emitted by an omnidirectionally radiating line source at several time instances, $0 < t_1 <$

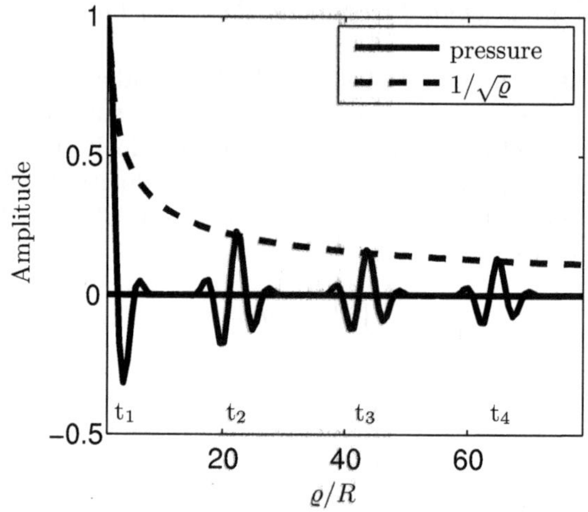

Fig. 2.6. Propagation of a cylindrically spreading wave-front at several time instances

$t_2 < t_3 < t_4$. These time instances are equivalent to the respective distance of the wavefield with respect to the radiating line source. The attenuation factor, $1/\sqrt{\varrho}$, is also shown as a dashed curve.

Note that Fig. 2.6 is valid only for a wave-front generated by a line source in the farfield since two-dimensional wave propagation, in general, not only results in an attenuation of the wave-front as it propagates but also in a modification of the shape of the wave. This can be understood by examining the inverse temporal Fourier transform of $iH_0^{(1)}(\omega\|\varrho - \varrho_0\|/c)/4$, the fundamental solution of the acoustic wave equation in two dimensions, cf. Eq. (2.36), which is [MF53],

$$\mathcal{F}_\omega\left\{\frac{i}{4}H_0^{(1)}(\omega\|\varrho - \varrho_0\|/c)\right\} = \begin{cases} 0, & t < \|\varrho - \varrho_0\|/c \\ \dfrac{1}{2\pi\sqrt{t^2 - (\|\varrho - \varrho_0\|/c)^2}}, & t > \|\varrho - \varrho_0\|/c \end{cases} . \tag{2.69}$$

As can be seen, this wave does not only have a sharp front but leaves a slowly diminishing 'wake' behind. In other words, Fig 2.6 only considers the wave-front and not the 'wake'.

The Cylindrical Radiator of Finite Length

The z-dependent component in Eq. (2.56) for cylindrical radiators of finite length is, in general, not spatially periodic. This means that k_z needs to be expressed in a continuous fashion, that is by a continuous distribution $F(k_z)$. Thus, the z-dependent part of the surface velocity, i.e. $f(z)$ in Eq. (2.56), can be represented by an inverse spatial Fourier transform with respect to k_z (see Section A.1), i.e.,

$$\dot{W}(\varrho,\phi,z) = \frac{1}{2\pi}\sum_{n=-\infty}^{\infty} \dot{W}_n e^{in\phi} \int_{-\infty}^{\infty} F(k_z) e^{ik_z z} dk_z, \tag{2.70}$$

where,

$$F(k_z) = \int_{-\infty}^{\infty} f(z)\, e^{-ik_z z}\, dz = \int_{-L}^{L} f(z)\, e^{-ik_z z}\, dz, \tag{2.71}$$

due to the finite length of the cylindrical radiator.

The pressure of a cylindrical radiator of finite length can, as for the case of infinite length, be derived by applying Euler's equation, Eq. (2.45), to Eq. (2.55) and Eq. (2.70) evaluated at $\varrho = R$. It follows that,

$$A_n(k_z) = \frac{i\varrho_0 ck\, \dot{W}_n}{\sqrt{k^2-k_z^2} H_n'(\sqrt{k^2-k_z^2} R)}. \tag{2.72}$$

Plugging this result into Eq. (2.55) finally yields,

$$P(\varrho,\phi,z) = \frac{i\varrho_0 ck}{2\pi}\sum_{n=-\infty}^{\infty} \dot{W}_n e^{in\phi} \int_{-\infty}^{\infty} \frac{H_n(\sqrt{k^2-k_z^2}\,\varrho)}{\sqrt{k^2-k_z^2} H_n'(\sqrt{k^2-k_z^2}\, R)} F(k_z) e^{ik_z z} dk_z, \tag{2.73}$$

where $F(k_z)$ is defined in Eq. (2.71).

Note that the boundary conditions introduced by the endcaps of the finite cylindrical rigid baffle have been neglected.

2.3.3 The Cylindrical Scatterer

In this section, the wavefield is not generated by the cylindrical surface but by a point source located sufficiently far from the cylinder. As a first approximation, an often quoted requirement for 'sufficiently far' is, $k\|\boldsymbol{\varrho} - \boldsymbol{\varrho}_0\| \gg \pi$. The incoming wavefield interacting with the cylinder can therefore be regarded as planar, see Section 2.2. The cylinder can be considered as an acoustic obstacle that modifies the incoming wavefield by its boundary condition. More specifically, the cylinder scatters the wavefield. Assume that a cylinder is placed into a cylindrical wavefield as shown in Fig. 2.7. Furthermore, assume that the wavefield is two-dimensional, making it independent of the z-coordinate. This assumption yields a symmetric setup that greatly simplifies further derivations.

Two types of boundary conditions are considered in this section. First and foremost, this book deals mainly with so-called rigid boundaries. A material is considered acoustically rigid if its locally reacting acoustic impedance, Z, is

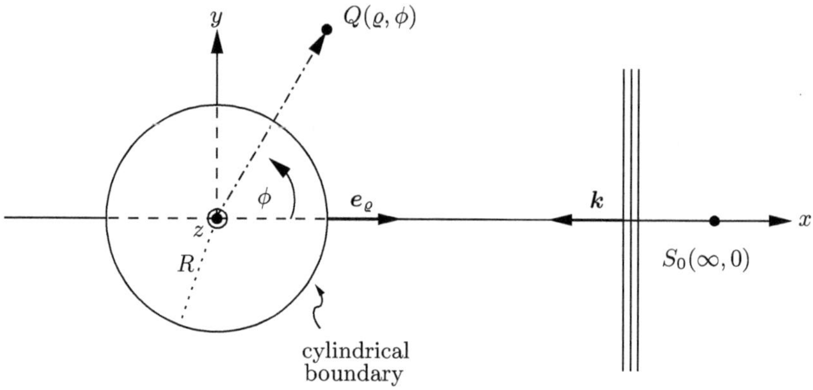

Fig. 2.7. Geometric model considered for the scattering problem for $z = 0$

much larger than the impedance of the transmitting medium. The frequency-dependent acoustic impedance leads to a boundary condition describing a surface and is defined as the ratio of the acoustic pressure, p or P, to the inward normal component of the particle velocity, $\dot{w}$ or $\dot{W}$. Assuming an airborne wavefield generated at a large distance from the cylinder, the impedance of the medium, i.e. air, is Z_0, where [JF93],

$$Z_0 = \varrho_0 c, \tag{2.74}$$

is called the *characteristic* acoustic impedance. A large acoustic impedance of a rigid scatterer means that the scattering medium is acoustically much 'harder' than the one in which the rigid scatterer is placed, i.e. its density (or speed of sound) is much higher. It can be shown that, for $kR > 1$, the pressure on the surface of the rigid scatterer where the plane-wave impinges, i.e. $(\varrho, \phi) = (R, 0)$ in Fig. 2.7, is twice its free-field value [JF93], cf. Fig. 2.8. Many structurally 'hard' materials placed in air can be considered rigid as a first approximation.

Materials that satisfy $Z \ll Z_0$ are considered acoustically soft, or 'pressure-release'. This terminology is explained by the fact that the acoustic pressure is zero at the pressure-release boundary. To a wave that is transmitted by a source located underwater, the air surface, for instance the ocean surface or bubbles in water, can be considered as acting as a pressure-release boundary. Therefore, pressure-release boundaries are mainly dealt with in the underwater literature. A more general discussion of scattering involving materials that can neither be classified as strictly acoustically rigid nor strictly pressure-release can be found in [And50].

The Infinite-Length Cylinder as a Rigid Scatterer

Figure 2.7 shows the geometry of the scattering problem under consideration where a point source, S_0, at $(\varrho, \varphi, z) = (\infty, 0, 0)$ emits a wavefield that is assumed to be planar at the surface of the rigid scatterer.

Because of the assumed symmetry of the problem – a plane-wave traveling perpendicularly to the z-axis – the incident wavefield at point $Q(\varrho, \phi)$ resulting in $k = k_\varrho$ can be written as,

$$P_{\text{inc}}(\varrho, \phi) = \hat{P}_{\text{inc}} e^{ik\varrho\cos\phi}. \tag{2.75}$$

See Section 3.1.1 for a detailed derivation of this relation. Without loss of generality the amplitude of the plane-wave, $\hat{P}_{\text{inc}}$, is assumed to equal unity in all of the subsequent discussions. The subscript 'inc' here and in the following denotes an incoming wave.

Since Eq. (2.75) is independent of z, it can be expanded in a Fourier series [AS72] as,

$$P_{\text{inc}}(\varrho, \phi) = e^{ik\varrho\cos\phi} = \sum_{n=-\infty}^{\infty} i^n J_n(k\varrho) e^{in\phi}, \tag{2.76}$$

which could have been obtained also by applying Eq. (2.44). $P_{\text{inc}}(\varrho, \phi)$ is the sound pressure at $Q(\varrho, \phi)$ if the rigid scatterer were not present. As will be shown below, this plane-wave by itself does not satisfy the boundary conditions at $\varrho = R$. Therefore, an additional wavefield, caused by the presence of the cylinder, must exist that satisfies the acoustic wave equation at that boundary. This wavefield is called the scattered wavefield, $P_{\text{scat}}(\varrho, \phi)$, that must consist of components traveling outward with respect to the scatterer. These outward traveling waves are described by Hankel functions of the first kind.

Rigid boundaries result in the particle velocity having a zero component in the direction of the surface normal, $\boldsymbol{e}_\varrho$, i.e. [Wil99],

$$\dot{W}_{\text{scat}}(\varrho, \phi)|_{\varrho=R} + \dot{W}_{\text{inc}}(\varrho, \phi)|_{\varrho=R} = 0, \tag{2.77}$$

where the subscript 'scat' here and in the following denotes the scattered wave. According to Eq. (2.45), it follows that,

$$i\varrho_0 ck\dot{W}_{\text{inc}}(\varrho, \phi)|_{\varrho=R} = \frac{\partial P_{\text{inc}}(\varrho, \phi)}{\partial \varrho}|_{\varrho=R}. \tag{2.78}$$

Keeping in mind the symmetry of the problem and evaluating Eq. (2.78) with Eq. (2.76) one obtains with Eq. (2.77),

$$\dot{W}_{\text{scat}}(R, \phi) = -\frac{1}{i\varrho_0 c} \sum_{n=-\infty}^{\infty} i^n J_n'(kR) e^{in\phi}. \tag{2.79}$$

Since this velocity distribution on the cylindrical surface is independent of z, it follows from Eq. (2.57) that,

$$\dot{W}_{\text{scat}_{mn}} = \begin{cases} -\frac{1}{i\varrho_0 c} i^n J_n'(kR), & m = 0 \\ 0, & m \neq 0 \end{cases}. \tag{2.80}$$

The scattered wavefield due to the rigid cylinder can be obtained by plugging Eq. (2.80) into Eq. (2.61) to yield,

$$P_{\text{scat}}(\varrho, \phi) = -\sum_{n=-\infty}^{\infty} i^n \frac{J_n'(kR) H_n(k\varrho)}{H_n'(kR)} e^{in\phi}. \tag{2.81}$$

Finally, one obtains for the total wavefield,

$$P_{\text{tot}}(\varrho, \phi) = P_{\text{inc}}(\varrho, \phi) + P_{\text{scat}}(\varrho, \phi) = \sum_{n=-\infty}^{\infty} i^n \left[J_n(k\varrho) - \frac{J_n'(kR) H_n(k\varrho)}{H_n'(kR)} \right] e^{in\phi}, \tag{2.82}$$

where the subscript 'tot' here and in the following denotes the resulting (total) wavefield, i.e. the superposition of the incoming and scattered wave.

Figure 2.8 is a pictorial representation of Eq. (2.82), the total wavefield in the proximity of an infinite-length rigid cylindrical scatterer with $R = 0.04$ m due to a plane-wave incidence, in the time-domain. The area of the rigid scatterer is shown as a white disk since it is assumed that no wavefield is generated inside the scatterer, which results from the fourth assumption stated in the introductory paragraph in this chapter. The upper left snapshot depicts the incoming bandlimited plane-wave according to the setup shown in Fig. 2.7 before hitting the scatterer. For generating the plots, the temporal sampling rate was chosen to be 48 kHz, corresponding to a bandwidth of the plane-wave of 24 kHz. The upper right part of the figure shows a snapshot where the plane-wave has just hit the scatterer. The cylindrically spreading reflected wave is clearly visible. Also, it can be seen that the pressure on the cylinder surface is higher than anywhere else which is due to the specular reflection off the surface. The lower left sub-figure shows a snapshot where the incoming plane-wave has just passed the cylinder. Visible are the reflected wave as well as the effect of the acoustic shadow formed by the cylinder. Due to the presence of the cylinder, the shape of the plane-wave has changed in the close proximity of the cylinder, the gap being partly filled in by diffraction from the incident plane-wave. As time proceeds further, diffraction continues to fill in the gap in the plane-wave, thereby 'healing' the wave-front [Feu04]. This phenomenon is indicated in the lower right sub-figure. It can also be seen that this diffraction phenomenon generates a circumferential wave that propagates along the surface of the cylinder and that eventually breaks off. Figure 2.8 shows the scattering effect for an impulsive plane-wave incidence, which corresponds to a broadband signal. Scattering, however, is a frequency-dependent phenomenon that exhibits peculiarities for certain ranges of frequencies compared to the

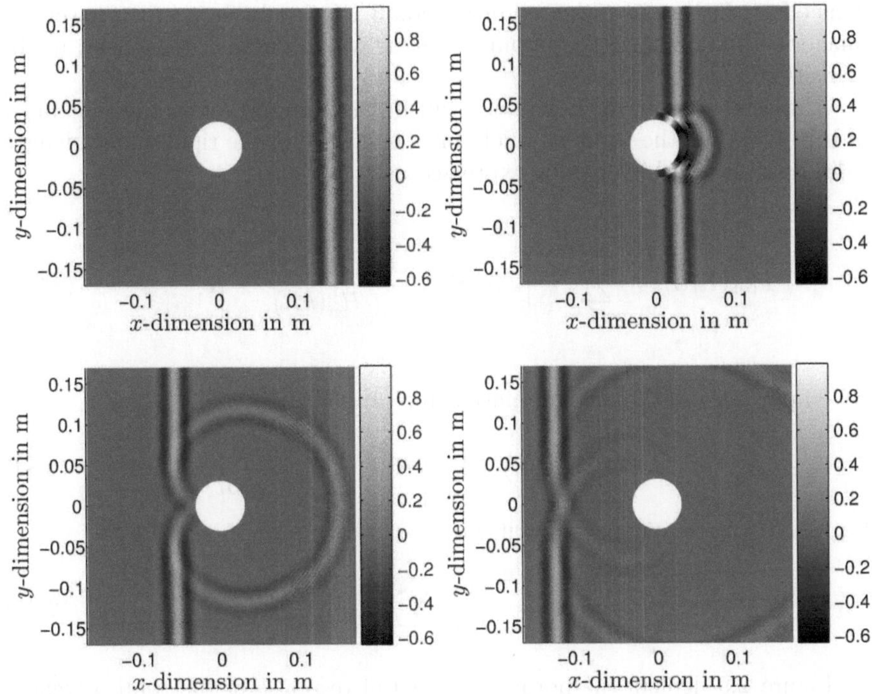

Fig. 2.8. Total wavefield in the proximity of an infinite-length rigid cylindrical scatterer due to an incident plane-wave

object's radius [ÜDM66] and, therefore, behaves in a more complicated manner than indicated in Fig. 2.8. The first frequency region satisfies $kR \ll 1$ and is known as the Rayleigh region. Here, the scattered wave propagates in all directions and no shadow behind the scatterer is formed. Only a few terms in Eq. (2.82) are necessary to describe these waves. Frequently, $n = 0, 1$ is deemed to be sufficient [JF93]. The second case is the region of geometrical optics, $kR \gg 1$. Put somewhat informally, some of the energy contained in the scattered wave spreads out uniformly from the scatterer while some of the energy interferes destructively behind the scatterer, forming an acoustic shadow. In the transition region, where $kR \sim 1$, it is shown in [FD52] that incident plane-waves are not only reflected in the geometrical optics sense, but also enter the shadow region of the scatterer along the surface and continue propagating along the surface with reduced phase velocity and decreasing amplitude. This decrease in velocity and amplitude is frequency-dependent. It is said that the waves *creep* along the surface of the scatterer. These tangentially radiating waves, caused by the zeros of the Hankel functions [HS69], result therefore in evanescent waves that attenuate quickly. It follows from the nature of creeping waves that the scatterer appears to be larger in diameter.

This effect, further investigated in Section 3.1.1, can be seen in the lower left image in Fig. 2.8 since the impulsive plane-wave contains frequencies in the transition region.

A special solution of Eq. (2.82) will be very important in the remainder of this work, i.e. the total wavefield on the surface of the rigid infinite-length cylinder, $\varrho = R$, which can be expressed as,

$$P_{\text{tot}}(R,\phi) = \sum_{n=-\infty}^{\infty} i^n \left[J_n(kR) - \frac{J_n^{'}(kR)H_n(kR)}{H_n^{'}(kR)} \right] e^{in\phi}. \tag{2.83}$$

By employing the Wronskian relationship [JF93],

$$J_n(kR)Y_n^{'}(kR) - Y_n(kR)J_n^{'}(kR) = \frac{2}{\pi kR}, \tag{2.84}$$

Eq. (2.83) can be compactly written as,

$$P_{\text{tot}}(R,\phi) = \frac{2i}{\pi kR} \sum_{n=-\infty}^{\infty} \frac{i^n}{H_n^{'}(kR)} e^{in\phi}. \tag{2.85}$$

Figure 2.9 depicts the normalized spatial response of the total wavefield due to a single plane-wave impinging from $\varphi = 0 \ldots 2\pi$ at a point-like omnidirectional receiver mounted at $(r, \phi, z) = (R, 0, 0)$ into the infinite-length rigid cylindrical baffle of Fig. 2.8. It can be seen that the spatial selectivity depends on the direction of the impinging plane-wave, in the extreme case resulting in the above discussed shadow zone corresponding to the minima in the spatial response. This phenomenon does not occur when the point-like receiver is not mounted into a rigid cylindrical baffle. The spatial selectivity is dependent on kR as well, as a result of the presence of creeping waves [Wie47]. Note that Fig 2.9(b) is merely an excerpt of Fig 2.9(a) covering a range of kR that will be of importance in later chapters.

Figure 2.10 shows the pressure distribution in the time-domain along a circular region in space, with $R = 0.04$ m, due to a bandlimited plane-wave incident from $\varphi = 0$. As before, the temporal sampling rate was chosen to be 48 kHz, corresponding to a bandwidth of the plane-wave of 24 kHz. Figure 2.10(a) depicts the case where no boundary conditions are imposed (case 1), while Fig. 2.10(b) depicts the case where a rigid boundary condition at $R = 0.04$ m is assumed (case 2). Figure 2.10(b) shows how the pressure distribution changes as the plane-wave travels along the rigid boundary, cf. Fig. 2.8. It can be seen that the pressure at the angle of incidence, $\phi = 0$, is higher in case 2 than in case 1, as a result of the amplitude increase due to specular reflection. Also, the amplitude decreases as the wave travels along the boundary in case 2. As expected, no amplitude decrease is observed in case 1. Both aforementioned effects of creeping waves, the presence of circumferential

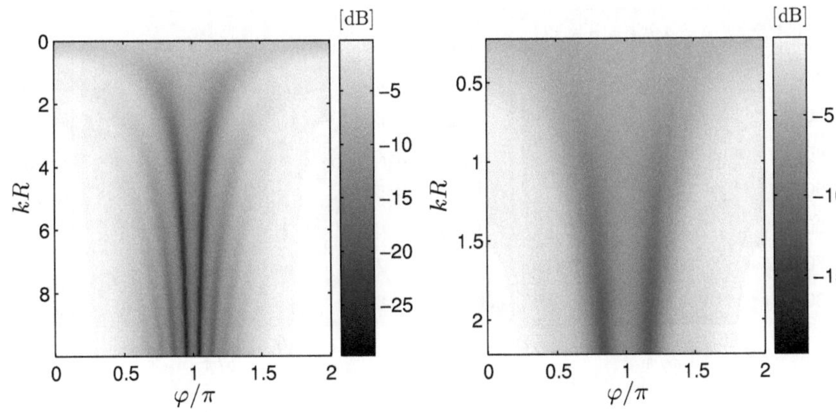

(a) $kR = 0 \ldots 10$ (b) $kR = 0.22 \ldots 2.2$

Fig. 2.9. Spatial response of the total wavefield due to a single plane-wave impinging from $\varphi = 0 \ldots 2\pi$ at a point-like receiver positioned at $(r, \phi, z) = (R, 0, 0)$ mounted in an infinite-length rigid cylindrical baffle

waves that eventually break off and the fact that the rigid boundary appears to be larger in diameter, can be verified in this figure. The circumferential waves correspond to the two 'lobes' that originate from $\phi = \pi$ in Fig. 2.10(b). The larger diameter can be interpreted as the longer response time (in terms of samples) the wave causes as it travels along the circular region in space.

The Finite-Length Cylinder as a Rigid Scatterer

It is again assumed that the wavefield impinges the finite-length rigid scatterer normally with respect to the cylindrical axis, cf. Fig. 2.7. Then, after neglecting the boundary conditions at the endcaps and setting $f(z) = 1$ in Eq. (2.71), the scattered field generated by a cylinder of finite length $2L$ can be approximated by restricting the velocity distribution to the length of the cylinder, i.e.,

$$\dot{W}_{\text{scat}}(R, \phi, z) = \begin{cases} \sum\limits_{n=-\infty}^{\infty} \dot{W}_n e^{in\phi}, & |z| < L \\ 0, & |z| > L \end{cases}, \tag{2.86}$$

where it follows from Eq. (2.80) that,

$$\dot{W}_n = -\frac{1}{i\varrho_0 c} i^n J_n'(kR). \tag{2.87}$$

The spatial Fourier transform of Eq. (2.86) yields,

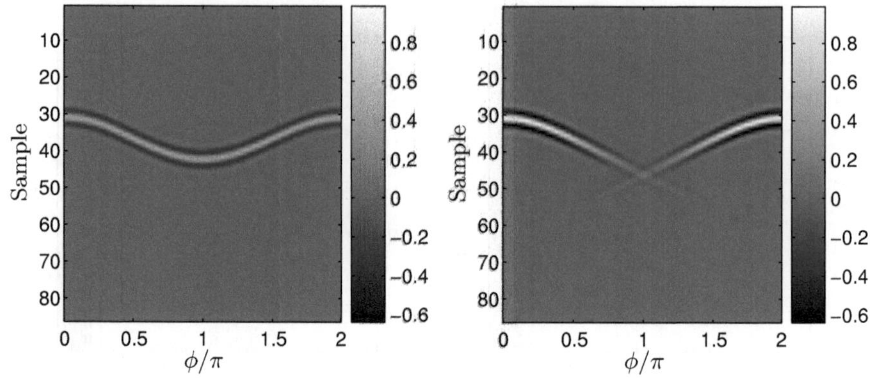

(a) no boundary

(b) rigid boundary condition at $R =$ 0.04 m

Fig. 2.10. Pressure distribution in the time-domain along a circular region in space, with $R = 0.04$ m, due to a bandlimited plane-wave incident from $\varphi = 0$

$$
\begin{aligned}
\dot{W}_{\mathrm{scat}}(R, \phi, k_z) &= \int_{-\infty}^{\infty} \dot{W}_{\mathrm{scat}}(R, \phi, z) e^{-ik_z z}\, dz \\
&= \sum_{n=-\infty}^{\infty} \dot{W}_n e^{in\phi} \int_{-L}^{L} e^{-ik_z z}\, dz \\
&= \sum_{n=-\infty}^{\infty} \dot{W}_n e^{in\phi} \underbrace{2L \,\mathrm{sinc}(k_z L)}_{F(k_z)},
\end{aligned}
\tag{2.88}
$$

where in the last step Eq. (2.70) has been used and where $\mathrm{sinc}(k_z L) = \sin(k_z L)/(k_z L)$. The scattered wavefield produced by a finite-length rigid cylinder is therefore with Eq. (2.73), Eq. (2.87), and Eq. (2.88),

$$
\begin{aligned}
P_{\mathrm{scat}}(\varrho, \phi, z) = &-\frac{kL}{\pi} \sum_{n=-\infty}^{\infty} i^n J_n'(kR) e^{in\phi} \\
&\times \int_{-\infty}^{\infty} \frac{H_n(\sqrt{k^2 - k_z^2}\,\varrho)\,\mathrm{sinc}(k_z L)}{\sqrt{k^2 - k_z^2} H_n'(\sqrt{k^2 - k_z^2}\, R)} e^{ik_z z}\, dk_z .
\end{aligned}
\tag{2.89}
$$

One can easily show that the scattered wavefield due to the presence of an infinite-length rigid cylinder, Eq. (2.81), is a special case of Eq. (2.89) by realizing that in Eq. (2.88),

$$
F(k_z) = \int_{-\infty}^{\infty} e^{-ik_z z}\, dz = 2\pi\delta(k_z), \tag{2.90}
$$

and by applying this result to Eq. (2.73).

Finally, one obtains for the total pressure,

$$P_{\text{tot}}(\varrho,\phi) = P_{\text{inc}}(\varrho,\phi) + P_{\text{scat}}(\varrho,\phi,z), \tag{2.91}$$

and therefore,

$$P_{\text{tot}}(\varrho,\phi,z) = \sum_{n=-\infty}^{\infty} i^n J_n(k\varrho)e^{in\phi} - \frac{kL}{\pi}\sum_{n=-\infty}^{\infty} i^n J_n'(kR)e^{in\phi}\int_{-\infty}^{\infty}\frac{H_n(\sqrt{k^2-k_z^2}\,\varrho)\,\text{sinc}(k_zL)}{\sqrt{k^2-k_z^2}H_n'(\sqrt{k^2-k_z^2}\,R)}e^{ik_zz}\,dk_z. \tag{2.92}$$

A special case that is of central importance in this book is the pressure on the surface of a finite-length rigid cylinder in the xy-plane, i.e. $z = 0$, which follows as,

$$P_{\text{tot}}(R,\phi) = \sum_{n=-\infty}^{\infty} i^n J_n(kR)e^{in\phi} - \frac{kL}{\pi}\sum_{n=-\infty}^{\infty} i^n J_n'(kR)e^{in\phi}\int_{-\infty}^{\infty}\frac{H_n(\sqrt{k^2-k_z^2}\,R)\,\text{sinc}(k_zL)}{\sqrt{k^2-k_z^2}H_n'(\sqrt{k^2-k_z^2}\,R)}\,dk_z. \tag{2.93}$$

The total wavefield in the proximity of a rigid cylindrical scatterer of finite length due to a normally incident plane-wave qualitatively exhibits the same behavior as the rigid cylindrical scatterer of infinite length in the xy-plane, see Fig. 2.8. However, this statement is true only if the cylinder is 'sufficiently long', i.e. the boundary conditions that are associated with the endcaps of the cylinder do not contribute significantly to the pressure in the xy-plane. It is assumed throughout this work that this condition is fulfilled. More discussion on 'sufficiently long' cylinders is provided in Section 3.1.1.

Note that Eq. (2.83) follows directly from Eq. (2.93) for infinitely long cylindrical scatterers since [Wei03],

$$\lim_{L\to\infty}\frac{L}{\pi}\,\text{sinc}(k_zL) = \delta(k_z). \tag{2.94}$$

Infinite-Length Cylindrical Scatterer with Pressure-Release Boundaries

Equation (2.77) describes the boundary conditions of materials that can be considered acoustically rigid. A second class of materials are considered pressure-release [Wil99], i.e. they require the pressure to be zero on its surface, i.e.,

$$P_{\text{tot}}(\varrho,\phi)|_{\varrho=R} = P_{\text{scat}}(\varrho,\phi)|_{\varrho=R} + P_{\text{inc}}(\varrho,\phi)|_{\varrho=R} = 0. \tag{2.95}$$

Now, consider the same setup as in Fig. 2.7 where the only difference is that the boundary satisfies the pressure-release boundary condition, Eq. (2.95). Therefore, it follows for the scattered wavefield, corresponding to Eq. (2.81),

$$P_{\text{scat}}(\varrho,\phi) = \sum_{n=-\infty}^{\infty} A_n(kR)H_n(k\varrho)e^{in\phi}. \tag{2.96}$$

Introducing Eq. (2.76) into Eq. (2.95) yields,

$$\sum_{n=-\infty}^{\infty} A_n(kR)H_n(kR)e^{in\phi} + \sum_{n=-\infty}^{\infty} i^n J_n(kR)e^{in\phi} = 0, \tag{2.97}$$

resulting in,

$$A_n(kR) = -i^n \frac{J_n(kR)}{H_n(kR)}. \tag{2.98}$$

Finally, the total wavefield due to an infinite-length cylinder satisfying the pressure-release boundary condition can be written as,

$$P_{\text{tot}}(\varrho,\phi) = \sum_{n=-\infty}^{\infty} i^n \left[J_n(k\varrho) - \frac{J_n(kR)}{H_n(kR)} H_n(k\varrho) \right] e^{in\phi}. \tag{2.99}$$

It follows immediately that $P_{\text{tot}}(R,\phi) = 0$, thus satisfying Eq. (2.95).

Figure 2.11 shows the total wavefield in the proximity of a pressure-release cylindrical scatterer due to an incident plane-wave. The snapshots were chosen such that a direct comparison with the rigid cylindrical boundary, presented in Fig. 2.8, is possible. An important difference is the fact that no strong surface waves are generated. However, it has been shown in [ÜDM66] that surface waves are expected exhibiting an amplitude that is much weaker than in the case of a rigid scatterer. Therefore, these waves are not visible in the linearly scaled plots in Fig. 2.11. Another difference with respect to rigid scatterers is the fact that the reflected cylindrical wavefield experiences a phase change due to the requirements of the boundary condition. Since the incoming wavefield has a positive peak, the scattered one must have a negative peak in order to cancel the total wavefield on the cylinder surface.

As stated before, conditions that require to model the scatterer as pressure-release are mainly present in underwater acoustics and are not further pursued in the following.

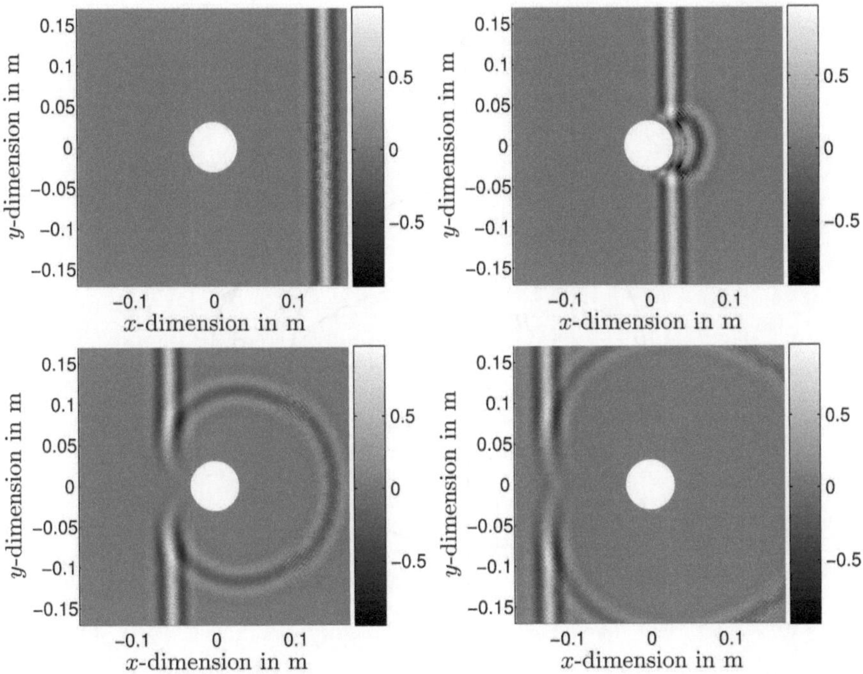

Fig. 2.11. Total wavefield in the proximity of a pressure-release cylindrical scatterer due to an incident plane-wave

2.4 Acoustic Wave Equation in Spherical Coordinates

This section examines wavefields in spherical coordinates. Both wavefields produced by a spherically shaped object and modifications of wavefields due to the presence of a spherical object in a wavefield are discussed. Section 2.4.1 presents a general solution of the homogeneous Helmholtz equation. Section 2.4.2 detail the wavefield of a spherical radiator and scattering phenomena of spherical objects are outlined in Section 2.4.3.

2.4.1 General Solution of the Acoustic Wave Equation in Spherical Coordinates

The spherical coordinate system and the geometric model of a spherical radiator/scatterer with radius R is shown in Fig. 2.12. The relation of the spherical coordinate system with respect to the Cartesian coordinate system was given in Eq. (2.3). The field point $Q(r, \theta, \phi)$, located on the spherical surface, is also depicted. Transferring the results from Section 2.1 to the particular case of spherical coordinate systems gives for Euler's equation, Eq. (2.13),

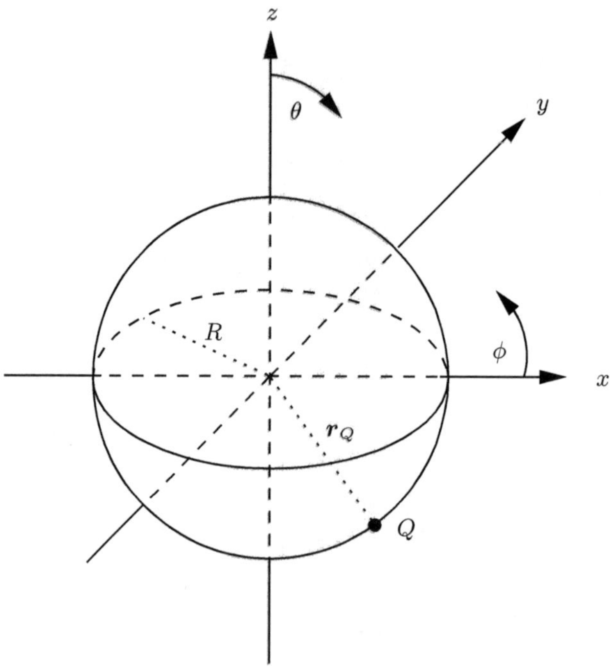

Fig. 2.12. Geometric model of a spherical radiator/scatterer

$$\dot{W}(r,\theta,\phi,\omega) = \frac{1}{i\varrho_0 ck}\boldsymbol{\nabla}P(r,\theta,\phi,\omega), \tag{2.100}$$

where with Eq. (2.7), Eq. (2.3), and Eq. (2.8),

$$\boldsymbol{\nabla} = \frac{\partial}{\partial r}\boldsymbol{e}_r + \frac{1}{r}\frac{\partial}{\partial\theta}\boldsymbol{e}_\theta + \frac{1}{r\sin\theta}\frac{\partial}{\partial\phi}\boldsymbol{e}_\phi. \tag{2.101}$$

For the acoustic wave equation in the frequency-domain, i.e. the homogeneous Helmholtz equation, cf. Eq. (2.16), it follows that,

$$\boldsymbol{\nabla}^2 P(r,\theta,\phi,\omega) + k^2 P(r,\theta,\phi,\omega) = 0, \tag{2.102}$$

with, cf. Eq. (2.15) and Eq. (2.8),

$$\boldsymbol{\nabla}^2 = \frac{1}{r^2}\frac{\partial}{\partial r}\left(r^2\frac{\partial}{\partial r}\right) + \frac{1}{r^2\sin\theta}\frac{\partial}{\partial\theta}\left(\sin\theta\frac{\partial}{\partial\theta}\right) + \frac{1}{r^2\sin^2\theta}\frac{\partial^2}{\partial\phi^2}. \tag{2.103}$$

Subsequently, steady-state conditions are assumed and the explicit time-dependence is dropped for notational convenience in further developments. The steady-state solution of the acoustic wave equation in spherical coordinates can be shown to be [Wil99],

$$P(r, \theta, \phi) = \sum_{n=0}^{\infty} \sum_{m=-n}^{n} \left[A_{mn} h_n^{(1)}(kr) + B_{mn} h_n^{(2)}(kr) \right] Y_n^m(\theta, \phi). \tag{2.104}$$

By defining the time-dependence to be $e^{-i\omega t}$ rather than $e^{+i\omega t}$ in the time-domain formulation of the displacement vector, cf. Section 2.1, the spherical Hankel function of the first kind corresponds to outgoing spherical waves while the spherical Hankel function of the second kind corresponds to incoming spherical waves with respect to the coordinate origin.

By expressing the spherical Hankel functions, $h_n^{(1,2)}(kr)$, with spherical Bessel, $j_n(kr)$, and spherical Neumann functions, $y_n(kr)$, according to Eq. (2.22), Eq. (2.104) can be written as,

$$P(r, \theta, \phi) = \sum_{n=0}^{\infty} \sum_{m=-n}^{n} \left[C_{mn} j_n(kr) + D_{mn} y_n(kr) \right] Y_n^m(\theta, \phi), \tag{2.105}$$

where,

$$C_{mn} = A_{mn} + B_{mn}, \tag{2.106}$$
$$D_{mn} = i(A_{mn} - B_{mn}). \tag{2.107}$$

The terms A_{mn} and B_{mn} are defined by the boundary values of the problem.

As in the cylindrical case, see Section 2.3.1, two important boundary value problems can be identified, i.e. the *interior* and the *exterior* boundary value problem. Following the same steps as presented for the discussion of Fig. 2.4 in cylindrical coordinates the general solution of the homogeneous wave equation in spherical coordinates for interior boundary value problems can be written as,

$$P(r, \theta, \phi) = \sum_{n=0}^{\infty} \sum_{m=-n}^{n} C_{mn} j_n(kr) Y_n^m(\theta, \phi), \tag{2.108}$$

and for exterior boundary value problems,

$$\begin{aligned} P(r, \theta, \phi) &= \sum_{n=0}^{\infty} \sum_{m=-n}^{n} A_{mn} h_n^{(1)}(kr) Y_n^m(\theta, \phi) \\ &\triangleq \sum_{n=0}^{\infty} \sum_{m=-n}^{n} A_{mn} h_n(kr) Y_n^m(\theta, \phi). \end{aligned} \tag{2.109}$$

Since most of the following discussions will not involve spherical Hankel functions of the second kind, the superscript $^{(1)}$ will be omitted in all further derivations, until noted otherwise.

2.4.2 The Spherical Radiator

Following the discussions of Section 2.3.2, most surface velocity distributions on spherical boundaries can be represented by double series as, cf. Section 2.2.2,

$$\dot{W}(\theta,\phi) = \sum_{n=0}^{\infty} \sum_{m=-n}^{n} \dot{W}_{mn} Y_n^m(\theta,\phi). \tag{2.110}$$

Applying this velocity distribution to exterior boundary value problems it can be shown that with Eq. (2.100) and Eq. (2.109),

$$A_{mn} = \frac{i\varrho_0 c \dot{W}_{mn}}{h'_n(kR)}, \tag{2.111}$$

and therefore for the pressure distribution due to a spherical radiator of radius $r = R$, it follows that,

$$P(r,\theta,\phi) = i\varrho_0 c \sum_{n=0}^{\infty} \sum_{m=-n}^{n} \frac{\dot{W}_{mn}\, h_n(kr)}{h'_n(kR)} Y_n^m(\theta,\phi). \tag{2.112}$$

Note that Eq. (2.112) is valid only for $r \geq R$.

The pressure field of a general spherical source in the θ- and ϕ-direction is determined by the strength of the respective spherical harmonic, $Y_n^m(\theta,\phi)$, in Eq. (2.112). The subscript n represents the number of nodal circles coaxial with the $\theta = 0$ axis. A nodal circle is a region on the sphere which does not move for a particular mode of vibration. The superscript m dictates the wavelength in the ϕ-direction and represents the number of nodal circles that are great circles – defined as circles having the same radius as the spherical source – intersecting the $\theta = 0$ axis [JF93]. Images of the spherical harmonics that should clarify these statements are provided in Section 3.2. Configurations with spherical symmetry, i.e. $n = m = 0$ are considered next.

Pressure Field from an Omnidirectionally Radiating Spherical Source

In the following, a special case of a spherical radiator of vanishing radius, i.e. a point source, is discussed. First, it is assumed that the surface velocity is independent of θ and ϕ. More complicated cases with surface velocity distributions other than of omnidirectional nature are discussed in [JF93]. In this simple case, the boundary condition on the spherical surface is,

$$\dot{W}_{mn} = \begin{cases} \dot{W}_{00}, & m = n = 0 \\ 0, & \text{else} \end{cases}. \tag{2.113}$$

Equation (2.112) then becomes,

$$P(r,\theta,\phi) = i\varrho_0 c \frac{\dot{W}_{00}\, h_0(kr)}{h'_0(kR)} Y_0^0(\theta,\phi). \tag{2.114}$$

Since $Y_0^0(\theta,\phi) = 1/\sqrt{4\pi}$, see Tab. B.2, Eq. (2.114) can be simplified to yield,

$$P(r,\theta,\phi) = \frac{i\varrho_0 c \dot{W}_{00}}{\sqrt{4\pi}} \frac{h_0(kr)}{h'_0(kR)} \sim h_0(kr). \tag{2.115}$$

In the farfield, $r \to \infty$, from Eq. (B.43) it follows for the spherical Hankel function of the first kind that,

$$h_0(kr) \approx \frac{-i}{kr}\, e^{ikr}, \tag{2.116}$$

and therefore,

$$P_P(r) \approx \frac{\varrho_0 c \dot{W}_{00}}{k\sqrt{4\pi} h_0'(kR)} \frac{1}{r}\, e^{ikr} \sim \frac{1}{r}\, e^{ikr}. \tag{2.117}$$

Equation (2.117) is an expression for the outgoing wavefield generated by an omnidirectionally radiating, i.e. pulsating, spherical source in spherical coordinates. The factor $1/r$ is responsible for an 6 dB attenuation of the wavefield when the point of observation is subsequently doubled in distance from the source. This is in contrast to the attenuation factor that occurs in cylindrical wavefields, see Section 2.3.2. Note that Eq. (2.117) describes the pressure field of a point source for small radius R, in accordance with the point source solution of the inhomogeneous Helmholtz equation, Eq. (2.19).

Figure 2.13 shows the position of a bandlimited impulse emitted by an omnidirectionally radiating spherical source at four time instances, $0 < t_1 < t_2 < t_3 < t_4$. These time instances are equivalent to the respective distance of the wavefield to the radiating source. The attenuation factor, $1/r$, is also shown as a dashed curve.

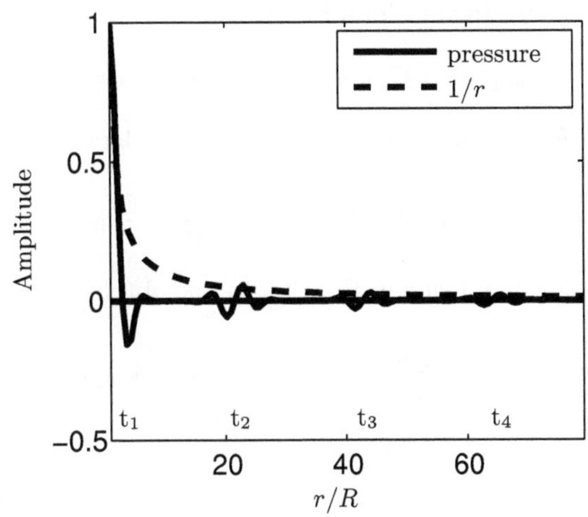

Fig. 2.13. Propagation of a spherically spreading wave-front at several time instances

The shape of the wave-fronts in Fig. 2.13 are, in contrast to the cylindrical radiator, valid for a wave-front generated by a point source in the nearfield as well as in the farfield since three-dimensional wave propagation only results in an attenuation of the wave-front as it propagates. This phenomenon can be understood by considering the inverse temporal Fourier transform of the fundamental solution of the acoustic wave equation in three dimensions, Eq. (2.19) or Eq. (2.20), which is a time-shifted Dirac function.

2.4.3 The Spherical Scatterer

The model under consideration for the spherical scatterer can be visualized using Fig. 2.7 for the special case of normal plane-wave incidence. A plane-wave coming from any direction (ϑ, φ) can be expanded into spherical harmonics as,

$$P_{\text{inc}}(r,\theta,\phi) = 4\pi \sum_{n=0}^{\infty} i^n j_n(kr) \sum_{m=-n}^{n} Y_n^m(\theta,\phi) Y_n^m(\vartheta,\varphi)^*, \tag{2.118}$$

see Eq. (2.35). For the total wavefield, the following relation applies,

$$P_{\text{tot}}(r,\theta,\phi) = P_{\text{inc}}(r,\theta,\phi) + P_{\text{scat}}(r,\theta,\phi). \tag{2.119}$$

The Rigid Spherical Scatterer

As in the cylindrical case, a rigid boundary requires the normal component of the particle velocity to vanish on the surface, i.e.,

$$\dot{W}_{\text{inc}}(r,\theta,\phi)|_{r=R} + \dot{W}_{\text{scat}}(r,\theta,\phi)|_{r=R} = 0. \tag{2.120}$$

Euler's equation for the incoming wave on the boundary is,

$$i\varrho_0 ck \dot{W}_{\text{inc}}(r,\theta,\phi)|_{r=R} = \frac{\partial P_{\text{inc}}(r,\theta,\phi)}{\partial r}|_{r=R}. \tag{2.121}$$

Evaluating the right-hand side of Eq. (2.121) one obtains with Eq. (2.118),

$$\frac{\partial P_{\text{inc}}(r,\theta,\phi)}{\partial r}|_{r=R} = 4\pi k \sum_{n=0}^{\infty} i^n j_n'(kR) \sum_{m=-n}^{n} Y_n^m(\theta,\phi) Y_n^m(\vartheta,\varphi)^*. \tag{2.122}$$

Therefore with Eq. (2.120) it follows that,

$$\begin{aligned} \dot{W}_{\text{scat}}(r,\theta,\phi)|_{r=R} &= -\dot{W}_{\text{inc}}(r,\theta,\phi)|_{r=R} \\ &= -\frac{4\pi}{i\varrho_0 c} \sum_{n=0}^{\infty} i^n j_n'(kR) \sum_{m=-n}^{n} Y_n^m(\theta,\phi) Y_n^m(\vartheta,\varphi)^*. \end{aligned} \tag{2.123}$$

Realizing that the scattered wavefield is equivalent to a wavefield produced by a spherical radiator it follows with Eq. (2.110) that,

$$\dot{W}_{\text{scat}_{mn}} = -i^n \frac{4\pi}{i\varrho_0 c} j_n^{'}(kR) Y_n^m(\vartheta, \varphi)^*, \tag{2.124}$$

and with Eq. (2.112),

$$P_{\text{scat}}(r, \theta, \phi) = -4\pi \sum_{n=0}^{\infty} i^n \frac{j_n^{'}(kR) h_n(kr)}{h_n^{'}(kR)} \sum_{m=-n}^{n} Y_n^m(\theta, \phi) Y_n^m(\vartheta, \varphi)^*. \tag{2.125}$$

The total wavefield due to the presence of a rigid spherical scatterer is therefore with Eq. (2.119),

$$P_{\text{tot}}(r, \theta, \phi) = 4\pi \sum_{n=0}^{\infty} i^n \left[j_n(kr) - \frac{j_n^{'}(kR) h_n(kr)}{h_n^{'}(kR)} \right] \sum_{m=-n}^{n} Y_n^m(\theta, \phi) Y_n^m(\vartheta, \varphi)^*. \tag{2.126}$$

Due to the symmetry of the spherical scatterer the wavefield *qualitatively* equals the one derived in Section 2.3.3 for the infinite-length cylindrical scatterer with normal plane-wave incidence. The total wavefield in the proximity of a rigid spherical scatterer looks like the one shown in Fig. 2.8 and, thus, is not reproduced here.

Again, this book is concerned with the total pressure *on* the spherical surface, i.e.,

$$P_{\text{tot}}(R, \theta, \phi) = 4\pi \sum_{n=0}^{\infty} i^n \left[j_n(kR) - \frac{j_n^{'}(kR) h_n(kR)}{h_n^{'}(kR)} \right] \sum_{m=-n}^{n} Y_n^m(\theta, \phi) Y_n^m(\vartheta, \varphi)^*. \tag{2.127}$$

This result can be expressed in a condensed way using the Wronskian relation [JF93],

$$j_n(kR) h_n^{'}(kR) - j_n^{'}(kR) h_n(kR) = \frac{i}{(kR)^2}, \tag{2.128}$$

to finally yield,

$$P_{\text{tot}}(R, \theta, \phi) = \frac{4\pi i}{(kR)^2} \sum_{n=0}^{\infty} \frac{i^n}{h_n^{'}(kR)} \sum_{m=-n}^{n} Y_n^m(\theta, \phi) Y_n^m(\vartheta, \varphi)^*. \tag{2.129}$$

The spatial response of the total wavefield due to a single plane-wave impinging from $\varphi = 0 \ldots 2\pi$ and $\vartheta = \pi/2$ at a point-like receiver positioned at $(r, \theta, \phi) = (R, \pi/2, 0)$ in a rigid spherical baffle is qualitatively similar to the result for infinite-length cylindrical baffles, see Fig. 2.9. Quantitatively, the difference is that plane-waves impinging from the opposite side w.r.t. the point-like receiver result in a higher-amplitude total response. This is due to the so-called *bright spot* exhibited by spherical scatterers [Wie47].

Note that all results presented in Section 2.3.3 can be qualitatively carried over to the spherical case and are therefore not further extended in more detail here.

The Pressure-Release Spherical Scatterer

As discussed in the cylindrical case, pressure-release boundaries require the pressure to be zero on the spherical surface in order to satisfy the boundary condition, i.e.,

$$P_{\text{scat}}(r,\theta,\phi)|_{r=R} + P_{\text{inc}}(r,\theta,\phi)|_{r=R} = 0. \tag{2.130}$$

By expressing the scattered wavefield using Eq. (2.109) and the incoming wavefield using Eq. (2.118) it can be easily shown that it follows for the total pressure due to the presence of a pressure-release spherical scatterer that,

$$P_{\text{tot}}(r,\theta,\phi) = 4\pi \sum_{n=0}^{\infty} i^n \left[j_n(kr) - \frac{j_n(kR)h_n(kr)}{h_n(kR)} \right] \sum_{m=-n}^{n} Y_n^m(\theta,\phi) Y_n^m(\vartheta,\varphi)^*, \tag{2.131}$$

which, again, vanishes on the spherical surface, as expected. The total wavefield close to the pressure-release spherical scatterer due to a plane-wave incidence looks like the one shown in Fig. 2.11 and is therefore not reproduced here.

3

Wavefield Decomposition

This chapter deals with wavefield decomposition making extensive use of the theoretical foundations of acoustic wavefields laid out in Chapter 2. Wavefield decomposition (WFD), as treated in this book, is a technique that decomposes a wavefield into spatially orthogonal eigen-solutions of the acoustic wave equation in a coordinate system that best suits the geometry of the aperture under consideration. Here, apertures are spatially distributed observation devices for analyzing the characteristics of the spatio-temporal nature exhibited by wavefields. This aperture may be a continuous processor in three-dimensional space. Alternatively, this continuous aperture may be sampled by discrete points in space. The resulting processor is called an *array*.

Widely used apertures include linear, planar, circular, and spherical apertures. For WFD methods, two apertures are of particular interest. The first geometry of interest in this context is the circular aperture. It is shown in Section 3.1 that circular apertures are particularly suitable for capturing and describing two-dimensional wavefields, since circular apertures decompose the wavefield into eigen-solutions of the acoustic wave equation in cylindrical coordinates. As an alternative to conventional circular apertures, circular apertures mounted into rigid cylindrical baffles are also investigated. It is shown that this configuration has several advantages over conventional circular apertures.

Section 3.2 details the wavefield decomposition capabilities offered by spherical apertures and spherical apertures mounted into rigid spherical baffles. It will be shown that spherical apertures decompose the wavefield into eigen-solutions of the acoustic wave equation in spherical coordinates, thus making this aperture attractive for capturing and describing wavefields in three spatial dimensions.

Two other types of apertures are briefly discussed in Section 3.3, namely linear apertures and cylindrical apertures, which, for reasons that will become clear in subsequent sections, only play subordinate roles in wavefield decomposition methods.

It may be worthwhile noting here that an advantage of spherical apertures over other geometries is its uniform spatial selectivity over the entire 4π-space, i.e. θ and ϕ. Note that circular apertures offer uniform spatial selectivity over 2π-space, i.e. ϕ, only. Most other types of apertures do not offer uniform spatial selectivity at all.

3.1 Wavefield Decomposition Using Circular Apertures and Arrays

In this section, wavefields are analyzed by circular continuous apertures and by sampled continuous apertures, i.e. circularly symmetric microphone arrays. Circular apertures can be successfully deployed in situations where the wavefield is invariant with respect to the z-axis, for examples in rooms where reflections off the ceiling and floor are sufficiently attenuated. Both omnidirectional and directional apertures are considered.

3.1.1 Continuous Circular Apertures

Figure 3.1 depicts the geometric model of a circular aperture of radius R. This aperture is optionally mounted into a (rigid) cylindrical baffle of length $2L$. The setup where the circular aperture is mounted into a *rigid* cylindrical baffle will be referred to as *baffled circular aperture*. Otherwise the setup will be called an *unbaffled circular aperture*.

According to Eq. (2.34) in the general case of three-dimensional wave propagation, the pressure of a plane-wave impinging on the circular aperture can be written as,

$$P_{\text{inc}}(\boldsymbol{k}^T \boldsymbol{r}) = e^{i\boldsymbol{k}^T \boldsymbol{r}}, \tag{3.1}$$

where $\boldsymbol{k}$ is the wave-vector of the incoming plane-wave with $k = \|\boldsymbol{k}\| = 2\pi f/c$, f denoting the frequency of the incoming plane-wave, and $\boldsymbol{r}$ is the position of the observation point, $Q_s(\boldsymbol{r})$. The time-dependence is assumed to be $e^{-i\omega t}$ and is dropped in the following for notational convenience. For general three-dimensional scenarios it holds in spherical coordinates that,

$$\boldsymbol{k} = k \begin{bmatrix} \sin\vartheta\cos\varphi \\ \sin\vartheta\sin\varphi \\ \cos\vartheta \end{bmatrix}, \tag{3.2}$$

and,

$$\boldsymbol{r} = r \begin{bmatrix} \sin\theta\cos\phi \\ \sin\theta\sin\phi \\ \cos\theta \end{bmatrix}. \tag{3.3}$$

Combining Eq. (3.2) and Eq. (3.3), as required by Eq. (3.1), one obtains,

$$\boldsymbol{k}^T \boldsymbol{r} = kr[\sin\theta\sin\vartheta\cos(\phi - \varphi) + \cos\theta\cos\vartheta]. \tag{3.4}$$

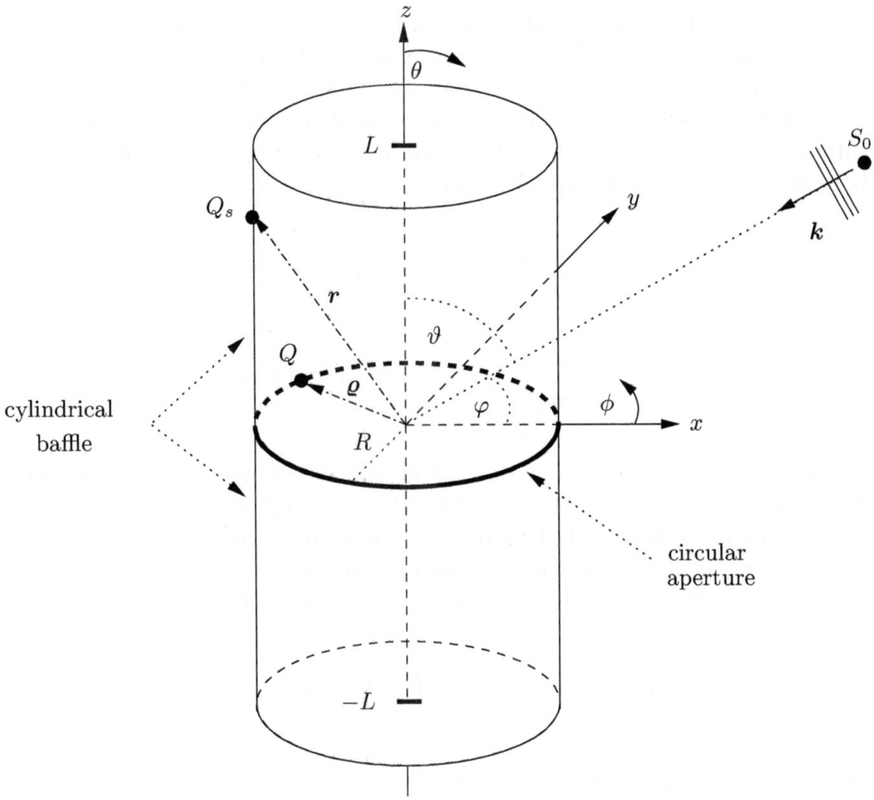

Fig. 3.1. Geometric model of a circular aperture mounted into a finite-length rigid cylindrical baffle

The Unbaffled Circular Aperture

In the following, the properties of unbaffled circular apertures are examined. As indicated in Fig. 3.1, only apertures that lie in the median plane, i.e. $\theta = \pi/2$, are considered. The observation point, therefore, is no longer $Q_s(\boldsymbol{r})$ but $Q(\boldsymbol{\varrho})$, since in the median plane it holds that $\boldsymbol{\varrho} = \boldsymbol{r}$. This restriction does not compromise generality of the following discussions.

As a first step, the pressure on the circular aperture, $\varrho = R$, due to an incoming plane-wave is required. This pressure follows from Eq. (3.1), using Eq. (3.4), and can be written as,

$$P(kR, \phi) \triangleq P_{\text{inc}}(\boldsymbol{k}^T\boldsymbol{r})\Big|_{\theta=\pi/2,\ \varrho=R} = e^{ikR\sin\vartheta\cos(\phi-\varphi)}. \tag{3.5}$$

A quantity of considerable interest in this book is the total response of a circular aperture due to a plane-wave impinging from (ϑ, φ) emitted by a source S_0

which is located in the farfield of the array in the sense of Appendix C. This quantity can be obtained by applying a spatial Fourier series expansion to the incident pressure of the, in general, weighted circular aperture, Eq. (3.5). Physically, this means that all contributions of the plane-wave on the aperture are weighted and integrated to yield the total response of the aperture. It therefore follows with Eq. (A.6) that,

$$\begin{aligned} F_n(kR,\varphi,\vartheta) &\triangleq \frac{1}{2\pi}\int_0^{2\pi} w(kR,\phi)P(kR,\phi)e^{-in\phi}d\phi \\ &= \frac{1}{2\pi}\int_0^{2\pi} w(kR,\phi)e^{ikR\sin\vartheta\cos(\phi-\varphi)}\,e^{-in\phi}\,d\phi, \end{aligned} \tag{3.6}$$

where $w(kR,\phi)$ is a frequency-dependent aperture weighting function. This weighting function can be chosen arbitrarily, although, in this section, it will be chosen as $w(kR,\phi)=1$. Different weighting functions will be considered when directional apertures are discussed in Section 3.1.2.

By setting the weighting function to unity, Eq. (3.6) becomes with $\gamma \triangleq \phi-\varphi$,

$$F_n(kR,\varphi,\vartheta) = \frac{1}{2\pi}e^{-in\varphi}\int_0^{2\pi} e^{ikR\sin\vartheta\cos\gamma}\,e^{-in\gamma}\,d\gamma, \tag{3.7}$$

which, by utilizing Eq. (B.2), simplifies to

$$F_n(kR,\varphi,\vartheta) = i^{-n}J_{-n}(kR\sin\vartheta)\,e^{-in\varphi}. \tag{3.8}$$

Applying Eq. (B.6) yields,

$$F_n(kR,\varphi,\vartheta) = i^{n}J_{n}(kR\sin\vartheta)\,e^{-in\varphi}. \tag{3.9}$$

Physically, the procedure of applying a spatial Fourier series expansion with respect to the polar coordinate to a circular aperture can be interpreted as a decomposition of a plane-wave into so-called *circular harmonics* on a circle with radius R. These circular harmonics are the Fourier coefficients of the series, cf. Eq. (A.5),

$$\begin{aligned} e^{i\boldsymbol{k}^T\boldsymbol{r}} = e^{ikR\sin\vartheta\cos(\phi-\varphi)} &= \sum_{n=-\infty}^{\infty} F_n(kR,\varphi,\vartheta)\,e^{in\phi} \\ &= \sum_{n=-\infty}^{\infty} i^n J_n(kR\sin\vartheta)\,e^{in(\phi-\varphi)}. \end{aligned} \tag{3.10}$$

This result is also known as the Jacobi-Anger expansion [CK98]. Note that this expansion has been used in Eq. (2.76) with $\vartheta=\pi/2$. Equation (3.10) is therefore the expansion of a plane-wave propagating in the free-field of an undisturbed medium on a circular aperture. It can be easily shown that this plane-wave expansion satisfies the acoustic wave equation by considering Eq. (2.54) and choosing $C_n(k_z) = 2\pi i^n e^{-in\varphi}\delta(k_z)$. In summary, restating Eq. (3.9),

$$\overset{\circ}{F}_n(kR, \varphi, \vartheta) = i^n J_n(kR\sin\vartheta)e^{-in\varphi}, \tag{3.11}$$

where $\overset{\circ}{F}_n(kR, \varphi, \vartheta)$ denotes the *modal* response of an *unbaffled* circular aperture of mode n.

The angular dependence of this response can be found in the exponential term while the frequency dependence is solely confined to the argument of the Bessel function. Therefore, one can expect the individual circular harmonics to be frequency independent along the angular axis. This observation is of paramount importance for many subsequent developments and serves as a foundation for the modal array signal processing methods presented in Chapter 5.

Figure 3.2 shows the normalized frequency-dependent *modal* magnitude response of the first four harmonics, $20\log|\overset{\circ}{F}_n(kR, \varphi, \vartheta)|, n = 0(1)3$, for $\theta = \vartheta = \pi/2$ and $\varphi = 0$. Two properties of the response are of particular interest here. First, for $n > 0$ all harmonics exhibit a highpass-like characteristic with a slope of $6n$ dB/octave up to the normalized frequency $kR \sim n$. This in turn means that a decomposed wavefield, at very small kR, only has the zero-order (omni-directional) component of considerable strength. As kR increases,

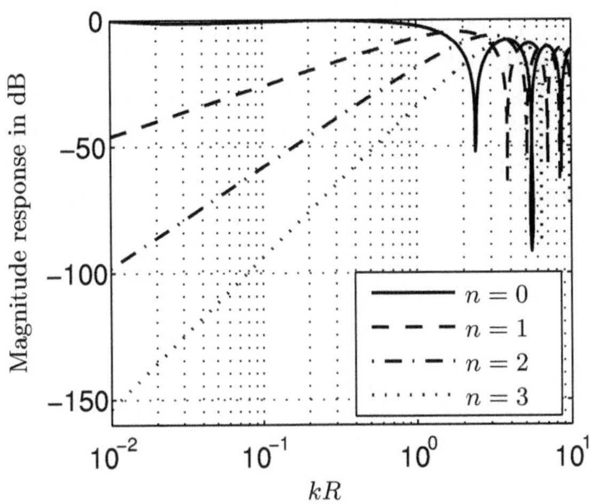

Fig. 3.2. Modal magnitude response, $20\log|\overset{\circ}{F}_n(kR, \varphi, \vartheta)|$, for $\theta = \vartheta = \pi/2$ and $\varphi = 0$

more and more harmonics are gaining in strength. Second, all harmonics show periodic dips in the magnitude response which correspond to the zeros of the respective Bessel functions $J_n(\cdot)$. As a consequence, signals that carry components in the vicinity of the zeros in the magnitude responses cannot be completely identified in the vicinity of these frequencies.

Figure 3.3 shows the azimuth-dependent component of the circular harmonics, $e^{-in\varphi}$, corresponding to Fig. 3.2. I.e. $|\mathrm{Re}\{e^{-in\varphi}\}| = |\cos(n\varphi)|$ is visualized. As can be seen, these harmonics correspond to multipoles, i.e. monopole, dipole, quadrupole, etc. These multipoles are mutually *orthonormal*, i.e.,

$$\frac{1}{2\pi}\int_0^{2\pi} e^{-in\varphi}e^{in'\varphi}\, d\varphi = \begin{cases} 1, \; n = n' \\ 0, \; n \neq n' \end{cases}. \tag{3.12}$$

This property, again, will be of importance for subsequent developments.

For illustration purposes, the full three-dimensional angular response due to a farfield source impinging from $\varphi \in [0 \ldots 2\pi]$ and $\vartheta = [0 \ldots \pi]$ is depicted in Fig. 3.4 with normalized magnitude. The dependence on the elevation of the source manifests itself in the argument of the n-th order Bessel function, see Eq. (3.11). The dependence on the azimuth of the source, however, can be found in the exponential term of Eq. (3.11). As a result, the angular response of the circular harmonics can be rotated in the azimuthal direction by introducing a phase offset in each harmonic. This rotation is not possible with respect to elevation. Therefore, a circular aperture offers flexibility only in the median plane, i.e.,

$$\vartheta = \pi/2. \tag{3.13}$$

Equation (3.13) is assumed to be valid throughout this book, unless noted otherwise. As a result, only two-dimensional wavefields are considered in the following. In practice, this requires all signals coming from the off-median plane to be sufficiently attenuated, if unambiguous discrimination of plane-wave incidence is desired.

The Jacobi-Anger expansion, Eq. (3.10), then becomes in two space dimensions,

$$e^{i\boldsymbol{k}^T\boldsymbol{\varrho}} = e^{ikR\cos(\phi-\varphi)} = \sum_{n=-\infty}^{\infty} i^n J_n(kR)\, e^{in(\phi-\varphi)}, \tag{3.14}$$

see also Eq. (2.44), and Eq. (3.11) finally simplifies to,

$$\overset{\circ}{F}_n\,(kR,\varphi) = i^n J_n(kR)e^{-in\varphi}. \tag{3.15}$$

Equation (3.15) is an expression for the modal response of an unbaffled circular aperture due to a plane-wave incidence from the aperture plane.

Multiple incoming plane-waves can be taken into account by a superposition of the respective individual harmonics, i.e.,

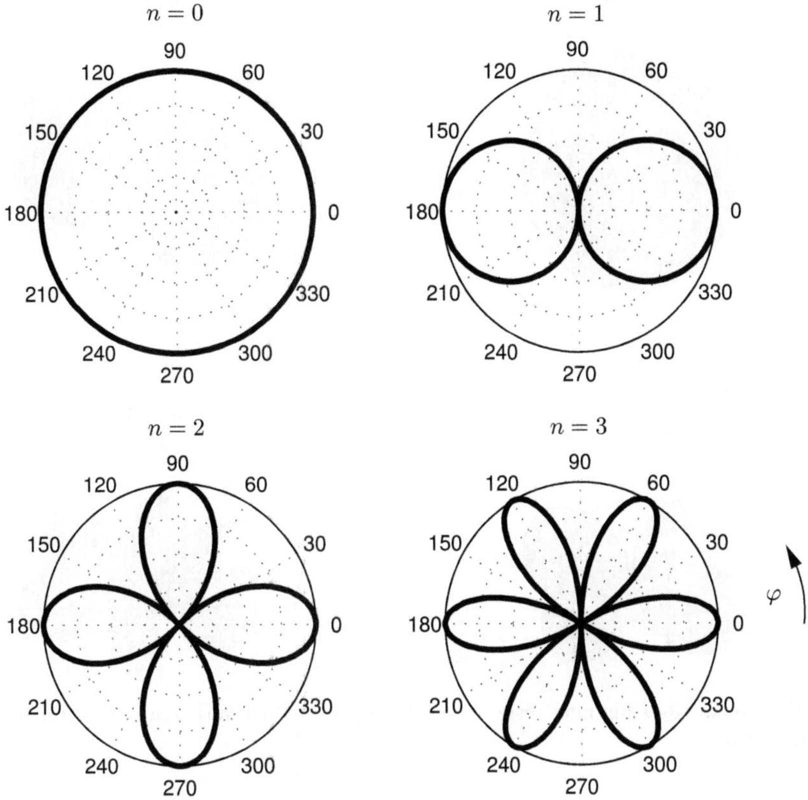

Fig. 3.3. Azimuthal response, $|\cos(n\varphi)|$, with respect to φ in degrees for $n = 0, 1, 2, 3$ and $\theta = \vartheta = \pi/2$

$$\mathring{F}_{n_{\text{sup}}} = \sum_{\nu=1}^{I} i^n J_n(k_\nu R) e^{-in\varphi_\nu}, \tag{3.16}$$

where I is the number of incident plane-waves.

By comparing the circular harmonics of Eq. (3.15) with the general solution of the homogeneous acoustic wave equation for interior boundary problems in cylindrical coordinates, Eq. (2.54), it becomes clear that the general boundary value problem is composed of the same individual harmonics as obtained by a decomposition of the cylindrical wavefields with circular apertures. These harmonics are therefore the eigen-solutions of the acoustic wave equation in cylindrical coordinates.

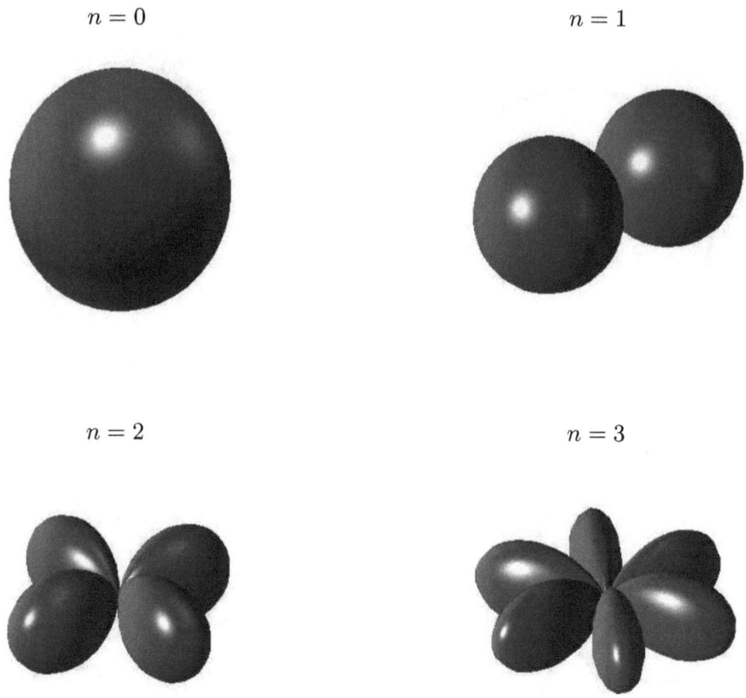

Fig. 3.4. 3-D angular response with respect to azimuth and elevation for $n = 0, 1, 2, 3$

The Baffled Circular Aperture

By analyzing a two-dimensional wavefield using circular apertures one has to consider dips in the magnitude responses of the decomposed harmonics, cf. Fig. 3.2. Depending on the radius of the aperture these dips may show up in the magnitude responses of the circular harmonics within the frequency range of interest. This problem can be avoided by mounting the circular aperture into a rigid cylindrical baffle, as in Fig. 3.1. Thereby, the scattering properties of rigid cylindrical baffles, see Section 2.3.3, can be applied advantageously. First, an infinite-length cylindrical baffle, i.e. $L \to \infty$, is considered. Then the more practical case of a finite-length cylindrical baffle is investigated.

The Infinite-Length Rigid Cylinder

The decomposition of an unperturbed wavefield into circular harmonics by a circular aperture is given by Eq. (3.10). In case of the presence of an infinite-length rigid cylindrical object in the wavefield, not only the incoming wavefield but the total wavefield, composed of the sum of the incoming and scattered wavefield, needs to be taken into account for this decomposition on the circular aperture, cf. Section 2.3.3. An expansion of the wavefield along the radius of a circular aperture mounted into a rigid cylindrical baffle is given in Eq. (2.83). By comparing Eq. (3.10) with Eq. (2.83) and Eq. (3.15) for $\vartheta = \pi/2$ one immediately obtains the circular harmonics decomposition for this setup as,

$$\overset{\bullet}{F}_n(kR,\varphi) = i^n\left[J_n(kR) - \frac{J_n'(kR)H_n(kR)}{H_n'(kR)}\right]e^{-in\varphi}, \tag{3.17}$$

where $\overset{\bullet}{F}_n(kR,\varphi)$ denotes the modal response of a baffled circular aperture of mode n.

Figure 3.5 shows the magnitude response of the circular aperture mounted into an infinite-length cylindrical baffle for $n = 0(1)3$. By comparing Fig. 3.5 with Fig. 3.2 one immediately recognizes a very desirable effect of the presence of a rigid cylindrical baffle. The additional components introduced by the scattered wavefield exactly cancel the zeros introduced by the Bessel functions. Mathematically, this can be explained by the fact that the two terms inside the brackets in Eq. (3.17) do not share any common zeros. This phenomenon makes this configuration more useful for wavefields containing wideband signals and/or when large apertures are considered. Also visible in Fig. 3.5 is that the slope of all harmonics is identically proportional to $1/\sqrt{kR}$ for approximately $n > kR$. This factor corresponds to the one that has been found in Section 2.3.2 as the attenuation of cylindrically spreading wave-fronts with respect to growing distance between source and receiver. It should be noted

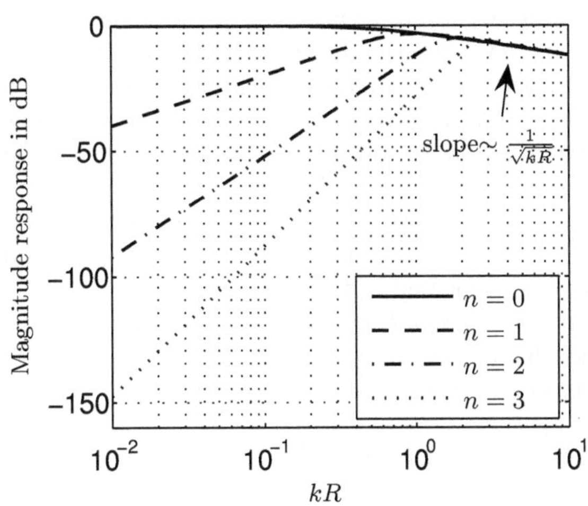

Fig. 3.5. Modal magnitude response, $20\log|\overset{\bullet}{F}_n(kR,0)|$ for $\varphi = 0$

that this attenuation is also present, though not obvious, in Fig. 3.2. As can be further deduced from Eq. (3.17), the cylindrical scatterer does not modify the azimuthal response of an unbaffled circular aperture, cf. Eq. (3.15). Therefore, Fig. 3.3 and Fig. 3.4 apply here, too.

The Finite-Length Rigid Cylinder

For actual implementations it is impossible to mount circular apertures into infinite-length cylindrical baffles. The length of the cylinder must be truncated to a reasonable size. As a result, the observable circular harmonics will exhibit magnitude responses that lie somewhat in between the ones that have been obtained for unbaffled and baffled circular apertures. An expression for a wavefield expanded on a circular aperture mounted into a rigid finite-length cylindrical baffle has been derived in Eq. (2.93). Unfortunately, this solution cannot be given in closed form. However, it will be shown by means of simulations that the finite-length case can be approximated by the infinite-length case for many practical cases. Figure 3.6 displays the effect of mounting circular apertures of radius R into rigid cylindrical baffles of length $2L$ for varying values for L/R. The result is shown for $n = 0(1)3$. It is clear that $L/R = 0$ corresponds to the case of unbaffled circular apertures while $L/R = \infty$ reflects the circular aperture mounted into an infinite-length cylindrical baffle. Two intermediate values for L/R are also shown. It can be seen that even a small ratio of $L/R = 0.1$, i.e. a relatively short cylindrical scatterer, is sufficient to effectively combat the zeros introduced by the respective Bessel functions. It can also be seen that the curves corresponding to $L/R = 1.4$ and $L/R = \infty$ already exhibit comparable properties. This means that for circular apertures of small radius, say $R < 0.1$ m, practical rigid cylindrical baffles can be constructed that model an infinite-length rigid baffle with satisfactory accuracy. Therefore, it seems reasonable to describe the circular harmonics decomposition of a circular aperture mounted into a finite-length rigid cylindrical baffle by the closed-form solution given in Eq. (3.17) in the following.

Note that the boundary conditions introduced by the endcaps of the finite cylindrical rigid baffle have been neglected in all developments. The results in this section have to be modified as soon as the circular aperture is located in the vicinity of the endcaps of the cylindrical baffle, since the endcap boundary conditions cannot be neglected in this case.

Note also that the case $L/R = 1.4$ corresponds to the actual system that has been realized for evaluating wavefield decomposition based array signal processing algorithms, see Section 6.

Additional Properties of Baffled Circular Apertures

An additional very interesting property of rigid cylindrical baffles can be deduced from Fig. 3.6 by comparing the results for $L/R = 0$ and $L/R = \infty$. For small kR and $n > 0$ the magnitude response due to the presence of a

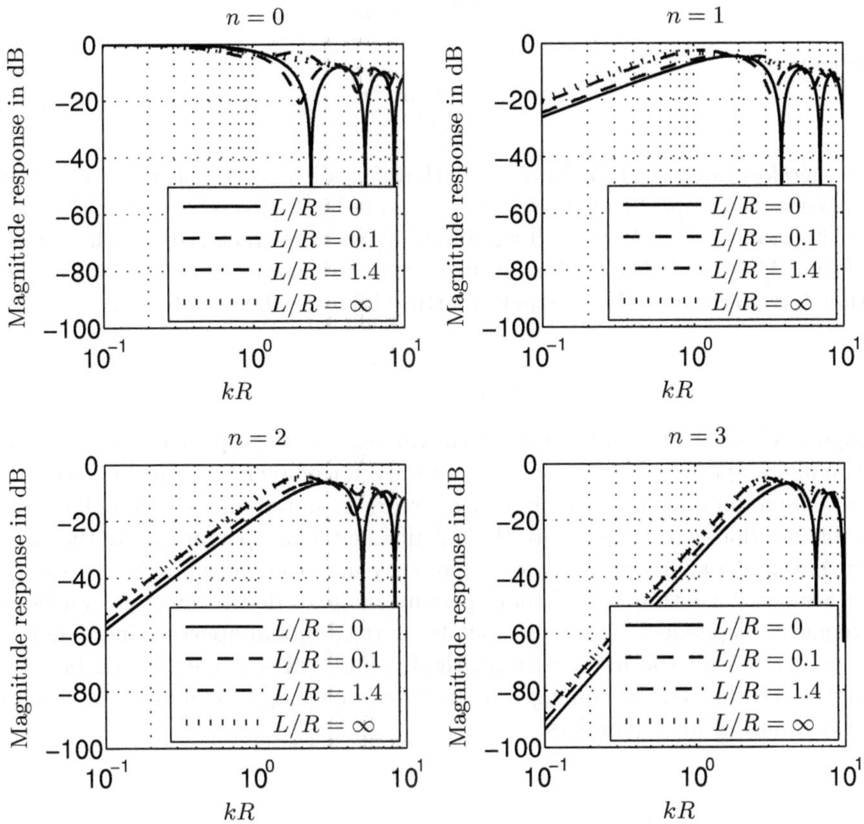

Fig. 3.6. Modal magnitude response for $\varphi = 0$ of a circular aperture mounted into a finite-length rigid baffle w.r.t. L/R

rigid baffle always seem to lie a few dB above the response of the unbaffled circular aperture. This effect can be interpreted as a virtual increase of the *modal aperture* due to the presence of the scatterer. This effect is examined further in the following. Expressing the relative magnitude response between the response of a circular aperture mounted into a rigid cylindrical baffle and an unbaffled circular aperture from Eq. (3.17) and Eq. (3.15), respectively, gives,

$$\Delta P(kR) \triangleq \frac{J_n(kR) - \dfrac{J_n'(kR)H_n(kR)}{H_n'(kR)}}{J_n(kR)} = \frac{2i}{\pi kR J_n(kR) H_n'(kR)}, \tag{3.18}$$

where Eq. (2.84) and Eq. (B.4) have been used. Since the aperture increase is only effective for small kR, it follows with Eq. (B.18) and Eq. (B.20) that,

$$\Delta P(kR) \approx \epsilon_n, \tag{3.19}$$

where,

$$\epsilon_n = \begin{cases} 1, \; n = 0 \\ 2, \; n > 0 \end{cases}. \tag{3.20}$$

It therefore follows from Eq. (3.19) that for all $n > 0$, the rigid scatterer increases the response of the circular aperture by $20 \cdot \log_{10}(2)$=6.02 dB. Since all harmonics of order $n > 0$ exhibit a highpass characteristic with a slope of $6n$ dB/octave it follows further from geometrical considerations in Fig. 3.6 that the effective modal (virtual) aperture increases by a factor of,

$$\frac{R_{\mathrm{virt}}}{R} = \sqrt[n]{\epsilon_n}. \tag{3.21}$$

Figure 3.7 shows the variability of this increase with respect to the mode n. Maximum aperture increase can be obtained for mode $n = 1$. In this case, the virtual modal aperture of a circular aperture mounted into a rigid cylindrical baffle has doubled compared to the unbaffled circular aperture. Note that this virtual *modal* aperture increase is frequency-independent while it was shown in Section 2.3.3 that the virtual aperture increase due to creeping waves is frequency-dependent. Note also that the virtual modal aperture increase has been derived for the infinite-length rigid cylindrical scatterer. It can be seen in Fig. 3.6 that the finite-length scatterer can only approximate these results in the limit of long cylinders.

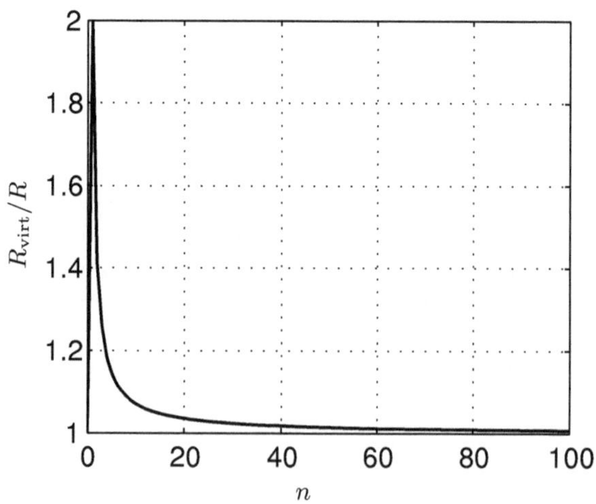

Fig. 3.7. Relative modal aperture increase due to an infinite-length rigid cylindrical scatterer

3.1.2 Directional Circular Apertures

In the previous sections, only circular apertures have been considered whose infinitesimal segments, $d\phi$, were modeled as being equally sensitive to wavefields impinging from all possible directions. I.e. they exhibit spatial characteristics that can be considered as omni-directional. This section examines infinitesimal segments that are more sensitive to specific directions of wave incidence than others. Assuming this dependence, $W(\phi - \varphi)$, to be confined to the azimuthal direction, the circular harmonics decomposed by a circular aperture of radius R can be written as [RD82],

$$\overset{\circ}{\mathscr{F}}_n(kR, \varphi, \vartheta) = \frac{1}{2\pi} \int_0^{2\pi} W(\phi - \varphi)\, e^{-i(n\phi - kR \sin\vartheta \cos(\phi - \varphi))}\, d\phi. \tag{3.22}$$

Of course, this expression can also be obtained directly by setting $w(kR, \phi) = W(\phi - \varphi)$ in Eq. (3.6). Note that for generality, ϑ is allowed to adopt values between 0 and π in this section. Since $W(\phi - \varphi)$ is periodic with period 2π, it can be expanded into a Fourier series as,

$$W(\phi - \varphi) = \sum_{l=-\infty}^{\infty} w_l e^{il(\phi - \varphi)}, \tag{3.23}$$

with Fourier coefficients w_l. Eq. (3.22) then becomes after changing summation and integration,

$$\overset{\circ}{\mathscr{F}}_n(kR, \varphi, \vartheta) = \frac{1}{2\pi} \sum_{l=-\infty}^{\infty} w_l \int_0^{2\pi} e^{i[l(\phi-\varphi) - n\phi + kR \sin\vartheta \cos(\phi - \varphi)]}\, d\phi. \tag{3.24}$$

Defining $\gamma \triangleq \phi - \varphi$ leads to,

$$\overset{\circ}{\mathscr{F}}_n(kR, \varphi, \vartheta) = \frac{1}{2\pi} \sum_{l=-\infty}^{\infty} w_l e^{-in\varphi} \int_0^{2\pi} e^{i[(l-n)\gamma + kR \sin\vartheta \cos\gamma]}\, d\gamma, \tag{3.25}$$

which finally simplifies with Eq. (B.2) to,

$$\begin{aligned} \overset{\circ}{\mathscr{F}}_n(kR, \varphi, \vartheta) &= e^{-in\varphi} \sum_{l=-\infty}^{\infty} w_l i^{l-n} J_{l-n}(kR \sin\vartheta) \\ &= e^{-in\varphi} \sum_{l=-\infty}^{\infty} w_l i^{n-l} J_{n-l}(kR \sin\vartheta). \end{aligned} \tag{3.26}$$

The following discussions detail two specific examples of directional apertures, i.e. the circular dipole aperture and the circular cardioid aperture.

Circular Dipole Aperture

The azimuthal dependence associated with continuous apertures having infinitesimal segments exhibiting dipole characteristics can be expressed as,

$$W(\phi - \varphi) = \cos(\phi - \varphi). \tag{3.27}$$

Note that this result assumes that the origin of the coordinate system lies on an infinitesimal segment of the aperture. Identifying the corresponding Fourier coefficients in Eq. (3.23) yields,

$$w_l = \begin{cases} 1/2, \, l = -1, 1 \\ 0, \quad \text{else} \end{cases}, \tag{3.28}$$

and give with Eq. (3.26),

$$\begin{aligned} \overset{\circ}{\mathscr{F}}_{nD}(kR, \varphi, \vartheta) &= \frac{1}{2}\left[i^{n-1} J_{n-1}(kR\sin\vartheta) + i^{n+1} J_{n+1}(kR\sin\vartheta)\right] e^{-in\varphi} \\ &= \frac{1}{2}\left[i^{n-1} J_{n-1}(kR\sin\vartheta) - i^{n-1} J_{n+1}(kR\sin\vartheta)\right] e^{-in\varphi} \\ &= i^{n-1} J_n'(kR\sin\vartheta) e^{-in\varphi}, \end{aligned} \tag{3.29}$$

where in the last step the recurrence relation for Bessel functions, Eq. (B.15), has been used.

It is often stated in the literature that apertures with dipole characteristics measure the velocity component of a wavefield. This is correct only up to a scaling factor as the following discussion shows. By applying Eq. (2.13) to Eq. (3.11), the circular harmonics decomposition using a circular aperture comprising radially oriented infinitesimal *velocity* segments, results with Eq. (3.11) and Eq. (3.29) in,

$$\begin{aligned} \overset{\circ}{\mathscr{F}}_{nV}(kR, \varphi, \vartheta) &= \frac{1}{i\varrho_0 ck} \nabla \overset{\circ}{F}_n(kr, \varphi, \vartheta)|_{r=R} \\ &= \frac{\sin\vartheta}{\varrho_0 c} i^{n-1} J_n'(kR\sin\vartheta) e^{-in\varphi} \\ &= \frac{\sin\vartheta}{\varrho_0 c} \overset{\circ}{\mathscr{F}}_{nD}(kR, \varphi, \vartheta) \\ &= \frac{\sin\vartheta}{Z_0} \overset{\circ}{\mathscr{F}}_{nD}(kR, \varphi, \vartheta), \end{aligned} \tag{3.30}$$

where in the last step the definition of the free-field impedance, Eq. (2.74), has been applied. Therefore, the decomposition of wavefields using apertures having dipole characteristics is identical to a decomposition of a wavefield using apertures whose segments measure the particle velocity up to a factor, $\sin\vartheta / Z_0$. Note that this simple relation is true only for plane-wave incidence.

Figure 3.8 shows the magnitude response of a decomposed wavefield using a circular dipole aperture for $\varphi = 0$ and $\vartheta = \pi/2$. It can be seen that the

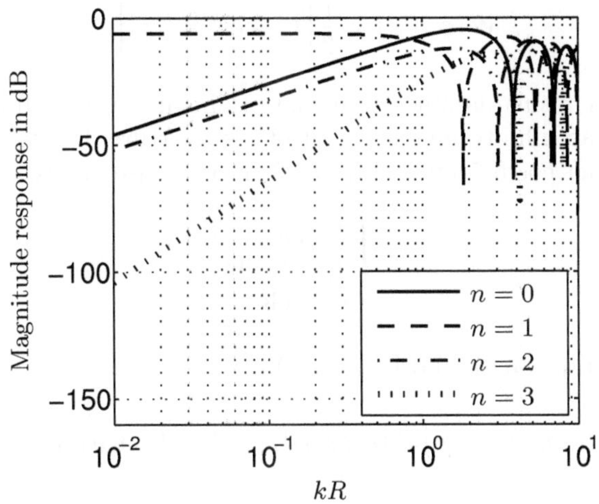

Fig. 3.8. Modal magnitude response of the decomposed wavefield using circular dipole aperture for $\varphi = 0$ and $\vartheta = \pi/2$

response differs significantly from the magnitude response of a decomposed wavefield using a circular omnidirectional aperture shown in Fig. 3.2. First, the first-order harmonic, i.e. the dipole, has a flat magnitude response for $kR < n$. This can be explained by the fact that $n = 1$, i.e. the dipole component, is the 'natural' harmonic of the decomposed wavefield using a circular dipole aperture, while $n = 0$ is the 'natural' harmonic of the decomposed wavefield using an omnidirectional aperture. Second, the dips in the magnitude response are now located according to the zeros of the first derivative of the Bessel function. Also, the slope of the highpass-like response is now $6(n-1)$ dB/octave up to $kR \sim 1$ for $n > 1$.

Note that the additional factor of $-i = e^{-i\pi/2}$ in Eq. (3.29) rotates the harmonics compared to the ones obtained by omnidirectional apertures by $\pi/2$, see Fig. 3.3.

It can be concluded that a circular dipole aperture measures the normal component of the particle velocity up to a scaling factor.

A circular dipole aperture mounted into a rigid cylindrical baffle yields no output at all since the applicable boundary condition, Eq. (2.77), forces the radial component of the particle velocity to vanish on the surface of the baffle.

Circular Cardioid Aperture

The azimuthal dependence associated with continuous apertures having infinitesimal segments exhibiting cardioid characteristics can be expressed as,

$$W(\phi - \varphi) = \frac{1}{2}\Big[1 + \cos(\phi - \varphi)\Big]. \tag{3.31}$$

Note that this result assumes that the origin of the coordinate system is identical with an infinitesimal segment of the aperture. Identifying the corresponding Fourier coefficients in Eq. (3.23) in this case yields,

$$w_l = \begin{cases} 1/4, \, l = -1, 1 \\ 1/2, \, l = 0 \\ 0, \quad \text{else} \end{cases}, \tag{3.32}$$

and give with Eq. (3.26), and Eq. (B.15),

$$\mathring{\mathscr{F}}_{nC}\,(kR, \varphi, \vartheta) = \frac{i^n}{2}\Big[J_n(kR\sin\vartheta) - iJ_n^{'}(kR\sin\vartheta)\Big]\, e^{-in\varphi}, \tag{3.33}$$

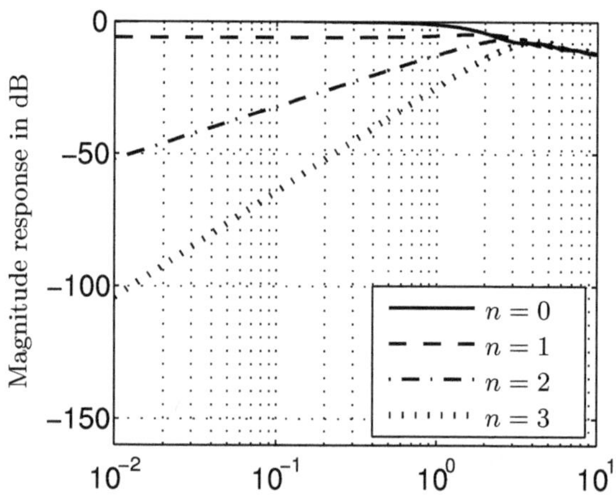

Fig. 3.9. Modal magnitude response of the decomposed wavefield using circular cardioid aperture for $\varphi = 0$ and $\vartheta = \pi/2$

Figure 3.9 shows the magnitude response of a decomposed wavefield using a circular cardioid aperture for $\varphi = 0$ and $\vartheta = \pi/2$. Again, the response differs significantly from the magnitude response of the decomposed wavefield using a circular omnidirectional aperture shown in Fig. 3.2. Here, both the zero-order harmonic as well as the first-order harmonic exhibit a flat response up to $kR \sim 1$. This behavior can be explained by realizing that both the omnidirectional as well as the dipole component comprise the 'natural' harmonics for a circular cardioid aperture. It can also be seen that there are no dips in the magnitude response at all. This results from the fact that there are no common zeros in the Bessel function and its derivative, cf. Eq. (3.33). The slope of the highpass-like response is $6(n-1)$ dB/octave up to $kR \sim 1$ for $n > 1$, as is exhibited by circular dipole apertures.

It can be gleaned from Eq. (3.33) that a circular cardioid aperture measures, in equal parts, the normal component of the particle velocity and the pressure of the incoming wavefield. Note that the additional factor of $-i = e^{-i\pi/2}$ in Eq. (3.33) again rotates the harmonics compared to the ones obtained by omnidirectional apertures. In this case, however, this rotation is only applied to the term that describes the normal component of the particle velocity since the factor $(-i)$ only appears in the second term of Eq. (3.33). There is no additional rotation of the component describing the pressure.

It is expected that mounting a circular cardioid aperture into a rigid cylindrical baffle will yield the same output as in Section 3.1.1 in the azimuthal direction. This can be explained by the fact that due to the discussion presented for circular dipole apertures, the normal component of the particle velocity vanishes on the surface of the baffle, thus only leaving the pressure of the wavefield as a non-zero component of the decomposition process.

3.1.3 Circularly Symmetric Microphone Arrays

For real-world applications, the circular aperture needs to be sampled by discrete points in space, i.e. microphones. In this section, it is assumed that this sampling is performed by ideal omnidirectional microphones. It is further assumed that one wishes to decompose an impinging wavefield into $\mathcal{N}+1$ circular harmonics, where $\mathcal{N}$ is the maximum order to be decomposed. For all examples that have been previously used for illustration purposes, $\mathcal{N}=3$. Following the same reasoning as for the sampling theorem of one-dimensional time signals [OS98], a minimum of,

$$M > 2\mathcal{N} \tag{3.34}$$

equidistant spaced microphones are necessary for this task [Tre02]. This requirement can easily be verified by noting that when considering $\mathcal{N}+1$ harmonics, $2\mathcal{N}+1$ Fourier coefficients of the *truncated* Fourier series, see Eq. (3.10) where the symbol '∞' is to be replaced by '$\mathcal{N}$', need to be identified by at least $2\mathcal{N}+1$ samples.

The highest order $\mathcal{N}$ that can be captured by a circular aperture depends on the largest wavenumber, $k = k_{\max}$, of the wavefield to be decomposed and on the radius R of the aperture. Therefore, the argument of the Bessel functions, $k_{\max}R$ is an indication of the strength of the respective harmonic. As a rule of thumb,

$$\mathcal{N} \approx k_{\max}R, \tag{3.35}$$

is often chosen as a first approximation [Tre02]. This choice is justified by looking at Fig. B.1 in Appendix B where it can be seen that the value of a particular Bessel function is small when the order $n > 0$ exceeds the argument.

Plugging Eq. (3.35) into Eq. (3.34) results in a spacing along the arc of the circular array that corresponds to less than half the smallest wavelength in the wavefield, λ_{min}, as,

$$d_{\text{arc}} \leq \lambda_{\text{min}}/2. \tag{3.36}$$

Note that for linear arrays a microphone spacing of $\lambda_{\text{min}}/2$ is sufficient to avoid *spatial* aliasing which are also denoted as grating-lobes in the array literature [Tre02].

Assuming $\mathcal{N}$ to be the highest order of the wavefield to be decomposed and $k_{\text{max}}R$ to be the highest spatial frequency, the sampling function is [Tre02],

$$\begin{aligned} S(\phi) &= \sum_{\nu=-\infty}^{\infty} \delta(\phi - \nu\phi_T) \\ &= \frac{1}{\phi_T} \sum_{q=-\infty}^{\infty} e^{iqM\phi} \\ &= \frac{1}{\phi_T}\left(1 + \sum_{q=1}^{\infty} e^{iqM\phi} + \sum_{q=1}^{\infty} e^{-iqM\phi}\right), \end{aligned} \tag{3.37}$$

where,

$$\phi_T = \frac{2\pi}{2k_{\text{max}}R} = \frac{\pi}{k_{\text{max}}R}, \tag{3.38}$$

is the sampling interval along the circle. Applying the sampling operation, Eq. (3.37), to the circular harmonics obtained by continuous apertures, Eq. (3.15), the circular harmonics of a sampled circular aperture is,

$$\overset{\circ}{F}{}^{s}_{n}(kR,\varphi) \sim \underbrace{i^n J_n(kR) e^{-in\varphi}}_{\overset{\circ}{F}_n(kR,\varphi)} + \underbrace{\sum_{q=1}^{\infty} i^g J_g(kR) e^{ig\varphi} + \sum_{q=1}^{\infty} i^h J_h(kR) e^{-ih\varphi}}_{\overset{\circ}{F}{}^{e}_{n}(kR,\varphi)}, \tag{3.39}$$

where $g = (Mq - n)$ and $h = (Mq + n)$. The superscript 's' denotes the sampled circular harmonics while the superscript 'e' denotes the component due to aliasing.

Similarly, introducing Eq. (3.37) into Eq. (3.17), the circular harmonics of a sampled continuous circular aperture mounted into an infinite-length cylindrical baffle can be written as,

$$\begin{aligned} \overset{\bullet}{F}{}^{s}_{n}(kR,\varphi) \sim\; & i^n \left[J_n(kR) - \frac{J_n'(kR) H_n(kR)}{H_n'(kR)} \right] e^{-in\varphi} \\ &+ \sum_{q=1}^{\infty} i^g \left[J_g(kR) - \frac{J_g'(kR) H_g(kR)}{H_g'(kR)} \right] e^{ig\varphi} \\ &+ \sum_{q=1}^{\infty} i^h \left[J_h(kR) - \frac{J_h'(kR) H_h(kR)}{H_h'(kR)} \right] e^{-ih\varphi}, \end{aligned} \tag{3.40}$$

The first term in Eq. (3.39), and Eq. (3.40), is equivalent to the respective circular harmonic of the continuous circular aperture in Eq. (3.15) and Eq. (3.17), respectively. The remaining terms are residuals due to the sampling operation applied to the circular aperture. Sampling, therefore, results in not only obtaining a harmonic of order n but also, inseparably superimposed, harmonics of order $g > n$ and $h > n$. In analogy to the error occurring due to sampling of one-dimensional time signals, this distortion is denoted as *modal aliasing*. However, there is one significant difference between one-dimensional time signals and two- or three-dimensional space-time signals. Time signals can be bandlimited in an effective way before performing any type of digital signal processing. This can be done by applying lowpass filters of high orders. Therefore, a strict rule can be provided of how aliasing can be avoided, i.e. the Nyquist frequency. It is very difficult, if not impossible, to perform effective band-limitation for space-time signals before applying digital signal processing algorithms since high-order spatial lowpass filters do not seem to exist. Hence, there is no strict rule guaranteeing an error-free sampling operation. As a consequence, modal aliasing cannot be entirely avoided. However, by judiciously choosing the number of microphones, M, and the maximum spatial frequency, $k_{\max}R$, these distortion modes can be made as small as desired. An expression that quantifies the relative error due to sampling, $\overset{\circ}{\mathcal{E}}_n(kR)$, can be given as the relation of the squared absolute value of the aliasing component with respect to the total energy recorded by the circular microphone array, i.e.,

$$\overset{\circ}{\mathcal{E}}_n(kR,\varphi) = \frac{\left|\overset{\circ}{F}{}^{e}_{n}(kR,\varphi)\right|^2}{\left|\overset{\circ}{F}{}^{s}_{n}(kR,\varphi)\right|^2}. \tag{3.41}$$

A similar expression can be provided for the sampled circular aperture mounted into a rigid cylindrical baffle, $\overset{\bullet}{\mathcal{E}}_n(kR,\varphi)$.

As an example, Fig. 3.10 displays the error due to modal aliasing by sampling a continuous baffled aperture with $M = 10$ and $M = 15$ microphones and $k_{\max}R = 10$. As can be seen the error is an increasing function with respect to kR. A second property is the fact that modal aliasing increases with the order of the circular harmonic. It can also be seen that the error decreases significantly with decreasing sampling interval. As stated before, there is no strict rule on how to determine M other than Eq. (3.34). A design procedure may include a maximum amount of error to be introduced due to modal aliasing of, say, -50 dB. In the example shown in Fig. 3.10, using $\mathcal{N} = 3$, the maximum spatial frequency that can be captured with this array is $k_{\max}R \approx 1.8$ and $k_{\max}R \approx 6$ for $M = 10$ and $M = 15$, respectively, both resulting in less than -50 dB of error due to modal aliasing.

As indicated in Fig. 3.3 and Fig. 3.4, circular harmonics can be regarded as multipoles or beam patterns. The effect of sampling a baffled circular aperture

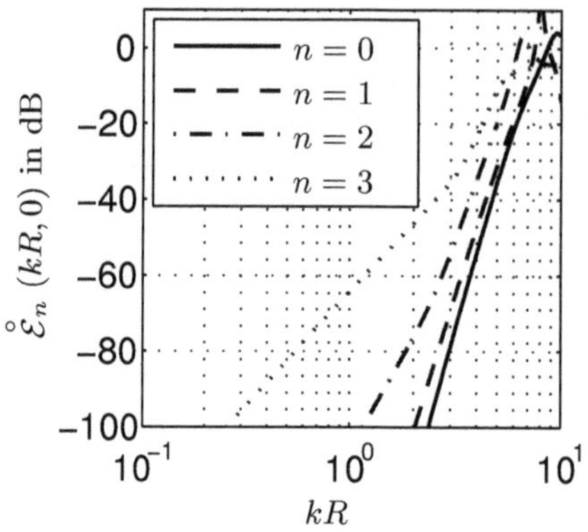

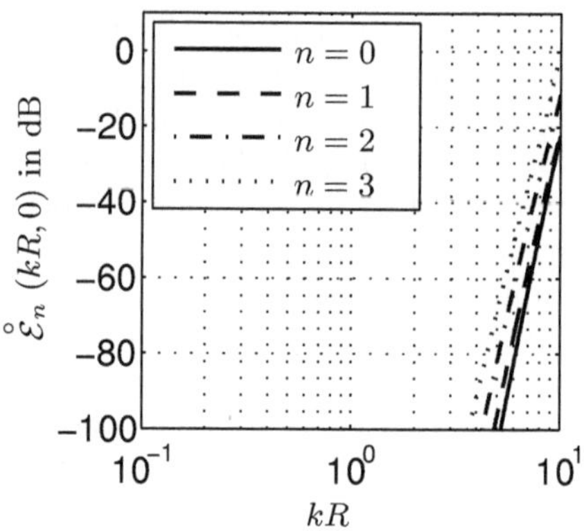

Fig. 3.10. Error due to sampling of a baffled circular aperture utilizing $M = 10$ (left) and $M = 15$ microphones (right) for $\varphi = 0$

with $M = 10$ on the resulting beam pattern is shown in Fig. 3.11. Circular harmonics of order zero (left column) and of order three (right column) are considered here. The first row shows the circular harmonic obtained by the continuous baffled circular aperture. One can identify the respective azimuthal response depicted in Fig. 3.3 by taking a horizontal slice through the figures in the first row and by wrapping the result around the polar axis. The second row depicts the circular harmonic obtained by the sampled circular aperture. In this example, severe distortion of the modes are obvious for $kR > 5$. The third row shows the error due to modal aliasing which is a monotonously increasing function of frequency (kR) and which is also a function of the azimuthal angle, φ. From Eq. (3.34) it follows that using M microphones a wavefield can be decomposed into a limited number of harmonics only, i.e.,

$$\mathcal{N} = \begin{cases} M/2 - 1, & M \text{ even} \\ (M-1)/2, & M \text{ odd} \end{cases} . \tag{3.42}$$

After applying Eq. (3.35) it becomes clear that, assuming ten microphones for sampling, the wavefield under observation should not contain spatial frequencies above $kR \approx 4$. This means that one is faced with the task of how to apply a spatial lowpass filter to an incoming wavefield *before* converting the signals into the digital domain. Analog lowpass filter of *low* order could be implemented by simply using microphones with greater diameter that effectively perform the filtering operation by averaging the pressure over its surface [ME02]. Analog lowpass filter of *high* order, however, are virtually impossible to realize. This makes efficient suppression of modal aliasing a very challenging, if not impossible, task.

Sampling the same aperture with 15 microphones results in reduced modal aliasing as shown in Fig. 3.12. In this example, severe distortion modes only come into play from about $kR \approx 7$ for $n = 3$. Very similar results are ob-

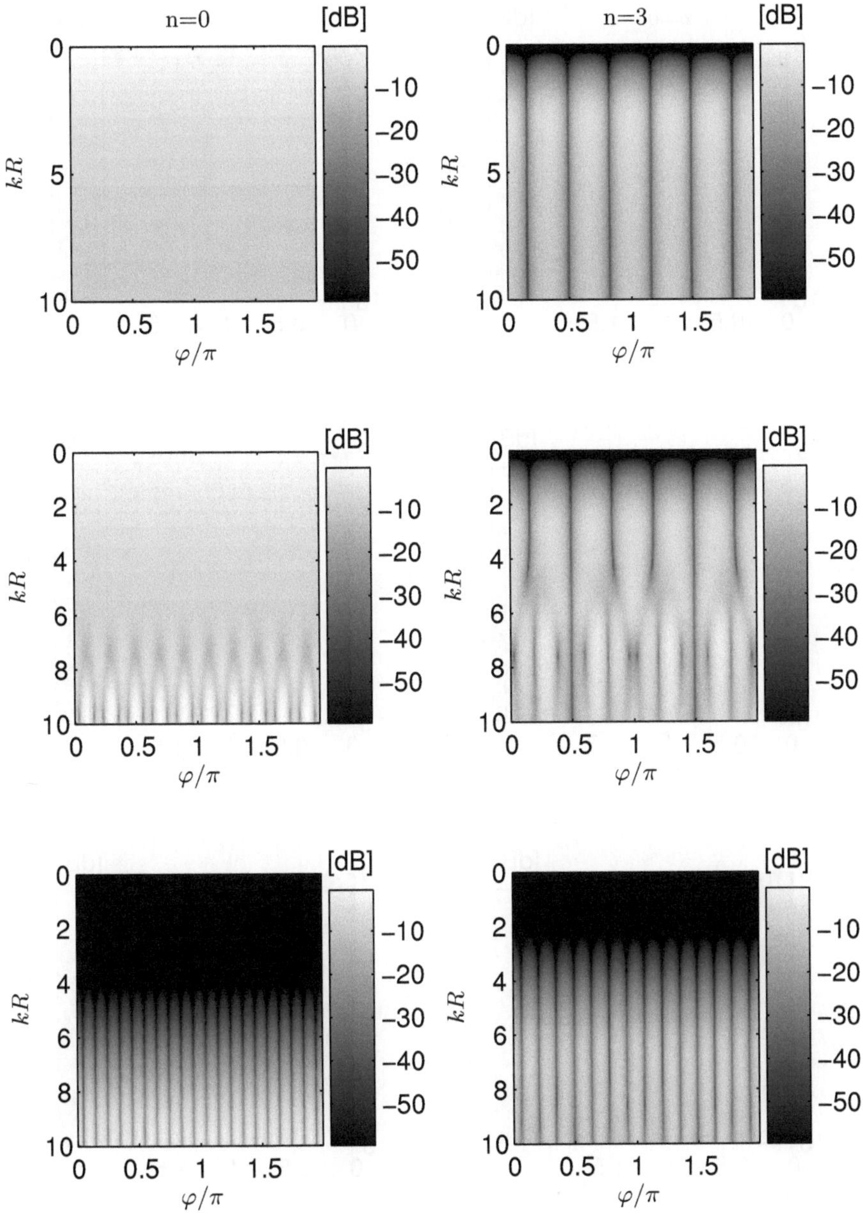

Fig. 3.11. Circular harmonics ($n = 0, 3$) obtained for the baffled circular aperture (upper row), for the sampled baffled aperture (middle row), and the error due to modal aliasing (lower row) where $M = 10$

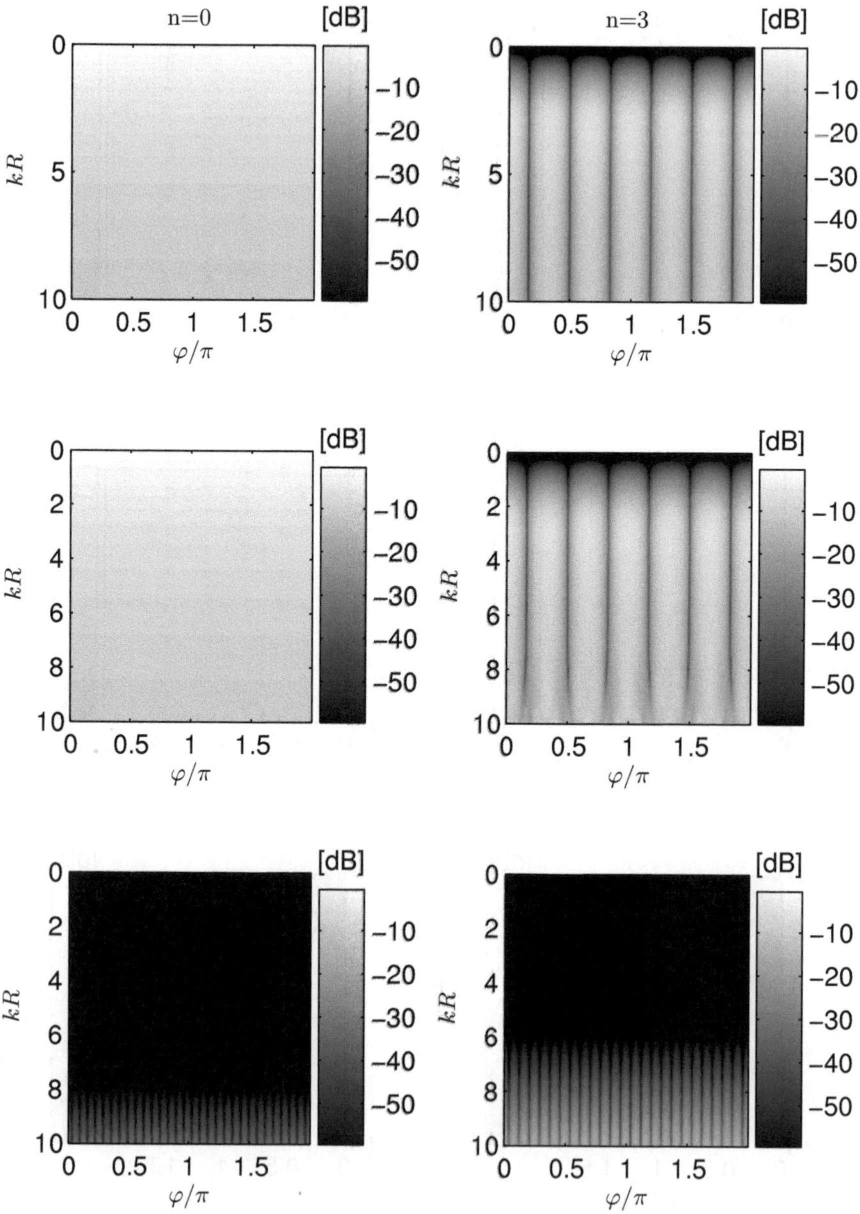

Fig. 3.12. Circular harmonics ($n = 0, 3$) obtained for the baffled circular aperture (upper row), for the sampled baffled aperture (middle row), and the error due to modal aliasing (lower row) where $M = 15$

tained by considering sampled unbaffled circular apertures. Thus, they are not repeated here. In summary, it can be stated somewhat trivially that an increasing number of microphones decreases the amount of modal aliasing.

All derivations for dipole and cardioid microphones are straightforward and do not provide any new insight. Therefore, their explicit expressions are omitted, especially as they do not offer any practical potential when mounted into rigid cylindrical baffles.

It should be noted that when an even number of microphones is used, one can consider applying the notion of critical sampling, just as in the sampling process of one-dimensional time signals [OS98]. Then, the sampling theorem given by Eq. (3.34) becomes,

$$M \geq 2\mathcal{N}. \tag{3.43}$$

When critical sampling is considered, one has to ensure, though, that the sampling operation does not sample the $\mathcal{N}$-th order harmonic, whose real part is equivalent to $\cos(\mathcal{N}\varphi)$, at its zero crossings as this would, of course, result in a zero output. This requirement can be met by ensuring that the $\mathcal{N}$-th order harmonic is *not* rotated by, $(2\nu + 1)\pi/\mathcal{N}$, where ν is an integer. It is assumed in the following that, whenever critical sampling is considered, this condition is satisfied.

It is further assumed that the effect of modal aliasing can be neglected due to proper array design. Therefore, the eigenbeams obtained by continuous circular apertures can be approximated by eigenbeams obtained by circular microphone arrays with sufficient accuracy in *all* subsequent discussions.

3.1.4 Representation of a 2D Wavefield Using a Finite Number of Harmonics

In Section 3.1.1 it was shown how a planar wavefield can be decomposed into circular harmonics using unbaffled as well as baffled circular apertures. Section 3.1.3 has shown that a wavefield of, in general, infinite dimensional space, $n \in [-\infty \ldots \infty]$, cannot be represented exactly by using a finite number of microphones. This section examines the error that occurs when a wavefield is represented by a finite number of circular harmonics $\mathcal{N}$, decomposed by a circular aperture. I.e., the question arises on how much information is lost by only decomposing a wavefield into $\mathcal{N}$ circular harmonics.

The wavefield truncation error can be quantified as,

$$\begin{aligned}\mathring{\mathcal{E}}_{trunc}(kR, \varphi) &= \sum_{n=-\infty}^{\infty} i^n B_n(kR) e^{in(\phi-\varphi)} - \sum_{n=-\mathcal{N}}^{\mathcal{N}} i^n B_n(kR) e^{in(\phi-\varphi)} \\ &= \sum_{|n|>\mathcal{N}} i^n B_n(kR) e^{in(\phi-\varphi)},\end{aligned} \tag{3.44}$$

where, cf. Eqs. (3.15) and (3.17),

$$B_n(kR) \triangleq \begin{cases} J_n(kR), & \text{unbaffled aperture} \\ J_n(kR) - \dfrac{J_n'(kR)}{H_n'(kR)} H_n(kR), & \text{baffled aperture} \end{cases} . \tag{3.45}$$

Parseval's theorem states that for a 2π-periodic function $f(\alpha)$ Fourier series, cf. Eq. (A.5),

$$f(\alpha) = \sum_{n=-\infty}^{\infty} a_n e^{in\alpha}, \tag{3.46}$$

it follows that [Wei03],

$$\frac{1}{2\pi} \int_0^{2\pi} |f(\alpha)|^2 \, d\alpha = \sum_{n=-\infty}^{\infty} |a_n|^2. \tag{3.47}$$

Therefore, one can obtain the mean square truncation error as,

$$\mathcal{E}^2_{trunc}(kR) \triangleq \frac{1}{2\pi} \int_0^{2\pi} \left| \sum_{|n|>\mathcal{N}} i^n B_n(kR) e^{in(\phi-\varphi)} \right|^2 d\phi = 2 \sum_{n=\mathcal{N}+1}^{\infty} B_n^2(kR), \tag{3.48}$$

which can also be written, using a finite sum [DD94] as,

$$\mathcal{E}^2_{trunc}(kR) = 1 - \sum_{n=0}^{\mathcal{N}} (2 - \delta_{n0}) B_n^2(kR), \tag{3.49}$$

where the Kronecker symbol $\delta_{n0} = 1$ for $n = 0$ and zero else. Figure 3.13 shows the wavefield truncation error for $\mathcal{N} = 0, 1, 2, 3, 10$ utilizing an unbaffled continuous circular aperture. It can be seen that for low spatial frequencies, kR, only a small number of circular harmonics are necessary to describe a wavefield with satisfactory accuracy. For example, if a circular aperture decomposes an incoming wavefield which only contains wavenumbers corresponding to $kR = 1$ into four circular harmonics, the mean square error is about -50 dB. The truncation error utilizing continuous circular apertures mounted into rigid cylindrical baffles are qualitatively identical and their marginally different numerical results do not warrant a separate discussion.

Of course, a practical realization utilizes microphones to sample the circular apertures and the error due to modal aliasing, as discussed in Section 3.1.3, needs to be added to the error due to wavefield truncation.

3.1.5 Circular Apertures and Nearfield Sources

Although not explicitly detailed in Chapter 2, the notion and effects of nearfield sources on circular apertures are introduced in this section since the results are of some importance for real-world implementations.

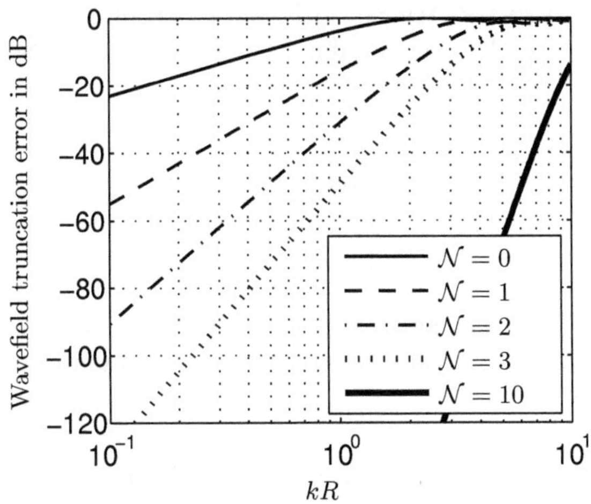

Fig. 3.13. Wavefield truncation error for $\mathcal{N} = 0, 1, 2, 3, 10$

A fundamental solution of the acoustic wave equation in two dimensions was given in Section 2.2.3, cf. Eq. (2.36), and is repeated here for convenience,

$$g_k(\boldsymbol{\varrho}|\boldsymbol{\varrho}_0) = \frac{i}{4} H_0^{(1,2)}(k\|\boldsymbol{\varrho} - \boldsymbol{\varrho}_0\|), \tag{3.50}$$

where the superscript $^{(1)}$ is used for outward traveling waves, and the superscript $^{(2)}$ is used for inward traveling waves. The geometric model is the one considered in Fig. 3.1, where the source S_0 is now located with respect to the aperture such that plane-wave incidence can no longer be assumed. Equation (3.50) with Eq. (3.3) and $\theta = \vartheta = \pi/2$ then becomes,

$$H_0^{(1,2)}(k\|\boldsymbol{\varrho} - \boldsymbol{\varrho}_0\|) = H_0^{(1,2)}\left(k\sqrt{\varrho^2 + \varrho_0^2 - 2\varrho\varrho_0 \cos(\phi - \varphi)}\right), \tag{3.51}$$

which can be expanded to yield [GR65],

$$H_0^{(1,2)}\left(k\sqrt{\varrho^2 + \varrho_0^2 - 2\varrho\varrho_0 \cos(\phi - \varphi)}\right) = \sum_{n=-\infty}^{\infty} J_n(k\varrho) H_n^{(1,2)}(k\varrho_0) e^{in(\phi-\varphi)}. \tag{3.52}$$

Note that this expansion is correct only for $\|\boldsymbol{\varrho}_0\| > \|\boldsymbol{\varrho}\|$. This relation is assumed to be valid here since this book is concerned with interior boundary problems only, see Section 2.3.1. Since the wavefield is impinging on the aperture, the wave is traveling inward. Therefore, the Hankel function of the second kind is used in the following. Note that,

$$H_n^{(2)}(k\varrho_0) = H_n^{*(1)}(k\varrho_0) \triangleq H_n^*(k\varrho_0). \tag{3.53}$$

As in the farfield scenario, see Eq. (3.7), an important quantity of interest is the total pressure on the aperture due to an impinging nearfield source. Here it follows that,

$$\begin{aligned} F_{\text{near}}(kR,\varphi) &= \frac{1}{2\pi}\int_0^{2\pi}\sum_{n=-\infty}^{\infty} J_n(kR)H_n^*(k\varrho_0)e^{in(\phi-\varphi)}e^{-im\phi}\,d\phi \\ &= J_m(kR)H_m^*(akR)e^{-im\varphi}, \end{aligned} \tag{3.54}$$

where $a > 1$ is a real constant relating ϱ_0 and R. Defining $n \triangleq m$ results in the modal decomposition of a wavefield due to a nearfield source on the unbaffled circular aperture

$$\overset{\circ}{F}_{n_{\text{near}}}(kR,\varphi) = J_n(kR)H_n^*(akR)e^{-in\varphi}. \tag{3.55}$$

This result can be directly compared with the modal decomposition of a wavefield due to a farfield source on the unbaffled circular aperture, Eq. (3.15). It can be seen that the decomposed wavefields due to nearfield and farfield sources result in identical azimuthal, i.e. angular, response. Deviations, however, are expected in the frequency response of the individual harmonics, which will be considered when the focus is on baffled circular apertures. First, it will be shown that Eq. (3.55) leads to Eq. (3.15) in the limit of large ϱ. For $\varrho \to \infty$, it holds that,

$$H_n^*(akR) \approx i^n H_0^*(akR), \tag{3.56}$$

where Eq. (B.25) has been used. Therefore,

$$\overset{\circ}{F}_{n_{\text{near}}}(kR,\varphi) \approx i^n J_n(kR)H_0^*(akR)e^{-in\phi} \sim i^n J_n(kR)e^{-in\varphi} = \overset{\circ}{F}_n(kR,\varphi). \tag{3.57}$$

Since all operations introduced to describe rigid scatterers in Section 2.3.3 do not affect the term indicating a nearfield source, $H_n^*(akR)$, the modal decomposition of a wavefield impinging on a baffled circular aperture can be written as, cf. Eq. (3.17),

$$\overset{\bullet}{F}_{n_{\text{near}}}(kR,\varphi) = H_n^*(akR)\left[J_n(kR) - \frac{J_n'(kR)}{H_n'(kR)}H_n(kR)\right]e^{-in\varphi}. \tag{3.58}$$

Again, the decomposed wavefields due to nearfield and farfield sources result in identical azimuthal response. Figure 3.14 shows the change of the modal magnitude response due to a nearfield source impinging on a baffled circular aperture decomposed into several harmonics with respect to the factor a. It can be seen that the higher the mode n the more sensitive the response

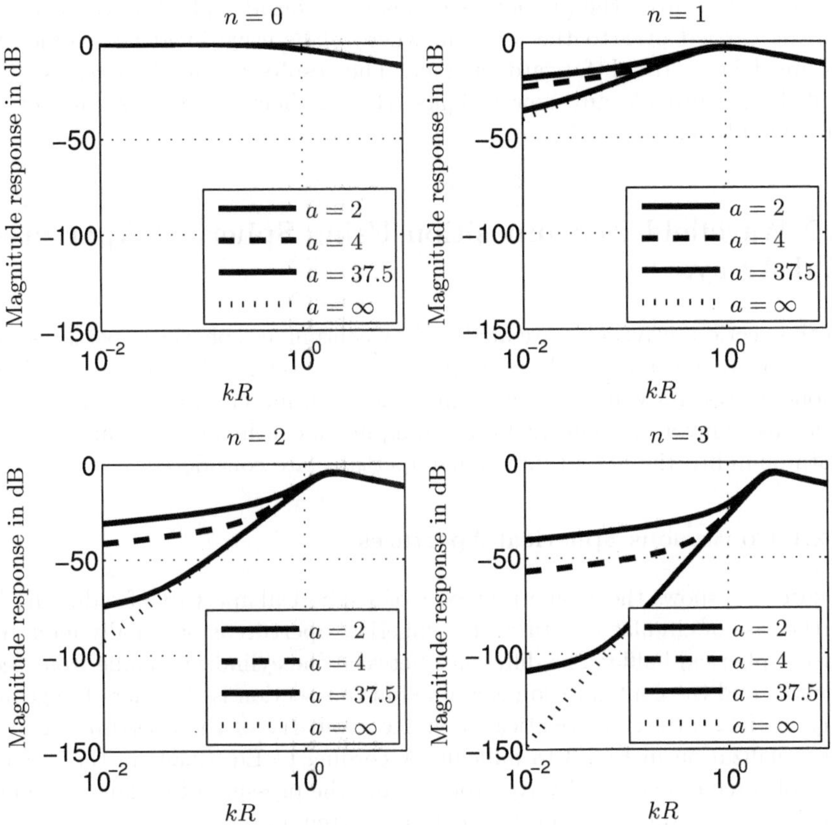

Fig. 3.14. Modal magnitude response due to a nearfield source impinging on a baffled circular aperture at $\varphi = 0$

to nearfield sources becomes, more so for low frequencies than for high frequencies. This strong dependency of the magnitude response on the distance from the source to the aperture poses a serious problem to signal processing algorithms that depend on a farfield assumption at low frequencies, see Chapter 5. It can be also seen, however, that by restricting the frequency range to be considered for further signal processing algorithms to, say $kR > 0.1$, the problems related to the varying magnitude response with respect to a can be minimized for many practical problems. It may be instructive at this point to try out a few numbers. For example, the system introduced in Chapter 6 has a radius of $R = 0.04$ m. This means that the curve for $a = 37.5$ in Fig. 3.14 corresponds to a distance from the source to the aperture of 1.5 m. This is turn means that for frequencies $f > 300$ Hz, or $kR > 0.22$, the source positioned at 1.5 m (or further away) can be approximated by a farfield source for $n \leq 3$ with satisfactory accuracy. This result will be used in Chapter 6.

All results from the previous sections that consider plane-wave incidence can be carried over to the scenario where plane-wave incidence cannot be assumed in a straightforward manner. The results are *qualitatively* similar and do not provide any new insights. Hence, their derivations will not be presented here.

3.2 Wavefield Decomposition Using Spherical Apertures and Arrays

In this section, wavefields are analyzed by continuous spherical apertures and by sampled continuous spherical apertures, i.e. spherically symmetric microphone arrays. As will be shown, sampling a continuous spherical aperture is much more involved than the task of sampling a continuous circular aperture. Both omnidirectional and directional apertures are considered.

3.2.1 Continuous Spherical Apertures

Figure 3.15 shows the geometric model of a spherical aperture of radius R. As in the case of circular apertures, the spherical aperture is optionally mounted into a spherical baffle. The difference between the cylindrical baffle from Section 3.1 and the configuration examined in this section is that here the geometry of the scatterer is identical with the geometry of the aperture, making its visualization in Fig. 3.15 difficult. According to Eq. (2.34) in the general case of three-dimensional wave propagation, the pressure of a plane-wave impinging on the spherical aperture can be written as,

$$P_{\text{inc}}(\boldsymbol{k}^T\boldsymbol{r}) = e^{i\boldsymbol{k}^T\boldsymbol{r}}, \tag{3.59}$$

where, see Section 3.1.1,

$$\boldsymbol{k}^T\boldsymbol{r} = kr[\sin\theta\sin\vartheta\cos(\phi-\varphi)+\cos\theta\cos\vartheta]. \tag{3.60}$$

The Unbaffled Spherical Aperture

In the following, the properties of unbaffled spherical apertures are considered. As in Section 3.1.1, an expression for the pressure on the spherical aperture, $r = \|\boldsymbol{r}\| = R$, due to an incoming plane-wave is required. This expression can be obtained by integrating the pressure field present on the spherical aperture. Mathematically, this is done by applying an inverse spherical harmonics transform, Eq. (A.8), to Eq. (3.59) with Eq. (3.60), i.e.,

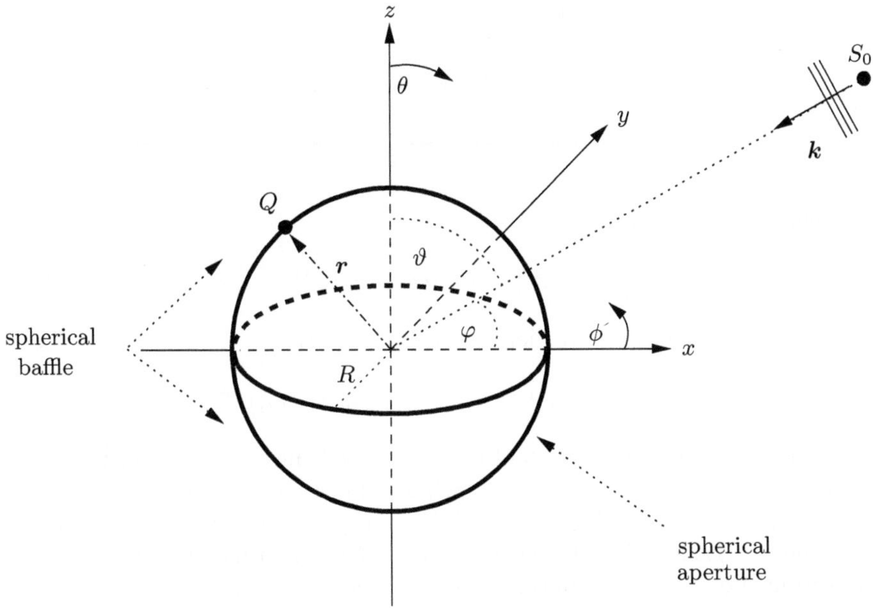

Fig. 3.15. Geometric model of a spherical aperture optionally mounted into a spherical baffle

$$
\begin{aligned}
G_n^m(kR,\varphi,\vartheta) &= \frac{1}{\sqrt{4\pi}} \int_0^{2\pi}\int_0^{\pi} e^{ikR[\sin\theta\sin\vartheta\cos(\phi-\varphi)+\cos\theta\cos\vartheta]} Y_n^m(\theta,\phi)^* \sin\theta d\theta d\phi \\
&= \frac{a_n^m}{\sqrt{4\pi}} \int_0^{\pi} e^{ikR\cos\theta\cos\vartheta} P_n^m(\cos\theta) \\
&\times \left[\int_0^{2\pi} e^{ikR\sin\theta\sin\vartheta\cos(\phi-\varphi)}\, e^{-im\phi}\, d\phi \right] \sin\theta\, d\theta,
\end{aligned}
\tag{3.61}
$$

where,

$$
a_n^m = \sqrt{\frac{2n+1}{4\pi}\frac{(n-m)!}{(n+m)!}}. \tag{3.62}
$$

The bracketed term in Eq. (3.61) was already discussed in Section 3.1.1. Therefore, Eq. (3.61) becomes,

$$G_n^m(kR,\varphi,\vartheta) = \sqrt{\pi} a_n^m i^m e^{-im\varphi} \times \underbrace{\int_0^\pi e^{ikR\cos\vartheta\cos\theta} J_m(kR\sin\vartheta\sin\theta) P_n^m(\cos\theta)\sin\theta\, d\theta}_{\triangleq I} . \quad (3.63)$$

It is shown in [MF53] that,

$$I = 2i^{n-m} P_n^m(\cos\vartheta) j_n(kR). \quad (3.64)$$

Therefore, Eq. (3.63) finally yields with Eq. (B.53),

$$G_n^m(kR,\varphi,\vartheta) = \sqrt{4\pi} i^n j_n(kR) Y_n^m(\vartheta,\varphi)^* . \quad (3.65)$$

It can be seen that the wavefield is decomposed into a set of spherical harmonics of order n and degree m, which are mutually orthonormal, as shown in Section B.3. This orthonormality will be of importance in subsequent developments. By applying these coefficients to the forward spherical harmonics transform, Eq. (A.7), one obtains the expansion for plane-waves in spherical coordinates,

$$e^{i\boldsymbol{k}^T\boldsymbol{r}}\Big|_{r=R} = P_{\mathrm{inc}}(kR,\theta,\phi) = 4\pi \sum_{n=0}^{\infty} i^n j_n(kR) \sum_{m=-n}^{n} Y_n^m(\theta,\phi) Y_n^m(\vartheta,\varphi)^* . \quad (3.66)$$

Note that this expression was already used in Section 2.2.1. In summary, an incoming plane-wave can be decomposed into several spherical harmonics of order n and degree m that, spatially, only depend on the direction-of-arrival (ϑ,φ). As in the case of wavefield decomposition using circular apertures, the frequency-dependence is decoupled from the spatial dependence. In the following, the decomposed wavefield due to an incoming plane-wave by an unbaffled spherical aperture is written as,

$$\overset{\circ}{G}{}_n^m(kR,\vartheta,\varphi) = \sqrt{4\pi} i^n j_n(kR) Y_n^m(\vartheta,\varphi)^* . \quad (3.67)$$

As before, the symbol $\overset{\circ}{(\cdot)}$ denotes an unbaffled aperture.

Figure 3.16 shows the frequency-dependent component of the decomposed wavefield by an unbaffled spherical aperture, for $n = 0(1)3$ and $\vartheta = \pi/2, \varphi = 0$, and $n = m$. Compared to Fig. 3.2, the frequency-dependent component of the decomposed wavefield by an unbaffled circular aperture, striking similarities are obvious. Again, for $n > 0$ all harmonics exhibit a highpass-like characteristic with a slope of $6n$ dB/octave up to approximately $kR < n$. Also, all harmonics have repetitive dips in their respective magnitude response, which correspond to the zeros in the spherical Bessel functions, $j_n(\cdot)$, cf. Appendix B.2.

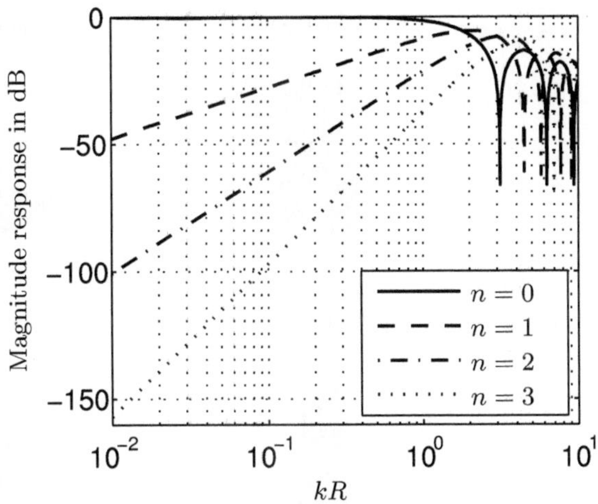

Fig. 3.16. Modal magnitude response of an unbaffled spherical aperture for $\vartheta = \pi/2, \varphi = 0$, and $n = m$

Figure 3.17 depicts the three-dimensional directional component of the spherical harmonics with respect to φ and ϑ for $n = 0, 1, 2$ and $m = 0, 1, 2$. This figure can, again, be compared to the respective plot for circular apertures, Fig. 3.4. As can be seen the wavefield can be decomposed into more components by using spherical apertures than by using circular apertures. There are not only variations in azimuth but also in elevation. Furthermore, it can be shown that the spherical harmonics can be rotated in both azimuth and elevation. This gives spherical apertures the capability of being able to decompose the wavefield in all three spatial dimensions.

The components comprising the decomposed wavefield are mutually orthogonal, i.e.,

$$\int_0^{2\pi}\int_0^{\pi} \mathring{G}_n^m \, \mathring{G}_{n'}^{m'*} \sin\vartheta d\vartheta d\varphi = \begin{cases} 4\pi(-1)^n j_n^2(kR), & n' = n, \; m' = m \\ 0, & \text{else} \end{cases} . \tag{3.68}$$

This very desirable property will be of importance in further developments.

The Baffled Spherical Aperture

As in the case of circular apertures, cf. Section 3.1.1, the dips in the magnitude response of unbaffled apertures can be 'equalized' by mounting the aperture into a rigid baffle, thereby taking advantage of its scattering properties, see Section 2.4.3. The expansion of the total pressure on the surface of a rigid spherical scatterer with radius R due to a plane-wave incidence has been

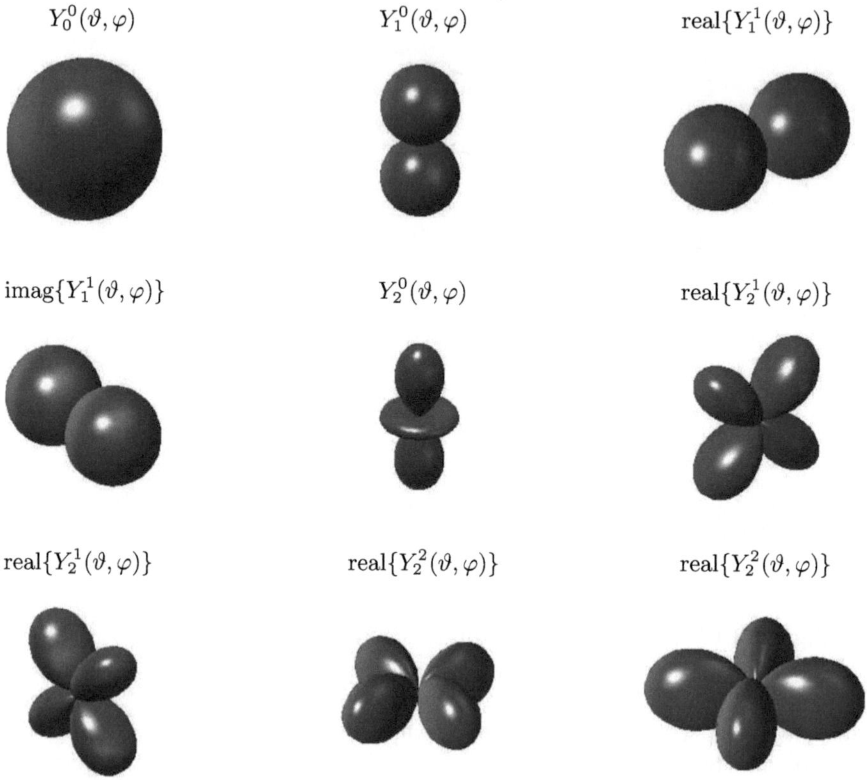

Fig. 3.17. Three-dimensional directional modal response with respect to φ and ϑ for $n = 0, 1, 2$ and $m = 0, 1, 2$

given in Eq. (2.127). The coefficients of this series expansion can be readily identified with the desired modal components of the wavefield decomposition with baffled spherical apertures as,

$$\overset{\bullet}{G}{}_n^m(kR,\vartheta,\varphi) = \sqrt{4\pi} i^n \left[j_n(kR) - \frac{j_n'(kR) h_n(kR)}{h_n'(kR)} \right] Y_n^m(\vartheta,\varphi)^* . \qquad (3.69)$$

The modal magnitude response of a baffled spherical aperture for $\vartheta = \pi/2$ and $\varphi = 0$ is reproduced in Fig. 3.18. With reference to the unbaffled spherical aperture, cf. Fig. 3.16, it can be seen that the additional components due to the rigid spherical baffle, as in the cylindrical case, exactly cancel the zeros introduced by the spherical Bessel functions, thus resulting in a smooth modal magnitude response. Also visible is the highpass character with a slope of

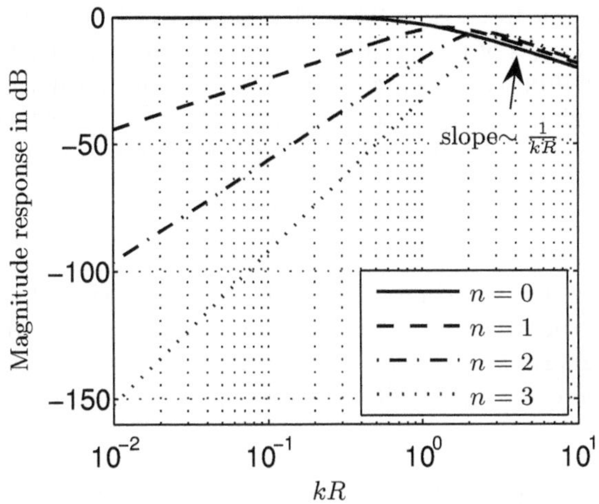

Fig. 3.18. Modal magnitude response of a baffled spherical aperture for $\vartheta = \pi/2, \varphi = 0$ and $n = m$

$6n$ dB/octave of the individual modal responses for $n > 0$ and $kR < n$. For approximately $kR > n$ the slope of all spherical harmonics is identically $1/(kR)$. This factor corresponds to the one that has been found in Section 2.4.2 for describing the attenuation of spherically propagating wave-fronts.

Another observation from Eq. (3.69) compared to Eq. (3.67) is that the rigid spherical baffle does not have any influence on the directional response of unbaffled spherical apertures. Therefore, the results shown Fig. 3.17 are not modified by baffled spherical apertures.

Additional Properties of Baffled Spherical Apertures

As in the case of a cylindrical scatterer, see Section 2.3.3, for small kR and $n > 0$ the magnitude response due to the presence of a rigid baffle always seem to lie a few dB above the setup where no rigid baffle is used. This effect can, again, be interpreted as an virtual increase of the *modal aperture*. Expressing the relative magnitude response between the response of a spherical aperture mounted into a rigid spherical baffle and an unbaffled spherical aperture from Eq. (3.69) and Eq. (3.67), respectively, gives,

$$\Delta P(kR) \triangleq \frac{j_n(kR) - \dfrac{j_n'(kR)h_n(kR)}{h_n'(kR)}}{j_n(kR)} = \frac{i}{(kR)^2 j_n(kR) h_n'(kR)}. \tag{3.70}$$

Since the aperture increase is only effective for small kR, it follows with Eq. (B.38) and Eq. (B.41) that,

$$\Delta P(kR) \approx \frac{(2n+1)!!}{(n+1)(2n-1)!!} = \frac{(2n+1)!}{(n+1)(2n)!} \triangleq f_n. \tag{3.71}$$

It therefore follows from Eq. (3.71) that for all $n > 0$ the rigid scatterer increases the response of the circular aperture by about $20 \log f_n$ dB, differing here from the cylindrical case, cf. Section 3.1.1. Since all harmonics of order $n > 0$ exhibit a highpass characteristic with a slope of $6n$ dB/octave it follows that the effective modal (virtual) aperture increases by a factor of,

$$\frac{R_{\text{virt}}}{R} = \sqrt[n]{f_n}. \tag{3.72}$$

Figure 3.19 shows the variability of the increase of the virtual modal aperture with respect to the mode n. Note that, interestingly, the maximum effective virtual modal aperture increase is 1.5 for $n = 1$, compared to a maximum increase by a factor of 2 for baffled circular apertures, see Fig. 3.7.

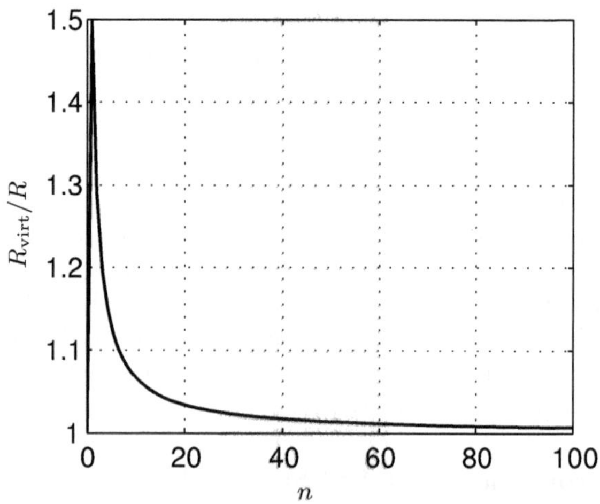

Fig. 3.19. Relative modal aperture increase due to a rigid spherical scatterer

3.2.2 Directional Spherical Apertures

A directional unbaffled spherical aperture can be modeled by applying a directional weighting factor to the inverse spherical harmonics transform of a plane-wave, i.e.,

$$\mathring{\mathcal{G}}_n^m(kR, \vartheta, \varphi) = \frac{1}{\sqrt{4\pi}} \int_0^{2\pi} \int_0^{\pi} W(\theta, \vartheta, \phi, \varphi) e^{i\boldsymbol{k}^T \boldsymbol{r}} Y_n^m(\theta, \phi)^* \sin\theta d\theta d\phi, \tag{3.73}$$

where $W(\theta, \vartheta, \phi, \varphi)$ describes the sensitivity of an infinitesimal point on the aperture. In order to find an expression for the weighting term pointing into the (ϑ, φ)-direction, an expression for the unrotated weighting term, $W(\theta, \phi)$, needs to be derived. Any sensitivity pattern that is rotationally symmetric with respect to the z-axis can be expressed as [ME02],

$$W(\theta, \phi) = \sum_{p=0}^{\infty} w_p Y_p^0(\theta, \phi), \tag{3.74}$$

with the appropriate coefficients defined as,

$$w_p = \int_0^{2\pi} \int_0^{\pi} W(\theta, \phi) Y_p^0(\theta, \phi)^* \sin\theta d\theta d\phi. \tag{3.75}$$

The rotation in azimuth and elevation of the pattern given by Eq. (3.74) is considered next. Let γ define the angle between two points on the unit sphere, (θ, ϕ) and (ϑ, φ), as,

$$\cos\gamma = \cos\theta\cos\vartheta + \sin\theta\sin\vartheta\cos(\phi - \varphi). \tag{3.76}$$

Then, using Eq. (B.55), Eq. (3.74) can be written as,

$$W(\gamma) = \sum_{p=0}^{\infty} w_p \sqrt{\frac{2p+1}{4\pi}} P_p(\cos\gamma), \tag{3.77}$$

and therefore, using the addition theorem for spherical harmonics Eq. (B.59),

$$W(\theta, \vartheta, \phi, \varphi) = \sum_{p=0}^{\infty} w_p \sqrt{\frac{4\pi}{2p+1}} \sum_{q=-p}^{p} Y_p^q(\theta, \phi) Y_p^q(\vartheta, \varphi)^*. \tag{3.78}$$

Plugging this result and the plane-wave expansion, Eq. (3.66), into Eq. (3.73) yields after a change of summation and integration,

$$\begin{aligned} \mathring{\mathcal{G}}_n^m(kR, \vartheta, \varphi) = \sqrt{4\pi} \sum_{n=0}^{\infty} i^n j_n(kR) \sum_{p=0}^{\infty} w_p \sqrt{\frac{4\pi}{2p+1}} \sum_{q=-p}^{p} Y_p^q(\vartheta, \varphi)^* \\ \times \sum_{m=-n}^{n} Y_n^m(\vartheta, \varphi)^* \int_0^{2\pi} \int_0^{\pi} Y_n^m(\theta, \phi) Y_p^q(\theta, \phi) Y_{n'}^{m'}(\theta, \phi)^* \sin\theta d\theta d\phi. \end{aligned} \tag{3.79}$$

The evaluation of the integral for arbitrary p and q is very complicated. The problem of integrating the product of three spherical harmonics occurs frequently in quantum mechanics and their solutions make extensive use of so-called Clebsch-Gordan coefficients [Ros57]. As with directional circular apertures, see Section 3.1.2, two special cases of particular interest are considered next in some detail. First, spherical apertures where each infinitesimal point exhibits spatial dipole characteristics. Second, spatial cardioid characteristics are considered.

Spherical Dipole Aperture

The spatial dependence associated with dipole apertures in spherical coordinates can be written as,

$$W(\theta,\vartheta,\phi,\varphi) = \sin\theta\sin\vartheta\cos(\phi-\varphi) + \cos\theta\cos\vartheta. \tag{3.80}$$

By comparing Eq. (3.80) with Eq. (3.78), it can be readily shown that the expansion coefficients in Eq. (3.78) are then given by,

$$w_p = \begin{cases} \sqrt{\frac{4\pi}{3}}, & p=1 \\ 0, & \text{else} \end{cases}. \tag{3.81}$$

Applying the specific values for the spherical harmonics in p and q given in Tab. B.2 yields for Eq. (3.79),

$$\begin{aligned}
\mathring{\mathcal{G}}^{\,m'}_{n'D}(kR,\vartheta,\varphi) &= \sqrt{4\pi}\sum_{n=0}^{\infty} i^n j_n(kR) \sum_{m=-n}^{n} Y_n^m(\vartheta,\varphi)^* \\
&\times \Bigg\{ \frac{1}{2}\sin\vartheta e^{i\varphi} \int_0^{2\pi}\int_0^{\pi} Y_n^m(\theta,\phi)\sin\theta\, e^{-i\phi} Y_{n'}^{m'}(\theta,\phi)^* \sin\theta d\theta d\phi \\
&+ \cos\vartheta \int_0^{2\pi}\int_0^{\pi} Y_n^m(\theta,\phi)\cos\theta\, Y_{n'}^{m'}(\theta,\phi)^* \sin\theta d\theta d\phi \\
&+ \frac{1}{2}\sin\vartheta e^{-i\varphi} \int_0^{2\pi}\int_0^{\pi} Y_n^m(\theta,\phi)\sin\theta\, e^{i\phi} Y_{n'}^{m'}(\theta,\phi)^* \sin\theta d\theta d\phi \Bigg\}.
\end{aligned} \tag{3.82}$$

Utilizing recurrence relations for spherical harmonics, Eqs. (B.61), and applying the orthonormality principle one obtains after cumbersome but straightforward calculations with $n \triangleq n'$ and $m \triangleq m'$,

$$\begin{aligned}
&\mathring{\mathcal{G}}^{\,m}_{nD}(kR,\vartheta,\varphi) = \sqrt{\pi} i^{n-1} \\
&\times \Big\{ \sin\vartheta e^{i\varphi} \left[a_1 j_{n-1}(kR) Y_{n-1}^{m+1}(\vartheta,\varphi)^* + a_2 j_{n+1}(kR) Y_{n+1}^{m+1}(\vartheta,\varphi)^*\right] \\
&+ 2\cos\vartheta \left[a_3 j_{n-1}(kR) Y_{n-1}^{m}(\vartheta,\varphi)^* - a_4 j_{n+1}(kR) Y_{n+1}^{m}(\vartheta,\varphi)^*\right] \\
&- \sin\vartheta e^{-i\varphi} \left[a_5 j_{n-1}(kR) Y_{n-1}^{m-1}(\vartheta,\varphi)^* + a_6 j_{n+1}(kR) Y_{n+1}^{m-1}(\vartheta,\varphi)^*\right] \Big\},
\end{aligned} \tag{3.83}$$

where,

$$a_1 = \sqrt{\frac{(n-m+1)(n-m+2)}{(2n+1)(2n+3)}}, \qquad a_2 = \sqrt{\frac{(n+m)(n+m-1)}{(2n-1)(2n+1)}},$$
$$a_3 = \sqrt{\frac{(n-m+1)(n+m+1)}{(2n+1)(2n+3)}}, \qquad a_4 = \sqrt{\frac{(n-m)(n+m)}{(2n-1)(2n+1)}}, \tag{3.84}$$
$$a_5 = \sqrt{\frac{(n+m+1)(n+m+2)}{(2n+1)(2n+3)}}, \qquad a_6 = \sqrt{\frac{(n-m)(n-m-1)}{(2n-1)(2n+1)}}.$$

Note that the case $n = m = 0$ is not covered by Eq. (3.83), however, since the spherical harmonics are not defined for negative orders. A solution in this case can be found by letting $n' = 0$ and $m' = 0$ in Eq. (3.82) and by applying Eq. (B.56) as well as Eqs. (B.61), resulting in,

$$\mathring{\mathcal{G}}\,{}^{0}_{0D}(kR, \vartheta, \varphi) = i j_1(kR). \tag{3.85}$$

The decomposition of a plane-wave by spherical dipole apertures is mathematically of much higher intricacy than the decomposition by a circular dipole aperture, see Section 3.1.2. However, all statements made in Section 3.1.2 remain qualitatively valid here, too, and are therefore not repeated here.

By applying Eqs. (B.61) to Eq. (3.83) it can be shown that for $n, m \neq 0$,

$$\begin{aligned}\mathring{\mathcal{G}}\,{}^{m}_{nD}(kR, \vartheta, \varphi) = \sqrt{\pi} i^{n-1} \Big\{ & j_{n-1}(kR) \left[b_1 Y_n^m(\vartheta, \varphi)^* + b_2 Y_{n-2}^m(\vartheta, \varphi)^* \right] \\ & + j_{n+1}(kR) \left[b_3 Y_n^m(\vartheta, \varphi)^* + b_4 Y_{n+2}^m(\vartheta, \varphi)^* \right] \Big\},\end{aligned} \tag{3.86}$$

where $b_\nu, \nu = 1(1)4$, are rather complicated expressions involving n and m. This representation was chosen only to show that the components of the wavefield decomposed by spherical dipole apertures are orthogonal. This statement can be readily verified by multiplying Eq. (3.86) by $\mathring{\mathcal{G}}\,{}^{m'}_{n'D}(kR, \vartheta, \varphi)^*$ and integrating over the unit sphere, here covering all angles (ϑ, φ).

Spherical Cardioid Aperture

The spatial dependence associated with cardioid apertures in spherical coordinates can be expressed as,

$$W(\theta, \vartheta, \phi, \varphi) = \frac{1}{2} \Big[1 + \sin\theta \sin\vartheta \cos(\phi - \varphi) + \cos\theta \cos\vartheta \Big]. \tag{3.87}$$

It can be readily shown that the expansion coefficients in Eq. (3.78) are then given by,

$$w_p = \begin{cases} \frac{1}{2}\sqrt{4\pi}, & p = 0 \\ \frac{1}{2}\sqrt{\frac{4\pi}{3}}, & p = 1 \\ 0, & \text{else} \end{cases}. \tag{3.88}$$

From Eq. (3.87), Eq. (3.67), and Eq. (3.83) it follows immediately that the components of the wavefield decomposed by a spherical cardioid aperture can be written as,

$$\mathring{\mathcal{G}}\,{}^{m}_{nC}(kR,\vartheta,\varphi) = \frac{1}{2}\left[\mathring{G}\,{}^{m}_{n}(kR,\vartheta,\varphi) + \mathring{\mathcal{G}}\,{}^{m}_{nD}(kR,\vartheta,\varphi)\right]. \tag{3.89}$$

Again, all statements made in Section 3.1.2, where the decomposition of wavefields by circular cardioid apertures are discussed, remain qualitatively valid in this section and are not detailed here.

3.2.3 Spherical Microphone Arrays

For practical applications, it is necessary to sample the continuous aperture at discrete microphone positions. However, sampling a function on a sphere is significantly more difficult than sampling a function on a circle since true equidistant spatial sampling is only possible for arrangements that are constructed according to the so-called Platonic solids or regular polyhedrons [Wei03]. In three dimensions, only five such geometries exist, i.e. tetrahedron, cube, octahedron, dodecahedron, and icosahedron. These consist of 4,6,8,12, and 20 faces, respectively. Therefore, if true equidistant sampling is to be applied, one is restricted in choices of possible arrangements. Moreover, the maximum number of sampling points is 20 by placing them in the center of the faces of an icosahedron. An arrangement that provides a sampling scheme that is very close to being uniform is the truncated icosahedron (soccer ball) which comprises 32 faces [EKM03]. See Fig. 3.21(a) for the arrangement of this sampling scheme on the unit sphere.

As will be shown in Chapter 5, most eigenbeam array processing algorithms rely on the orthogonality property of the eigen-solutions of the acoustic wave equation in circular/cylindrical and spherical coordinates. Based on this requirement, it can be concluded that spherical sampling can be cast into the problem of finding a discretization of the orthonormality relation for spherical harmonics Eq. (B.56), i.e.,

$$\sum_{\ell=1}^{M} w_\ell Y_n^m(\theta_\ell,\phi_\ell) Y_{n'}^{m'}(\theta_\ell,\phi_\ell)^* = \delta_{nn'}\delta_{mm'}, \tag{3.90}$$

where M denotes the total number of sampling points on the sphere and w_ℓ, $\ell = 1(1)M$, is a possibly non-uniform weighting factor.

Therefore, spherical sampling can be performed by utilizing non-equidistantly positioned sampling points with non-uniform weighting. Research has come up with a vast variety of different sampling schemes for a sphere that do not provide equidistant coverage of the sphere but are optimum in some other respect. References discussing various schemes of sampling on a sphere include [SHS] and [FM96].

In this book, two sampling structures are considered that have proven useful for the application of wavefield decomposition, namely the so-called t-design method and the ϕ_{equ}-Gaussian method. The latter one is particularly interesting because it allows for fast implementations and closed-form analysis of the error due to modal aliasing.

The ϕ_{equ}-Gaussian Method for Spherical Sampling

First, it is noted that the expansion of a plane-wave on the surface of a sphere, Eq. (3.66), is equivalent to a spherical harmonics transform as defined in Eq. (A.7). Similarly, the wavefield decomposition of a plane-wave on the surface of a sphere, Eq. (3.67), is equivalent to an inverse spherical harmonics transform, Eq. (A.8). The task now is to find a discrete representation of the inverse spherical harmonics transform of a three-dimensional signal on the unit sphere. From Eq. (A.8) it follows that for a function $f(\cos\theta, \phi)$, the inverse spherical harmonics transform can be written as,

$$\begin{aligned} f_{mn} &= \frac{a_n^m}{\sqrt{4\pi}} \int_0^{2\pi}\int_0^{\pi} f(\cos\theta,\phi) P_n^m(\cos\theta) e^{-im\phi} \sin\theta d\theta d\phi \\ &= \frac{a_n^m}{\sqrt{4\pi}} \int_0^{\pi} P_n^m(\cos\theta) \left[\int_0^{2\pi} f(\cos\theta,\phi) e^{-im\phi} d\phi \right] \sin\theta d\theta, \end{aligned} \tag{3.91}$$

where a_n^m is defined in Eq. (3.62). The term within the brackets can be readily identified as the Fourier transform of an individual circle surrounding the unit sphere at elevation θ. This is the same situation as in the discussion for circular apertures and the same arguments that are brought forward in Section 3.1.3 apply here as well. In particular, the analysis of the modal sampling error can be readily carried over. Therefore, the horizontal sampling can be described the same way as in Eq. (3.39), i.e. by identifying a term corresponding to a continuous aperture and another term corresponding to modal aliasing. Hence,

$$f_{mn}^{s\phi} = \sqrt{\pi} a_n^m \int_0^{\pi} P_n^m(\cos\theta)[f_m(\cos\theta) + f_m^e(\cos\theta)] \sin\theta d\theta, \tag{3.92}$$

where $f_m^e(\cos\theta)$ is the additional component due to modal aliasing. The $f_{mn}^{s\phi}$ are the resulting coefficients after applying a discrete Fourier transform along a circle at elevation θ. After a change of variables, $\xi \triangleq \cos\theta$,

$$f_{mn} = \sqrt{\pi} a_n^m \int_{-1}^{1} [f_m(\xi) + f_m^e(\xi)] \, P_n^m(\xi) d\xi. \tag{3.93}$$

This integral is also known as a Legendre transform [DJ94] of the term within the brackets. The question now arises on how to discretize this integral with minimum error and the minimum number of evaluation points. An answer to this question was already given by Gauss in 1814 [Gau14] who discovered that a polynomial $f(\xi)$ of *limited* order $2O-1$ can be integrated *exactly* by O discrete evaluation points, i.e.,

$$\int_{-1}^{1} f(\xi)d\xi = \sum_{l=1}^{O} w_l f(\xi_l), \tag{3.94}$$

where ξ_l are the evaluation points with appropriate weighting w_l. It is shown in [DR84] that ξ_l corresponds to the l-th zero of $P_O(\xi)$ and that the weights are given by,

$$w_l = \frac{2}{(1-\xi_l^2)[P_O'(\xi_l)]^2}, \qquad l = 1(1)O. \tag{3.95}$$

This type of numerical integration is also known as the Gauss-Legendre quadrature formula. Values for (w_l, ξ_l) are tabulated in [AS72]. It can be readily verified that the integrand of Eq. (3.93) is well-behaved and that it can be represented by polynomials. These polynomials are, however, not order-limited which makes the integration in Eq. (3.94) approximate,

$$\int_{-1}^{1} f(\xi)d\xi = \sum_{l=1}^{O} w_l f(\xi_l) + R_O, \tag{3.96}$$

where [AS72],

$$R_O = \frac{2^{2O+1}(O!)^4}{(2O+1)[(2O)!]^3} \frac{d^{2O}}{du^{2O}} f(u) \quad (-1 < u < 1). \tag{3.97}$$

Applying the principles of numerical integration of arbitrary functions, $f(\cos\theta, \phi)$, on the unit sphere to the problem of decomposing a three-dimensional sound field into spherical harmonics using spherical microphone arrays yields with Eq. (3.91) and Eq. (3.63),

$$\begin{aligned} \overset{\circ}{G}{}_n^{ms}(kR,\vartheta,\varphi) = \sqrt{\pi} a_n^m \int_0^{\pi} e^{ikR\cos\vartheta\cos\theta} \Big[i^m J_m(kR\sin\vartheta\sin\theta) e^{-im\varphi} \\ + \overset{\circ}{F}{}_m^e(kR,\vartheta,\varphi,\theta) \Big] P_n^m(\cos\theta)\sin\theta d\theta, \end{aligned} \tag{3.98}$$

where the bracketed expression is the DFT of the output of the individual circular microphone arrays comprising M elements at elevation angle θ and, according to Eq. (3.39),

$$\begin{aligned} \overset{\circ}{F}{}_m^e(kR,\vartheta,\varphi,\theta) = \sum_{q=1}^{\infty} & i^{Mq-m} J_{Mq-m}(kR\sin\vartheta\sin\theta) e^{i(Mq-m)\varphi} \\ + \; & i^{-(Mq+m)} J_{Mq+m}(kR\sin\vartheta\sin\theta) e^{-i(Mq+m)\varphi}. \end{aligned} \tag{3.99}$$

A change of variables, $\xi \triangleq \cos\theta$, yields,

$$\overset{\circ}{G}{}_n^{ms}(kR,\vartheta,\varphi) = \sqrt{\pi} a_n^m \int_{-1}^{1} e^{ikR\xi\cos\vartheta} \left[i^m J_m \left(kR\sin\vartheta\sqrt{1-\xi^2}\right) e^{-im\varphi} + \overset{\circ}{F}{}_m^e \left(kR,\vartheta,\varphi,\sqrt{1-\xi^2}\right) \right] P_n^m(\xi) d\xi. \tag{3.100}$$

The output of a spherical microphone array using M equidistantly positioned microphones in azimuth and O microphones in elevation based on ϕ_{equ}-Gaussian sampling, cf. Fig. 3.21(b), is therefore with Eq. (3.96),

$$\overset{\circ}{G}{}_n^{ms}(kR,\vartheta,\varphi) = R_O + \sqrt{\pi} a_n^m \sum_{l=1}^{O} w_l e^{ikR\xi_l\cos\vartheta} \left[i^m J_m \left(kR\sin\vartheta\sqrt{1-\xi_l^2} \right) e^{-im\varphi} + \overset{\circ}{F}{}_m^e \left(kR,\vartheta,\varphi,\sqrt{1-\xi_l^2} \right) \right] P_n^m(\xi_l). \tag{3.101}$$

For large O, the first term in the sum of Eq. (3.101) approaches Eq. (3.63) and is therefore a representation for the modal decomposition of a continuous spherical aperture. The second term in the sum is then the additional component due to modal aliasing. The term R_O can be neglected in the following since the factor in Eq. (3.97) becomes very small quickly. For $O = 1$ the factor is $R_O = 1/3$ and for $O = 4$, $R_O \approx 2.9e^{-7}$. Additionally, the $2O$–th derivation of $f(u)$ is bounded by a number on the order of one.

With this assumption, an expression similar to Eq. (3.41) can readily be determined from Eq. (3.101) and Eq. (3.67) as,

$$\overset{\circ}{\mathcal{E}}{}_n^m(kR,\vartheta,\varphi) \triangleq \frac{|\overset{\circ}{G}{}_n^{me}(kR,\vartheta,\varphi)|^2}{|\overset{\circ}{G}{}_n^{ms}(kR,\vartheta,\varphi)|^2} = \frac{|\overset{\circ}{G}{}_n^{ms}(kR,\vartheta,\varphi) - \overset{\circ}{G}{}_n^{m}(kR,\vartheta,\varphi)|^2}{|\overset{\circ}{G}{}_n^{ms}(kR,\vartheta,\varphi)|^2}. \tag{3.102}$$

It therefore follows that modal aliasing is dependent on both azimuth and elevation for sampled continuous spherical apertures.

As in the case of sampling the circular aperture, it is impossible to derive strict criteria regarding the control of modal aliasing. Instead, by supplying a maximum amount of modal aliasing allowable for a given application one needs to check each component of the decomposed wavefield utilizing $M \cdot O$ microphones separately. As an example, Fig. 3.20 shows the relative modal aliasing error, cf. Fig. 3.10, with respect to elevation and kR for $n = m = 3$ and $\varphi = 0$.

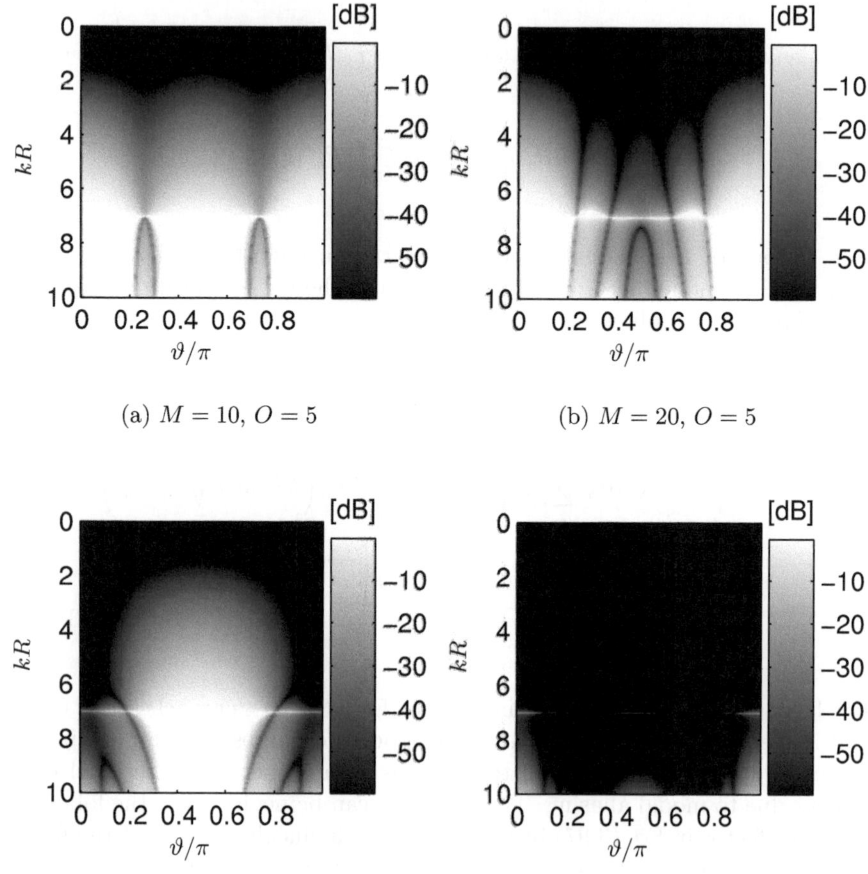

(a) $M = 10$, $O = 5$

(b) $M = 20$, $O = 5$

(c) $M = 10$, $O = 10$

(d) $M = 20$, $O = 10$

Fig. 3.20. Modal aliasing error, $\overset{\circ}{\mathcal{E}}{}_n^m\,(kR, \vartheta, \varphi)$, due to spherical sampling using ϕ_{equ}-Gaussian nodes for $n = m = 3$ and $\varphi = 0$

For certain applications, the distribution of sampling points on the sphere as described in this section is of particular interest since it allows for a fast implementation. The sampling in the horizontal direction can be performed by applying individual FFTs. The sampling in elevation involves a Legendre transform where fast implementations exist that make extensive use of the recurrence relations of the Legendre functions, see [HRM96] or [Moh99].

Using this type of spatial sampling, one is able to predict the minimum number of microphones required in order to decompose a wavefield into spherical harmonics of maximum order $\mathcal{N}$ and maximum degree $\mathcal{M} = \mathcal{N}$. Since the horizontal sampling is equidistant, Eq. (3.34), the condition for sampling circular apertures, holds here as well, i.e. $M \geq 2\mathcal{N} + 1$ microphones are necessary in order to extract $\mathcal{N}$ *circular* harmonics. In elevation, it can be shown [Moh99] that $O \geq \mathcal{M} + 1$ positioned on Gaussian nodes are required. Therefore, for example, a total number of at least 6, 15, and 28 microphones are necessary to decompose the wavefield up to order 1, 2, and 3, respectively. Of course, modal aliasing decreases if more than the minimum required number of microphones are used.

The *t*-Design Method for Spherical Sampling

A potential disadvantage of placing microphones on a sphere according to the ϕ_{equ}-Gaussian method for spherical sampling is the fact that the microphones are more densely packed near the poles of the sphere. They are not evenly distributed on the sphere's surface. Furthermore there is no uniform weighting of the sampling points. A discretization method that circumvents both problems is referred to as a so-called t-design [1] and is described in [HS96]. A set of M points $(T_1, \ldots, T_M)$ on the unit sphere forms a spherical t-design if,

$$\int_0^{2\pi}\int_0^{\pi} f(\cos\theta, \phi)\cos\theta d\theta d\phi = \frac{1}{M}\sum_{\ell=1}^{M} f(T_\ell), \tag{3.103}$$

holds for all polynomials $f(\cos\theta, \phi)$ described on the unit sphere of degree less than t. Only by numerical optimization methods it is possible to find a set of points on the sphere for which Eq. (3.103) is satisfied up to a given error. Solutions for M and t are tabulated in [HS96] and the coordinates are distributed electronically at [SHS]. The t-design sampling scheme for 32 evaluation points on the unit sphere is visualized in Fig. 3.21(c).

The optimization procedure for obtaining the integration nodes does not allow for *direct* closed-form error analysis expressions as presented earlier.As a consequence, the modal aliasing introduced by sampling using the t-design method can be obtained via simulations only. However, mathematical expressions can be derived by an *indirect* approach utilizing the discrete orthonormality criteria, Eq. (3.90), see [Raf05]. In the following, the discussion is restricted to results obtained via simulations. Fig. 3.22 shows the result of sampling the continuous spherical aperture at 32 points using three different sampling schemes. All three schemes, employing 32 sampling points, are applied to the decomposition of a plane-wave of order $n = 3$ and degree $m = 3$. The figures were created similarly as Fig. 3.20.

Unfortunately, it is impossible for sampling schemes based on the t-design method to provide a general rule determining the minimum number of microphones required to extract *all* harmonics up to a given order. Hardin and

[1] Note that the continuous time variable and the symbol t used for describing this method are fundamentally different.

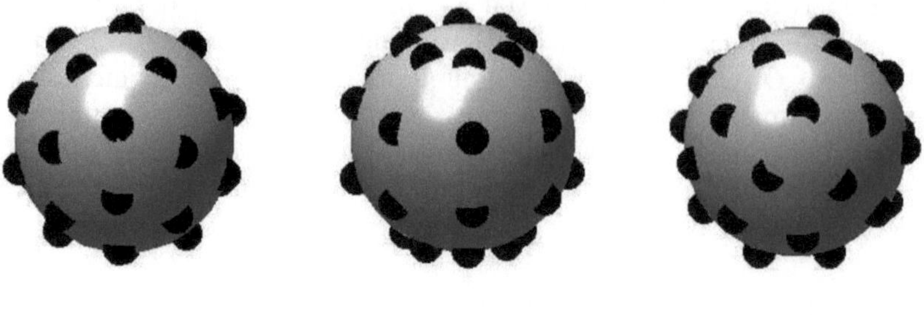

(a) truncated icosahedron (b) ϕ_{equ}-Gauss, $M = 8, O = 4$ (c) t-design

Fig. 3.21. Schemes for sampling a continuous spherical aperture using 32 sensors

Sloan [HS96] have found criteria, either in closed-form or numerically, on how many sampling points are required for numerically integrating a polynomial of a given order on a sphere. This problem is equivalent to finding the minimum number of microphones required for extracting harmonics on the spherical aperture of all degrees corresponding to a given order. However, rules of general validity are not given in [HS96]. Simulations have provided evidence that at least 6, 12, 22, and 30 microphones are required in order be able to decompose a wavefield into *all* spherical harmonics up to order 1, 2, 3, and 4, respectively.

3.2.4 Representation of a 3D Wavefield Using a Finite Number of Harmonics

In analogy to the definitions given for circular apertures, see Section 3.1.4, the wavefield truncation error can be quantified as,

$$\mathcal{E}_{\text{trunc}}(kR, \vartheta, \varphi) = 4\pi \sum_{n=\mathcal{N}+1}^{\infty} i^n b_n(kR) \sum_{m=-n}^{n} Y_n^m(\theta, \phi) Y_m^n(\vartheta, \varphi)^*, \tag{3.104}$$

where,

$$b_n(kR) \triangleq \begin{cases} j_n(kR), & \text{unbaffled aperture} \\ j_n(kR) - \dfrac{j_n'(kR)}{h_n'(kR)} h_n(kR), & \text{baffled aperture} \end{cases} . \tag{3.105}$$

Parseval's theorem states that if a function $f(\theta, \phi)$ can be expanded into a generalized Fourier series based on spherical harmonics as, Eq. (A.7),

$$f(\theta, \phi) = \sqrt{4\pi} \sum_{n=0}^{\infty} \sum_{m=-n}^{n} f_{mn} Y_n^m(\theta, \phi) \tag{3.106}$$

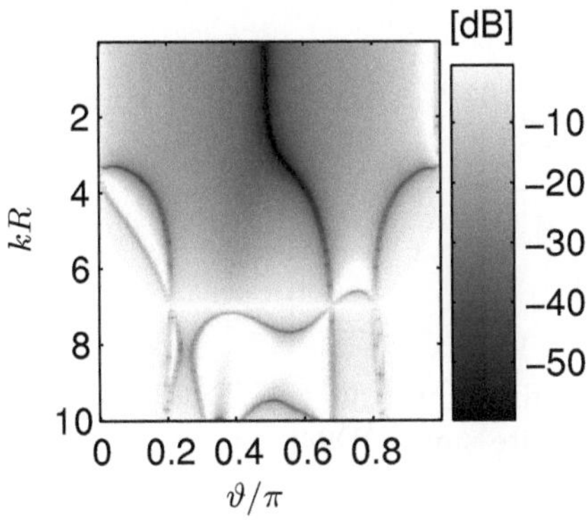

(a) truncated icosahedron sampling

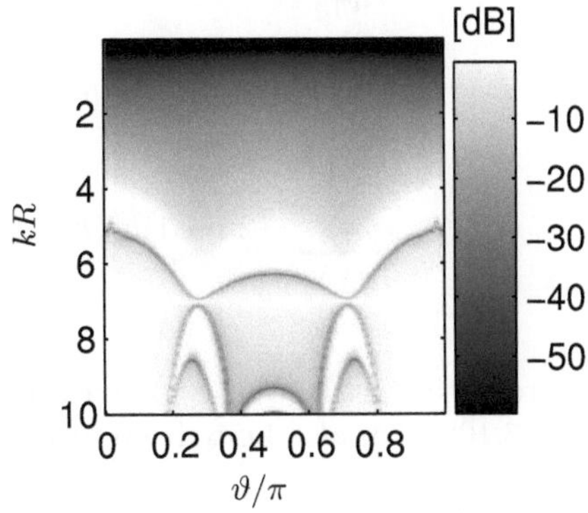

(b) ϕ_{equ}-Gaussian

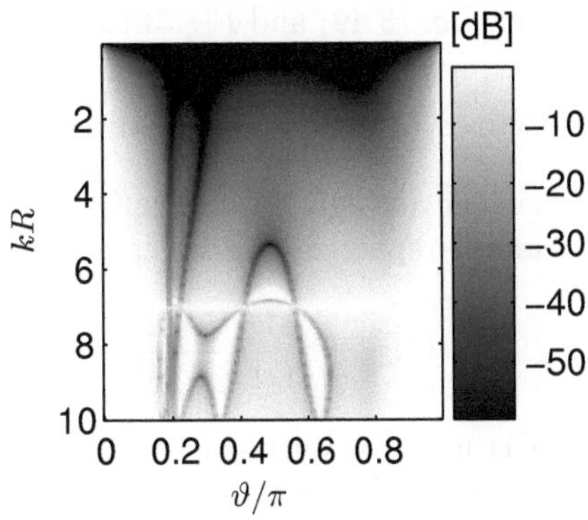

(c) t-design sampling

Fig. 3.22. Error due to modal aliasing by sampling a continuous spherical aperture at 32 points utilizing different sampling schemes where $n = m = 3$ and $\varphi = 0$

then it follows that [Wei03],

$$\frac{1}{4\pi}\int_0^{2\pi}\int_0^{\pi} |f(\theta,\phi)|^2 \sin\theta d\theta d\phi = \sum_{n=-\infty}^{\infty}\sum_{m=-n}^{n} |f_{mn}|^2. \tag{3.107}$$

Therefore, one can obtain the squared truncation error as,

$$\mathcal{E}^2_{\text{trunc}}(kR) \triangleq \frac{1}{4\pi}$$
$$\times \int_0^{2\pi}\int_0^{\pi}\left|4\pi \sum_{n=\mathcal{N}+1}^{\infty} i^n b_n(kR) \sum_{m=-n}^{n} Y_n^m(\theta,\phi)Y_n^m(\vartheta,\varphi)^*\right|^2 \sin\theta d\theta d\phi$$
$$= \sum_{n=\mathcal{N}+1}^{\infty} (2n+1)b_n^2(kR), \tag{3.108}$$

where in the last step, the spherical harmonics sum rule, Eq. (B.60) was used. Eq. (3.108) can also be written, using a finite sum [DD94], as,

$$\mathcal{E}^2_{\text{trunc}}(kR) = 1 - \sum_{n=0}^{\mathcal{N}}(2n+1)b_n^2(kR), \tag{3.109}$$

Numerically, this result differs only slightly from the one obtained for two-dimensional wavefields, cf. Eq. (3.49) and Fig. 3.13.

3.2.5 Spherical Apertures and Nearfield Sources

A fundamental solution of the acoustic wave equation in three dimensions was given in Section 2.2.2, and is repeated here for convenience,

$$g_k(\boldsymbol{r}|\boldsymbol{r}_0) = \frac{e^{\pm ik\|\boldsymbol{r}-\boldsymbol{r}_0\|}}{4\pi\|\boldsymbol{r}-\boldsymbol{r}_0\|}, \tag{3.110}$$

where the positive sign is used for outward traveling waves, and the negative sign is used for inward traveling waves. The geometric model is the one considered in Fig. 3.15, where the source S_0 is now located with respect to the aperture such that plane-wave incidence can no longer be assumed. An expansion of Eq. (3.110) has been given in Eq. (2.21) and Eq. (2.24). Since the wavefield is impinging on the aperture, the wave is traveling inward. Therefore, the expansion corresponding to the minus-sign in Eq. (3.110) has to be used here. With the definition,

$$h_n^{(2)}(kr_0) = h_n^{*(1)}(kr_0) \triangleq h_n^*(kr_0), \tag{3.111}$$

it follows for the modal response of a spherical aperture with radius $r = R$ and $a > 1$,

$$G_{\text{near}}(kR,\vartheta,\varphi) = -\sqrt{4\pi}ik \int_0^{2\pi}\int_0^{\pi} \sum_{n=0}^{\infty} j_n(kR)h_n^*(akR)$$
$$\times \sum_{m=-n}^{n} Y_n^m(\theta,\phi)Y_n^m(\vartheta,\varphi)^* Y_{n'}^{m'}(\theta,\phi)^* \sin\theta d\theta d\phi, \tag{3.112}$$

which finally simplifies to the modal decomposition of a spherical aperture due to a nearfield source,

$$\overset{\circ}{G}{}^{m}_{n_{\text{near}}}(kR,\vartheta,\varphi) = -\sqrt{4\pi} i k j_n(kR) h_n^*(akR) Y_n^m(\vartheta,\varphi)^*. \tag{3.113}$$

As in the case of modal decomposition by a circular aperture due to a nearfield source it can be easily shown that in the large-argument limit of the spherical Hankel function, Eq. (3.113) approaches Eq. (3.67).

Since all operations introduced to describe rigid spherical scatterers, see Section 2.4.3, do not affect the term indicating a nearfield source, $h_n^*(akR)$, the modal decomposition of a wavefield impinging on a baffled spherical aperture can be written as, cf. Eq. (3.69),

$$\overset{\bullet}{G}{}^{m}_{n_{\text{near}}}(kR,\vartheta,\varphi) = -\sqrt{4\pi} i k h_n^*(akR) \left[j_n(kR) - \frac{j_n'(kR)}{h_n'(kR)} h_n(kR) \right] Y_n^m(\vartheta,\varphi)^*. \tag{3.114}$$

Qualitatively, the consequences for nearfield plane-wave incidence on a spherical aperture are comparable to the situation of nearfield plane-wave incidence on a circular apertures, see Section 3.1.5, and, in particular, Fig. 3.14.

3.3 Wavefield Decomposition Using Other Types of Apertures

Although this work mainly restricts wavefield decomposition methods to circular and spherical apertures, this section introduces two other geometries that are, in principle, suitable for wavefield decomposition methods. First, the focus will be on linear continuous apertures and linear microphone arrays. Then, cylindrical apertures will be briefly considered.

3.3.1 Linear Apertures

Figure 3.23 shows a linear continuous aperture of length $2L$ coinciding with the z-axis. A planar wave-front with wavenumber vector, $\boldsymbol{k}$, impinges on the aperture, arriving from (ϑ, φ). At point $Q(\boldsymbol{r})$ on the aperture, the plane-wave is,

$$P_{\text{inc}}(\boldsymbol{k}^T \boldsymbol{r}) = e^{i\boldsymbol{k}^T \boldsymbol{r}}, \tag{3.115}$$

where, cf. Fig. 3.23, $\boldsymbol{k}$ is given by Eq. (3.2) and $\boldsymbol{r} = [0, 0, z]^T$.

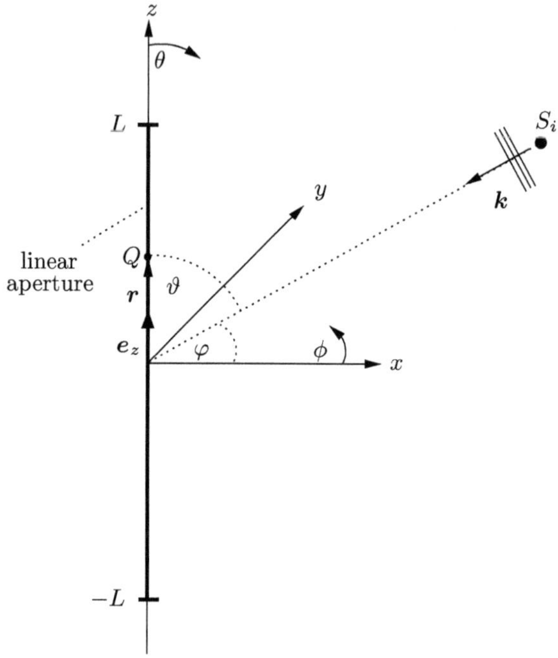

Fig. 3.23. Geometric model of a linear aperture

Consequently, Eq. (3.115) becomes,

$$P_{\text{inc}}(k, z) = e^{ikz\cos\vartheta} = e^{ik_z z}, \tag{3.116}$$

where $k_z = k\cos\vartheta$. It can be seen that Eq. (3.116) is not a function of φ. Therefore, a linear aperture cannot discriminate between wave-fronts coming from different azimuthal directions, φ. Moreover, linear apertures lack the ability to discriminate between plane-waves impinging from $0 \leq \vartheta \leq \pi$ and $\pi < \vartheta < 2\pi$. Therefore, it is obvious that linear apertures provide only limited support for three-dimensional wavefield decomposition methods.

Letting $L \to \infty$, the output of an infinite-length linear aperture due to an impinging plane-wave can be written as,

$$F(k_z) = \frac{1}{2\pi} \int_{-\infty}^{\infty} w(z) e^{ik_z z}\, dz, \tag{3.117}$$

where $w(z)$ is an aperture weighting function. This equation can be regarded as a spatial Fourier transform of this aperture weighting function, see Section A.1. In the particular case of omnidirectional spatial selectivity characteristics of an infinitesimal segment of the linear aperture, $w(z) = 1$, Eq. (3.117) becomes,

$$F(k_z) = \delta(k_z). \tag{3.118}$$

Therefore, the response of an uniformly weighted infinite linear aperture to an incoming plane-wave corresponds to an impulse in the direction of the plane-wave arrival.

Now, considering finite linear aperture results in,

$$F(k_z) = \frac{1}{2L} \int_{-L}^{L} w(z) e^{ik_z z} \, dz, \tag{3.119}$$

which simplifies for uniform weighting to,

$$F(k_z) = \operatorname{sinc}(k_z L) = j_0(k_z L), \tag{3.120}$$

where $\operatorname{sinc}(\zeta) \triangleq \sin\zeta/\zeta$. One can immediately see that Eq. (3.118) follows from Eq. (3.120) in the limit $L \to \infty$.

Two plane-wave expansions of Eq. (3.116) are [Wil99],

$$\begin{aligned} P_{\text{inc}}(k,z) &= e^{ikz\cos\vartheta} \\ &= \sum_{n=0}^{\infty} i^n (2n+1) j_n(kz) P_n(\cos\vartheta) \\ &= \sum_{n=0}^{\infty} \epsilon_n i^n J_n(kz\cos\vartheta). \end{aligned} \tag{3.121}$$

In practice, continuous linear apertures are very difficult, if not impossible, to realize. Hence, continuous apertures are usually sampled by discrete sensors. Mathematically, this sampling can most easily be described by expressing the aperture weighting function by a weighted impulse train evaluated at the positions of each of the M sensors, $\boldsymbol{r}_\ell$, as,

$$w(z) = \sum_{\ell=0}^{M-1} w_\ell \delta(z - \boldsymbol{r}_\ell^T \boldsymbol{e}_z), \tag{3.122}$$

where $\boldsymbol{e}_z$ is the unit vector in z-direction. Substituting $w(z)$ in Eq. (3.117) results in,

$$\begin{aligned} F(k_z) &= \frac{1}{2\pi} \int_{-\infty}^{\infty} \sum_{\ell=0}^{M-1} w_\ell \delta(z - \boldsymbol{r}_\ell^T \boldsymbol{e}_z) e^{ik_z z} \, dz \\ &= \sum_{\ell=0}^{M-1} w_\ell e^{ik_z \boldsymbol{r}_\ell^T \boldsymbol{e}_z}. \end{aligned} \tag{3.123}$$

For a complex scalar w_ℓ, Eq. (3.123) can be interpreted as the standard delay-and-sum beamformer [Tre02], while for a frequency-dependent w_ℓ, the so-called filter-and-sum beamformer is obtained. There is a vast amount of literature dedicated to determining w_ℓ in order to obtain the desired spatial characteristics, for example constant directivity beamforming, or superdirective beamformer. See Section 4.3 and [BW01] for an overview of these designs.

A simple technique that yields a spatial harmonic structure, much in the same way as circular and spherical microphone arrays discussed in previous sections, is the differential microphone array [Elk04]. The basic idea here is to simulate the acoustic pressure differentials of order n on the axis of the array by finite-difference approximations. One fundamental requirement that limits generality here is that the microphones need to be closely-spaced. More discussions on differential microphone arrays are deferred until Section 4.3.3.

3.3.2 Cylindrical Apertures

The geometry for continuous infinite-length cylindrical apertures can be taken from Fig. 3.1 for $L \to \infty$, where now the cylindrical baffle is part of the cylindrical aperture. By multiplying Eq. (3.66) with $Y_n^m(\vartheta, \varphi)$ and integrating over ϑ and φ gives,

$$Y_n^m(\theta, \phi) j_n(kr) = \frac{1}{4\pi i^n} \int_0^{2\pi} \int_0^{\pi} e^{i\boldsymbol{k}^T \boldsymbol{r}} \, Y_n^m(\vartheta, \varphi) \sin\vartheta \, d\vartheta \, d\varphi. \tag{3.124}$$

In cylindrical coordinates, $\varrho = r \sin\theta$ and $z = r\cos\theta$. Then,

$$e^{i\boldsymbol{k}^T \boldsymbol{r}} = e^{ik[z\cos\vartheta + \varrho \sin\vartheta \cos(\phi - \varphi)]}, \tag{3.125}$$

and therefore with Eq. (3.124) and Eq. (B.2),

$$P_n^m(\cos\theta) j_n(kr) = \frac{1}{2} i^{m-n} \int_0^{\pi} e^{ikz\cos\vartheta} J_m(k\varrho \sin\vartheta) P_n^m(\cos\vartheta) \sin\vartheta \, d\vartheta. \tag{3.126}$$

Expressing the spherical harmonics using modified Legendre functions, Eq. (3.66) can be written as,

$$e^{i\boldsymbol{k}^T \boldsymbol{r}} = \sum_{n=0}^{\infty} i^n (2n+1) \sum_{m=-n}^{n} \frac{(n-m)!}{(n+m)!} e^{im(\phi-\varphi)} P_n^m(\cos\vartheta) P_n^m(\cos\theta) j_n(kr), \tag{3.127}$$

and therefore in cylindrical coordinates using Eq. (3.126) as,

$$e^{i\boldsymbol{k}^T \boldsymbol{r}} = P_{\text{inc}}(k, \varrho, \phi, z) = \frac{1}{2} \sum_{n=0}^{\infty} (2n+1) \sum_{m=-n}^{n} i^m \frac{(n-m)!}{(n+m)!} e^{im(\phi-\varphi)} P_n^m(\cos\vartheta) \times \int_0^{\pi} e^{ikz\cos\vartheta'} J_m(k\varrho \sin\vartheta') P_n^m(\cos\vartheta') \sin\vartheta' \, d\vartheta'. \tag{3.128}$$

This expression is the plane-wave expansion in cylindrical coordinates, cf. Eq. (3.66) for the equivalent expression in spherical coordinates, and cf. Eq. (3.10) for the equivalent expression in polar coordinates.

Performing the integration with respect to the z-coordinate on the boundary of an infinite cylindrical aperture of radius $\varrho = R$, Eq. (3.128) becomes,

$$F(kR, \phi, \vartheta, \varphi) = \frac{1}{4\pi} \sum_{n=0}^{\infty} (2n+1) \sum_{m=-n}^{n} i^m \frac{(n-m)!}{(n+m)!} e^{im(\phi-\varphi)} P_n^m(\cos\vartheta) \times \int_0^{\pi} \int_{-\infty}^{\infty} e^{ikz\cos\vartheta'} J_m(kR\sin\vartheta') P_n^m(\cos\vartheta') \sin\vartheta' \, dz \, d\vartheta'. \tag{3.129}$$

Since it holds that,

$$\int_{-\infty}^{\infty} e^{ikz\cos\vartheta'} \, dz = 2\pi\delta(k\cos\vartheta'), \tag{3.130}$$

it follows that,

$$F(kR, \phi, \vartheta, \varphi) = \frac{1}{2} \sum_{n=0}^{\infty} (2n+1) \sum_{m=-n}^{n} i^m \frac{(n-m)!}{(n+m)!} e^{im(\phi-\varphi)} P_n^m(\cos\vartheta) \times \int_0^{\pi} \delta(kR\cos\vartheta'/R) J_m(kR\sin\vartheta') P_n^m(\cos\vartheta') \sin\vartheta' \, d\vartheta'. \tag{3.131}$$

Performing the substitution $t \triangleq kR\cos\vartheta'/R$ yields,

$$F(kR, \phi, \vartheta, \varphi) = \frac{R}{2kR} \sum_{n=0}^{\infty} (2n+1) \sum_{m=-n}^{n} i^m \frac{(n-m)!}{(n+m)!} e^{im(\phi-\varphi)} P_n^m(\cos\vartheta) \times \int_{-k}^{k} \delta(t) J_m(kR\sqrt{1-(tR/kR)^2}) P_n^m(tR/kR) \, dt. \tag{3.132}$$

Applying the sifting property of the Dirac delta function results in,

$$F(kR, \phi, \vartheta, \varphi) = \frac{R}{2kR} \sum_{n=0}^{\infty} (2n+1) \times \sum_{m=-n}^{n} i^m \frac{(n-m)!}{(n+m)!} e^{im(\phi-\varphi)} P_n^m(\cos\vartheta) J_m(kR) P_n^m(0). \tag{3.133}$$

Applying a spatial Fourier transform with respect to the ϕ-coordinate results in,

$$F_l(kR,\vartheta,\varphi) = \frac{R}{4\pi kR}\sum_{n=0}^{\infty}(2n+1)\sum_{m=-n}^{n} i^m \frac{(n-m)!}{(n+m)!}\, e^{-im\varphi} P_n^m(\cos\vartheta) \times J_m(kR) P_n^m(0) \int_0^{2\pi} e^{i(m-l)\phi}\, d\phi. \tag{3.134}$$

The last integral is simply,

$$\int_0^{2\pi} e^{i(m-l)\phi}\, d\phi = 2\pi\delta_{lm}, \tag{3.135}$$

where δ_{lm} denotes the Kronecker symbol which is unity for $m = l$ and zero for all other values. Consequently, Eq. (3.134) reduces to,

$$F_l(kR,\vartheta,\varphi) = \frac{R}{2kR}\, i^l J_l(kR) e^{-il\varphi} \sum_{n=l}^{\infty} \frac{(2n+1)(n-l)!}{(n+l)!} P_n^l(\cos\vartheta) P_n^l(0). \tag{3.136}$$

Expressing the associated Legendre functions by spherical harmonics, it finally follows that,

$$F_l(kR,\vartheta,\varphi) = \frac{2\pi R}{kR}\, i^l J_l(kR) \sum_{n=0}^{\infty} Y_n^l(\vartheta,\varphi)^* Y_n^l(\pi/2, 0). \tag{3.137}$$

Defining $n \triangleq l$, and using the completeness relation of the spherical harmonics, see Eq. (B.57), Eq. (3.137) can be further simplified to yield,

$$F_n(kR,\vartheta,\varphi) \sim i^n J_n(kR)\, e^{-in\varphi} \delta(k\cos\vartheta) = i^n J_n(kR)\, e^{-in\varphi} \cdot \delta(k_z), \tag{3.138}$$

where the relation $\delta(\cos\vartheta)/k = \delta(k\cos\vartheta)$ has been used.

This is an illustrative example of the *pattern multiplication theorem* which, applied to the present problem, states that the angular modal response of an infinite-length cylindrical aperture is equal to the product of the angular response of a continuous circular aperture, Eq. (3.15), multiplied by the angular response of a continuous infinite-length linear aperture, Eq. (3.118).

The pattern multiplication theorem enables the direct derivation of the modal output of a finite-length cylindrical aperture which reads with Eq. (3.120) and Eq. (3.138),

$$F_n(kR, \vartheta, \varphi) \sim i^n J_n(kR)\, e^{-in\varphi} \cdot j_0(k_z L), \tag{3.139}$$

where $2L$ is the length of the cylinder.

Although not examined any further, it is claimed here that cylindrical apertures do not offer any advantages over spherical apertures in terms of being able to decompose three-dimensional wavefields. Thus, cylindrical apertures and arrays will not be part of further developments and investigations, and detailed derivations corresponding to the ones performed for circular and spherical apertures are omitted.

4

Acoustic Scene Analysis Using Classical Array Signal Processing

In this chapter, the path of classical acoustics, illuminated by the previous two chapters, is abandoned for introducing the concepts of the second pillar necessary for understanding the fruitful combination of classical acoustics with classical array signal processing to be presented in Chapter 5. Array signal processing is the science of signal processing where more than one sensor is used to obtain an additional dimension of freedom for the solution of signal processing tasks. In most cases, this additional dimension is the spatial dimension. Since this book deals with acoustic wavefields propagating in air, the sensors utilized are microphones. However, most techniques developed here are also applicable to array signal processing in the fields of radar (antennas), sonar (hydrophones), and seismic imaging (accelerometers). Indeed, most of the techniques presented in this chapter have been originally developed in the radar and sonar community. As a tribute to this development, the more general term 'sensor' is used in most of the following discussions.

Many illustrations and specific examples in this chapter are based on the mathematically easily tractable linear sensor array since the fundamental concepts are most easily understood using this popular array geometry. The extension to other array geometries such as circular and spherical arrays is, in most cases, straightforward and therefore not treated separately.

In this book, acoustic scene analysis comprises two major tasks, namely waveform estimation (WE) and parameter estimation (PE), see Chapter 1 for an introduction of this terminology. Section 4.1 explains these terms, as used in this book, in more detail. Section 4.2 develops the signal models used and states major assumptions. Section 4.3 studies WE by introducing the concept of beamforming followed by commonly used performance measures and examples for array designs applied to WE. Section 4.4 details PE methods used for acoustic source localization and detection of the number of sources present in the wavefield under observation.

4.1 Overview

Figure 4.1 shows a general multiple-input, multiple-output (MIMO) system for acoustic array signal processing. An array of M sensors is placed in an enclosure at positions $\boldsymbol{r}_{\mathcal{M}_\ell}, \ell = 1(1)M$, possibly observing multipath propagation [1]. The center of the array, denoted as 'o', is assumed to be coinciding with the center of gravity of the sensor array, cf. Appendix C. The wavefield to be analyzed by array signal processing procedures comprises I, possibly simultaneously active, desired sources, $s_{\mathcal{D}_\iota}(\boldsymbol{r}_{\mathcal{D}_\iota}, t), \iota = 1(1)I$, and O interfering sources, $s_{\mathcal{I}_\nu}(\boldsymbol{r}_{\mathcal{I}_\nu}, t), \nu = 1(1)O$. Each microphone signal is corrupted by additive noise, $n_\ell(t)$, $\ell = 1(1)M$, produced by the microphones themselves and by the associated electronics.

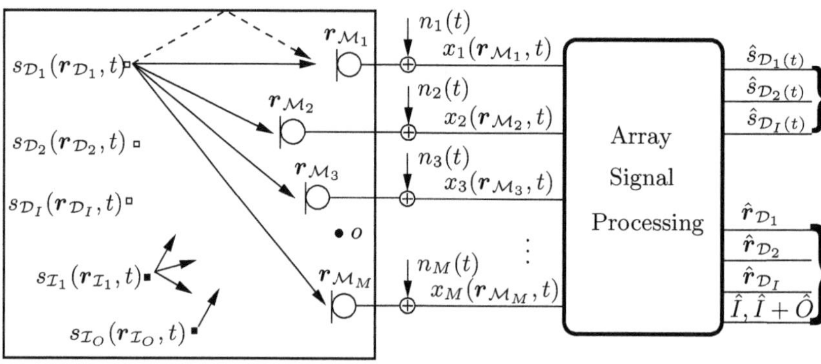

Fig. 4.1. Array signal processing for acoustic scene analysis

As stated earlier, array signal processing tasks are, in this book, divided into two major categories, namely waveform estimation and parameter estimation.

Waveform estimation algorithms try to extract the desired signal(s) from sensor array measurements that are corrupted by additional interferers, reverberation and additive noise, see Fig. 4.1. This task is usually performed by applying the paradigm of a spatial filter, or beamformer. Here, data-independent beamformers, exhibiting time-invariant spatial characteristics are considered, see Section 4.3.3. A second class of beamformers treated here are the data-dependent beamformers whose spatial characteristics are determined continuously from the statistics of the recorded data, see Section 4.3.4. These beamformers share the property that their specific geometry is an integral part of the design procedure. A rather new and increasingly important class of beamformer is the so-called *blind* beamformer, whose spatial characteristics are solely based on the statistics of the recorded wavefield. Knowledge of

[1] Note that the calligraphic symbol $\mathcal{M}$ used as a *subscript* in this section should not be confused with the maximum order of decomposition using spherical arrays.

the particular array geometry is not required. This class of beamformers is not treated in this book. The interested reader is referred to the literature, e.g. [CA02, PS00, BAK04] and references therein, for an in-depth treatment of blind beamforming and the related paradigm of blind source separation (BSS).

Parameter estimation as considered here deals with the extraction of additional information carried by a wavefield and observed by an array of sensors. The parameters that can be estimated include the position of the desired sources as well as the number of desired sources in the wavefield under observation, cf. Fig. 4.1. Here, only the directions-of-arrivals (DOAs) of the desired sources, see Section 4.4.2 and Section 4.4.3, and the number of active sources (NOS) in a wavefield, Section 4.4.4, are considered. The actual three-dimensional position of the sources, $\hat{\boldsymbol{r}}_{\mathcal{D}_\iota}$, can be obtained by estimating several DOAs using spatially distributed sensor arrays followed by application of geometric considerations. For an in-depth treatment of three-dimensional acoustic source localization, the interested reader is referred to the literature, e.g. [Bra95, Hua01].

4.2 Signal Models and Assumptions

In the previous chapters, only deterministic signals that can be described exactly in space and time as functions of $\exp[i(\boldsymbol{k}^T\boldsymbol{r} - \omega t)]$, were considered. Now, the wavefield impinging on the aperture is, in general, considered as a spatio-temporal random processes, or random field, with unknown second-order statistics.

All signals and interferers that can be modeled as point sources are assumed to be zero-mean temporally wide-sense stationary (WSS) Gaussian random processes. *Temporally* WSS random processes have the property that their second-order statistics, i.e. their mean and variance are independent of the time variable and that their correlation only depends on the time difference of observation [Pap65].

All signals and interferers that cannot be modeled as point sources are assumed to be zero-mean *spatio-temporally* WSS Gaussian processes. Spatio-temporal WSS random wavefields only depend on the Euclidean distance between two observation points along the aperture. Thus, spatio-temporally WSS random fields are spatially homogeneous, or isotropic, temporally WSS space-time processes [Her05]. Furthermore, they are considered to be statistically independent with respect to each other and with respect to all point-like sources.

As a consequence, all early distinct reflections due to multipath propagation of the desired signal(s) are modeled as point sources while the late reverberation is modeled as isotropic noise. In Section 4.4.2, however, the effect of reverberation is modeled explicitly. There, the signal model differs from the one presented here and its introduction is deferred until Section 4.4.2.

As long as not explicitly stated otherwise, all signals present in Fig. 4.1 are assumed to be located in the farfield of the aperture in the sense of Appendix C.

4.2.1 Sensor-Related Assumptions

In the following, several basic sensor-related assumptions are stated.

AS–1 All sensors of the array exhibit identical characteristics in terms of magnitude and phase response. They are therefore assumed to be perfectly *calibrated.*

AS–2 The impulse response describing an individual sensor is assumed to be a Dirac impulse, $\delta(t)$, i.e. each sensor adds unity gain and zero phase to the measurement.

AS–3 All sensors are assumed to be point-like, i.e. they do not alter in any way the wavefield they are measuring.

AS–4 The sensor array is compact, i.e. its maximum dimension, or aperture, is much smaller than the distance from the source(s) to the center-of-gravity of the array. Hence, the farfield assumption holds, cf. Appendix C.

Many discussions in this chapter deal with uniformly-spaced linear sensor arrays (ULAs). The geometric model for linear apertures has been used in Section 3.3.1. The geometric model for an ULA with a single incident plane-wave is depicted in Fig. 4.2. Note that only for an odd number of sensors M,

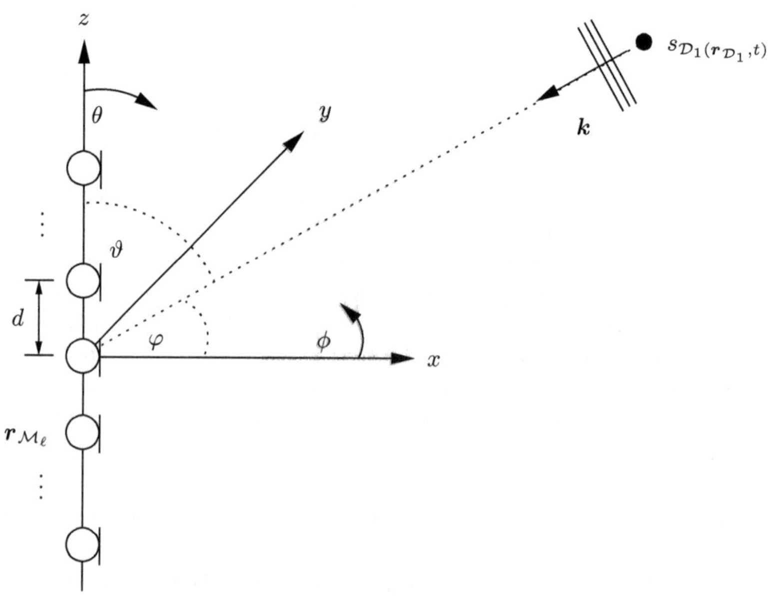

Fig. 4.2. Geometric model of an ULA using M sensors

there is a sensor coinciding with the phase center of the array, see Eq. (C.1). Note also that the array is not able to discriminate signals whose DOAs only differ in the ϕ coordinate.

A special and widely used array structure is the so-called standard ULA [Tre02]. The term 'standard' here corresponds to half-wavelength-spaced sensors comprising the array.

4.2.2 Time-Domain Signal Model

It is assumed that all time-domain signals are real-valued. In many of the subsequent developments, especially for the parameter estimation methods, strong interfering sources are not modeled separately but are treated as 'desired' signals. Additional algorithms, which will not be treated in this book, are then required to help separating desired sources from the undesired strong interferers. This task could be performed, for instance, by blind source separation techniques, see e.g. [CA02, BAK04], followed by signal classification algorithms, see e.g. [DHS00].

Weak directional interferers are, for simplicity, modeled as part of the noise signals, $n_\ell(t), \ell = 1(1)M$. Based on these preliminaries, the assumptions listed in Section 4.2.1, and after dropping the explicit spatial dependence, the time-domain signal model for an observation time interval $t \in [0, T]$ is,

$$x_\ell(t) = \sum_{\iota=1}^{I} s_{\iota,\ell}(t) + n_\ell(t) = \sum_{\iota=1}^{I} s(t - \boldsymbol{e}_{\boldsymbol{k}_\iota}^T \boldsymbol{r}_{\mathcal{M}_\ell}/c) + n_\ell(t), \qquad (4.1)$$

where, due to the assumed plane-wave incidence, $\boldsymbol{e}_{\boldsymbol{k}_\iota}^T \boldsymbol{r}_{\mathcal{M}_\ell}/c$ denotes the delay corresponding to each signal and sensor element with respect to an arbitrary reference point. Here, this reference point is chosen to be the center-of-gravity of the receiving array. $\boldsymbol{e}_{\boldsymbol{k}_\iota}$ is the unit vector pointing to the direction of the wavenumber vector associated with the ι-th desired source and $\boldsymbol{r}_{\mathcal{M}_\ell}$ is the position vector of the ℓ-th sensor, cf. Fig 3.23 and Fig. 4.2.

Note that the model, Eq. (4.1), does not take the attenuation of the plane-wave signals due to wave propagation from sensor to sensor into account. This is due to assumption AS-4. Employing vector notation, the signal received at an array of M sensors can be denoted compactly as,

$$\boldsymbol{x}(t) = [x_1(t), x_2(t), \ldots, x_M(t)]^T. \qquad (4.2)$$

An important quantity used in many subsequent discussions is the spatio-temporal correlation matrix which is defined as,

$$\underline{\boldsymbol{c}}_{\boldsymbol{x}\boldsymbol{x}}(\tau) = E\{\boldsymbol{x}(t) \cdot \boldsymbol{x}^T(t-\tau)\} \triangleq \begin{bmatrix} c_{x_1x_1}(\tau) & c_{x_1x_2}(\tau) & \cdots & c_{x_1x_M}(\tau) \\ c_{x_2x_1}(\tau) & c_{x_2x_2}(\tau) & \cdots & c_{x_2x_M}(\tau) \\ \vdots & \vdots & \ddots & \vdots \\ c_{x_Mx_1}(\tau) & c_{x_Mx_2}(\tau) & \cdots & c_{x_Mx_M}(\tau) \end{bmatrix}, \qquad (4.3)$$

where,

$$c_{\nu_1\nu_2}(\tau) = E\{x_{\nu_1}(t)x_{\nu_2}(t-\tau)\}, \quad \nu_1,\nu_2 = 1(1)M. \tag{4.4}$$

The spatio-temporal correlation matrix of the desired signals, $\underline{\boldsymbol{c}}_{\boldsymbol{ss}}(\tau)$ and the one of the noise signals, $\underline{\boldsymbol{c}}_{\boldsymbol{nn}}(\tau)$, are defined similarly. In practice, this spatio-temporal correlation matrix needs to be estimated from the output signals of the sensor array. For wide-sense ergodic WSS random processes, the expectation operator, $E\{\cdot\}$, can be replaced by a temporal averaging operation which, for finite observation time interval T, can be approximated as,

$$\hat{\underline{\boldsymbol{c}}}_{\boldsymbol{xx}}(\tau) \approx \frac{1}{T}\int_0^T \boldsymbol{x}(t)\boldsymbol{x}^T(t-\tau)\,dt. \tag{4.5}$$

4.2.3 Frequency-Domain Signal Model

A signal model that will be extensively used in the following sections is formulated in the frequency-domain. The signal vector $\boldsymbol{x}(t)$, cf. Eq. (4.2), is observed during a time interval of T seconds. The signal $\boldsymbol{x}(t)$ is assumed to be a zero-mean bandpass process centered around the temporal-radian frequency ω_c with bandwidth W. The so-called frequency-domain snapshot model [Tre02] is obtained by applying a (short-time) Fourier transform to the observed signal in the time-domain as [Dav70],

$$\boldsymbol{X}(\omega_\mu,\kappa) = \frac{1}{T}\int_{-T/2}^{T/2} \boldsymbol{x}(t+\kappa T)e^{i\omega_\mu t}\,dt, \tag{4.6}$$

where $\omega_\mu = \omega_c + \mu\omega_\Delta$, μ being an integer, and where $\omega_\Delta \triangleq 2\pi/T$ is the resolution of the transform. The quantity ω_μ can be interpreted as the center frequency of a frequency bin with width ω_Δ. Note that when $\mu = 0$, then a narrowband frequency-domain snapshot model is obtained. In addition to (short-time) stationarity, it is assumed that the different frequency bins can be regarded as asymptotically uncorrelated, see [SM93] and Assumption A-2 below. Then, the short-time Fourier transform of the time-domain signal model, Eq. (4.1), becomes for snapshot κ and temporal-radian frequency ω_μ [Tre02],

$$\boldsymbol{X}(\omega_\mu,\kappa) = \underline{\boldsymbol{V}}(\omega_\mu,\boldsymbol{\Theta})\boldsymbol{S}(\omega_\mu,\kappa) + \boldsymbol{N}(\omega_\mu,\kappa), \tag{4.7}$$

where,

$$\boldsymbol{X}(\omega_\mu,\kappa) = [X_1(\omega_\mu,\kappa), X_2(\omega_\mu,\kappa),\ldots,X_M(\omega_\mu,\kappa)]^T, \tag{4.8a}$$

$$\boldsymbol{S}(\omega_\mu,\kappa) = [S_1(\omega_\mu,\kappa), S_2(\omega_\mu,\kappa),\ldots,S_I(\omega_\mu,\kappa)]^T, \tag{4.8b}$$

$$\boldsymbol{N}(\omega_\mu,\kappa) = [N_1(\omega_\mu,\kappa), N_2(\omega_\mu,\kappa),\ldots,N_M(\omega_\mu,\kappa)]^T, \tag{4.8c}$$

and where the uppercase letters correspond to the Fourier transform of the respective lowercase letters. The matrix $\underline{\boldsymbol{V}}(\omega_\mu,\boldsymbol{\Theta})$ is commonly referred to as the *array manifold matrix*, which depends on the array geometry and the DOAs of the desired sources, $\boldsymbol{\Theta} = [\Theta_1, \Theta_2, \ldots, \Theta_I]^T$, where Θ_ι, $\iota = 1(1)I$, denotes the angle between the sources and the center-of-gravity of the sensor array. The array manifold matrix is defined as,

$$\underline{\boldsymbol{V}}(\omega_\mu, \boldsymbol{\Theta}) \triangleq [\boldsymbol{V}(\omega_\mu, \Theta_1) \mid \boldsymbol{V}(\omega_\mu, \Theta_2) \mid \ldots \mid \boldsymbol{V}(\omega_\mu, \Theta_I)]. \tag{4.9}$$

The *array manifold vector*, $\boldsymbol{V}(\omega_\mu, \Theta_\iota)$, is defined in Section 4.3.1 and given by Eq. (4.23). The array manifold vector describes the transfer characteristics from the source ι to the M sensors.

The formulation of the frequency-domain snapshot model is facilitated by the following common assumptions concerning the noise and the desired signals.

A–1 The columns of $\underline{\boldsymbol{V}}(\omega_\mu, \boldsymbol{\Theta})$ are linearly independent. This mild restriction on the array geometry ensures that all sources are observable by the array.

A–2 The length of the observation interval T is long with respect to the correlation times, i.e. the range of non-zero correlation, of the signals and noise. Therefore, individual snapshots are assumed to be uncorrelated [SM93]. Furthermore, T is assumed to be much larger than the maximum propagation time of the signal across the array. Therefore, it is assumed that the frequency bins in Eq. (4.6) are uncorrelated [HN76]. This requirement can also be stated by assuming a large time-bandwidth product. According to [WW83], the requirement is $TW \gg 2\pi$.

A–3 The noise is a spatially and temporally ergodic, zero-mean stochastic process.

A–4 The noise is uncorrelated from sensor to sensor and with respect to the desired signals. Note that this assumption is idealized and will, in general, not hold in diffuse noise fields.

A–5 The number of desired sources, I, does not exceed the number of sensors of the array. This assumption is further discussed below.

As in the time-domain model, a correlation matrix will be of paramount importance for many subsequent discussions. Here, the correlation matrix is a spatio-spectral matrix which is a short-time estimate of the Fourier transform of the spatio-temporal correlation matrix, Eq. (4.3). For each frequency ω_μ and snapshot κ, the $M \times M$ spatio-spectral matrix of the sensor outputs can then be written as,

$$\begin{aligned}\underline{\boldsymbol{S}}_{\boldsymbol{XX}}(\omega_\mu, \kappa) &= E\{\boldsymbol{X}(\omega_\mu, \kappa)\boldsymbol{X}^H(\omega_\mu, \kappa)\} \\ &= \underline{\boldsymbol{V}}(\omega_\mu, \boldsymbol{\Theta})\underline{\boldsymbol{S}}_{\boldsymbol{SS}}(\omega_\mu, \kappa)\underline{\boldsymbol{V}}^H(\omega_\mu, \boldsymbol{\Theta}) + \underline{\boldsymbol{S}}_{\boldsymbol{NN}}(\omega_\mu, \kappa),\end{aligned} \tag{4.10}$$

where the noise spectral matrix is assumed to be $\underline{\boldsymbol{S}}_{\boldsymbol{NN}}(\omega_\mu, \kappa) = \mathcal{F}_\tau\{\underline{\boldsymbol{c}}_{\boldsymbol{nn}}(\tau, \kappa)\}$ $= \sigma_w^2 \underline{\boldsymbol{I}}$ which complies with Assumptions A-3 and A-4. σ_w^2 denotes the variance of the noise. Note that Eq. (4.10) can be used for both narrowband as well as wideband signals, if, as is assumed here, the frequency bins ω_μ are mutually uncorrelated, see Assumption A-2. For notational convenience, where possible, the dependence on κ will be dropped and $\omega \triangleq \omega_\mu$.

The spatio-spectral matrix of the source signals, $\underline{\boldsymbol{S}}_{\boldsymbol{SS}}(\omega) = \mathcal{F}_\tau\{\underline{\boldsymbol{c}}_{\boldsymbol{ss}}(\tau)\}$ is diagonal – having the power spectral densities of the individual source signals as entries of the main diagonal – only if the signals are spatially uncorrelated. If the signals are spatially partly correlated, $\underline{\boldsymbol{S}}_{\boldsymbol{SS}}(\omega)$ is non-diagonal but non-singular. Fully correlated, i.e. fully coherent, signals result in a singular spatio-spectral matrix $\underline{\boldsymbol{S}}_{\boldsymbol{SS}}(\omega)$ due to the presence of linearly dependent columns.

An estimate of the spatio-spectral correlation matrix, Eq. (4.10) can, as for the spatio-temporal correlation matrix, be performed by a temporal averaging process, i.e.,

$$\hat{\underline{\boldsymbol{S}}}_{\boldsymbol{XX}}(\omega, \kappa) = \frac{1}{K} \sum_{\Lambda=1}^{K} \boldsymbol{X}(\omega, \kappa - \Lambda) \boldsymbol{X}^H(\omega, \kappa - \Lambda). \tag{4.11}$$

A comment on assumption A-5 is in order. This assumption was first quantified by Wax and Ziskind [WZ89] as the maximum number of sources that can be unambiguously resolved by an array of sensors at infinitely high signal-to-noise ratio at each sensor and infinitely long observation time as,

$$I < \frac{\text{rank}\{\underline{\boldsymbol{S}}_{\boldsymbol{SS}}(\omega)\}}{\text{rank}\{\underline{\boldsymbol{S}}_{\boldsymbol{SS}}(\omega)\} + 1} M, \tag{4.12}$$

under the assumption that the array manifold matrix is known and that the columns of the array manifold matrix are linearly independent. This result leads to the often quoted requirement that the number of sources need to be less than the number of sensors.

4.3 Waveform Estimation

In this section, the notion of waveform estimation, and in particular space-time filtering and beamforming, is introduced. Popular performance measures and the paradigms of data-independent beamforming as well as data-dependent beamforming are illustrated.

4.3.1 Space-Time Filtering and Beamforming

The concepts of space-time filtering and beamforming are discussed in this section. For the most part, the line of argumentation follows [Tre02].

Here it is assumed that a *single*, possibly deterministic, signal – e.g. $s_{\mathcal{D}_1}(\boldsymbol{r}_{\mathcal{D}_1}, t)$ from Fig. 4.1 – impinges on an M-element sensor array, as shown in Fig. 4.3. It is assumed that neither interfering signals nor other additive

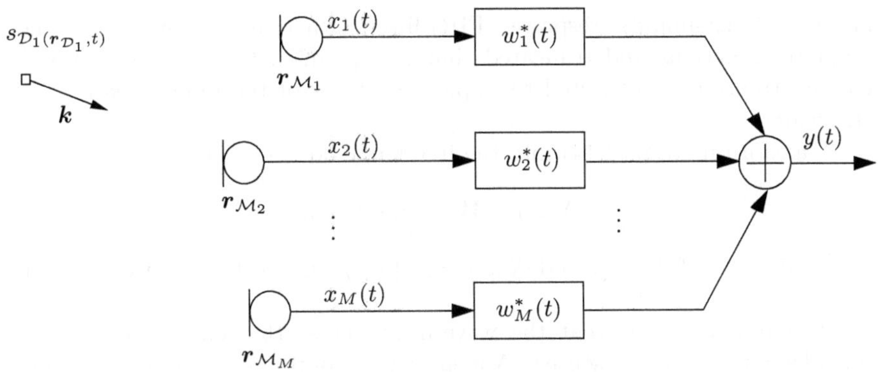

Fig. 4.3. A spatio-temporal filter

noise signals are present. Let the output of the sensor ℓ be processed by a linear filter with impulse response $w_\ell^*(t), \ell = 1(1)M$. [2] Then, assuming infinite observation time, the output of the structure shown in Fig. 4.3 is given by a superimposed convolution as,

$$y(t) = \sum_{\ell=1}^{M} w_\ell^*(t) * x_\ell(t), \tag{4.13}$$

where the symbol '$*$' relating two signals denotes a convolution operation. Using vector notation, this result can be written as,

$$y(t) = \boldsymbol{w}^H(t) * \boldsymbol{x}(t) \tag{4.14}$$

where,

$$\boldsymbol{w}(t) = \left[w_1(t), w_2(t), \ldots, w_M(t)\right]^T, \tag{4.15a}$$

$$\boldsymbol{x}(t) = \left[x_1(t), x_2(t), \ldots, x_M(t)\right]^T. \tag{4.15b}$$

Most actual implementations of the processor depicted in Fig. 4.3 require a finite-length impulse response $w_\ell(t)$. This can be achieved, in a very general way, by expanding the impulse response into a finite set of Q basis functions, $\chi_q(t)$, $q = 1(1)Q$, with the expansion coefficients for each sensor, $w_{\ell,q}$ [SK92]. Then,

$$w_\ell(t) = \sum_{q=0}^{Q-1} w_{\ell,q}\, \chi_q(t). \tag{4.16}$$

By selecting $\chi_q(t) = \delta(t - q\Delta t)$, where Δt is a delay usually related to the sampling interval T_s in digital systems as $\Delta t = T_s$, the filter $w_\ell(t)$ becomes

[2] Note that for real-valued coefficients, $w_\ell(t)$, it is merely a notational convention to use $w_\ell^*(t)$ instead of $w_\ell(t)$ [vVB88, WA01].

an Q-tap finite-impulse response (FIR) filter. Other choices for $\chi_q(t)$ include Laguerre functions and truncated sinusoids [Son67]. Here, however, the focus of attention is restricted to implementations of the filters yielding FIR structures.

Transforming Eq. (4.14) into the frequency-domain yields,

$$Y(\omega) = \boldsymbol{W}^H(\omega) \cdot \boldsymbol{X}(\omega), \tag{4.17}$$

where $\boldsymbol{W}(\omega) = \mathcal{F}_t\{\boldsymbol{w}(t)\}$ and $\boldsymbol{X}(\omega) = \mathcal{F}_t\{\boldsymbol{x}(t)\}$ are vectors of size $MQ \times 1$.

It is now assumed that the wave-front across the sensor array can be considered planar in the sense of Appendix C. In particular, it is again assumed that the origin of the coordinate system coincides with the center of gravity of the sensor array.

If the signal measured at the origin is denoted as $x(t)$, then the signal measured at an individual sensor is,

$$\boldsymbol{x}(t, \boldsymbol{\tau}) = \left[x(t-\tau_1), x(t-\tau_2), \ldots, x(t-\tau_M)\right]^T, \tag{4.18}$$

where $\boldsymbol{\tau} = \left[\tau_1, \tau_2, \ldots, \tau_M\right]^T$ and,

$$\tau_\ell = \frac{1}{c} \boldsymbol{e}_k^T \boldsymbol{r}_{\mathcal{M}_\ell}, \quad \ell = 1(1)M, \tag{4.19}$$

where $\boldsymbol{e}_k \triangleq \boldsymbol{k}/\|\boldsymbol{k}\| = c\boldsymbol{k}/\omega$.

It follows for the ℓ-th component of Eq. (4.18) in the frequency-domain that,

$$X_\ell(\omega) = \mathcal{F}_t\left\{x(t-\tau_\ell)\right\} = X(\omega) e^{i\omega\tau_\ell}, \tag{4.20}$$

where with Eq. (4.19),

$$\omega\tau_\ell = \frac{\omega}{c} \boldsymbol{e}_k^T \boldsymbol{r}_{\mathcal{M}_\ell} = \boldsymbol{k}^T \boldsymbol{r}_{\mathcal{M}_\ell}. \tag{4.21}$$

Therefore, Eq. (4.20) can be written as,

$$X_\ell(\omega) = X(\omega) \cdot e^{i\boldsymbol{k}^T \boldsymbol{r}_{\mathcal{M}_\ell}}. \tag{4.22}$$

By defining the so-called *array manifold vector*, which represents the spatial characteristics of a sensor array as,

$$\boldsymbol{V}(\omega, \Theta) \triangleq \left[e^{i\boldsymbol{k}^T \boldsymbol{r}_{\mathcal{M}_1}}, e^{i\boldsymbol{k}^T \boldsymbol{r}_{\mathcal{M}_2}}, \ldots, e^{i\boldsymbol{k}^T \boldsymbol{r}_{\mathcal{M}_M}}\right]^T, \tag{4.23}$$

where the dependency on ω and Θ is implicitly given by the scalar product in the exponential functions, Θ being the angle between $\boldsymbol{k}$ and $\boldsymbol{r}_{\mathcal{M}_\ell}$, Eq. (4.18) can finally be expressed in the frequency-domain as,

$$\boldsymbol{X}(\omega, \Theta) = \boldsymbol{V}(\omega, \Theta) \cdot X(\omega). \tag{4.24}$$

In summary, the output of the spatio-temporal processor shown in Fig. 4.3 reads for plane-wave incidence with Eq. (4.17),

$$Y(\omega, \Theta) = \boldsymbol{W}^H(\omega)\boldsymbol{V}(\omega, \Theta)X(\omega). \tag{4.25}$$

Let τ_0 denote the processing delay due to the structure shown in Fig. 4.3, a particular choice for the filters is,

$$w_\ell(t) = \delta(t + \tau_\ell - \tau_0)/M, \quad \ell = 1(1)M, \tag{4.26}$$

which compensates for the propagation delays present in the different sensor signals that are caused by the impinging plane-wave. In the time-domain, the output of the space-time filter, cf. Eq. (4.14), is then $y(t) = x(t - \tau_0)$. Therefore, by applying appropriate delays, the output of the space-time processor can be *steered* toward the emitting source. For this reason, a space-time processor is also known as a *beamformer.* The main objective of a beamformer as defined by Eqs. (4.25) and (4.26) is to add the desired signal present in each sensor output coherently, and to add the noise in each sensor output incoherently. If the filters in Fig. 4.3 are chosen according to Eq. (4.26), including a possible sensor-specific scalar weighting factor, the resulting structure is then called a *delay-and-sum* (DSB) beamformer. If FIR filters with frequency-dependent magnitude characteristics are applied, the structure is also known as a *filter-and-sum* (FSB) beamformer.

Beamformers are also classified into signal-independent, or fixed, beamformers utilizing time-invariant filters $\boldsymbol{W}(\omega)$. The other class of beamformers adaptively control the filters according to the statistics of the recorded wavefield, therefore yielding time-variant filters.

As can be deduced from Eq. (4.25), any beamformer design method has two parameter sets for optimization, i.e. the sensor positions, $\boldsymbol{r}_{\mathcal{M}_\ell}$, and the filters, $\boldsymbol{W}(\omega)$. Specific examples are discussed in the following sections.

4.3.2 Performance Measures

This section introduces the main performance measures used for designing sensor arrays applied to the problem of waveform estimation. Several performance measures are introduced for general sensor array geometries. For simplicity, specific examples are presented for the ULA, which is widely used in practice. For ULAs, where $\Theta = \vartheta$, see Fig. 4.2. Most of the discussion follows [Tre02].

Beampattern

An important quantity describing the performance of a beamformer is the *beampattern* which quantifies the spatial selectivity of a beamformer with respect to a plane-wave impinging from direction $\Theta \to (\vartheta, \varphi)$. The beampattern is defined from Eq. (4.25) as,

$$B(\omega, \vartheta, \varphi) \triangleq Y(\omega, \Theta)|_{X(\omega)=1}. \tag{4.27}$$

A graphical representation of the beampattern is obtained by varying the DOA of the incoming plane-wave as $0 \leq \vartheta \leq \pi$ and $0 \leq \varphi \leq 2\pi$ and then plotting the squared magnitude of Eq. (4.27). The resulting graph represents the so-called *directivity pattern* of an array.

For a linear array, $B(\omega, \vartheta, \varphi) = B(\omega, \vartheta)$, since linear sensor arrays, as depicted in Fig. 4.2, cannot discriminate in azimuthal direction. It was shown in Section 3.3.1 that a plane-wave arriving at a linear aperture can be written as,

$$e^{i\boldsymbol{k}^T \boldsymbol{r}} = e^{ik_z z} = e^{ikz \cos \vartheta}, \tag{4.28}$$

see also Fig. 4.2. Therefore, with $\|\boldsymbol{k}\| = k = \omega/c$, the array manifold vector, Eq. (4.23), for linear sensor arrays can be expressed as,

$$\boldsymbol{V}(\omega, \vartheta) = \left[e^{i\omega z_1 \cos \vartheta / c}, e^{i\omega z_2 \cos \vartheta / c}, \ldots, e^{i\omega z_M \cos \vartheta / c} \right]^T. \tag{4.29}$$

The spatio-temporal frequency response, $B(\omega, \vartheta)$, of a linear sensor array is then,

$$B(\omega, \vartheta) = \sum_{\ell=1}^{M} W_\ell^H(\omega) e^{i\omega z_\ell \cos \vartheta / c} = \boldsymbol{W}^H(\omega) \boldsymbol{V}(\omega, \vartheta). \tag{4.30}$$

For a ULA, $z_\ell = (\ell - 1) \cdot d$, $\ell = 1(1)M$, where d is the sensor spacing.

In the following, the beampattern of a uniformly weighted ULA is derived. The beamformer is steered to the direction of the desired source by letting $\boldsymbol{W}^H(\omega) = \boldsymbol{V}^H(\omega, \vartheta_s)/M$ in Eq. (4.30). Therefore,

$$B(\omega, \vartheta) = \frac{1}{M} \sum_{\ell=1}^{M} e^{i\omega(\ell-1)d(\cos \vartheta - \cos \vartheta_s)/c}. \tag{4.31}$$

The geometric series in Eq. (4.31) can be written in closed-form as,

$$B(\omega, \vartheta) = \frac{1}{M} \frac{\sin \left[M \frac{\omega d}{2c} (\cos \vartheta - \cos \vartheta_s) \right]}{\sin \left[\frac{\omega d}{2c} (\cos \vartheta - \cos \vartheta_s) \right]} e^{i \frac{(M-1)\omega d}{2c} (\cos \vartheta - \cos \vartheta_s)}. \tag{4.32}$$

At this point, it is instructive to consider a specific example. Figure 4.4 depicts the directivity pattern of a uniformly weighted standard ULA, i.e. $d = \lambda/2$, with $M = 9$. A plane-wave signal with a wavelength of $\lambda = 2d$ impinges from $\vartheta = \pi/2$. The main-lobe of the beamformer, is steered toward the impinging plane-wave, i.e. $\vartheta_s = \pi/2$. Visible are the main-lobe and several side-lobes with decreasing amplitude for $0 \leq \vartheta \leq \pi$. Also shown are various performance parameters that are widely used for the description of beamformers [Tre02], such as,

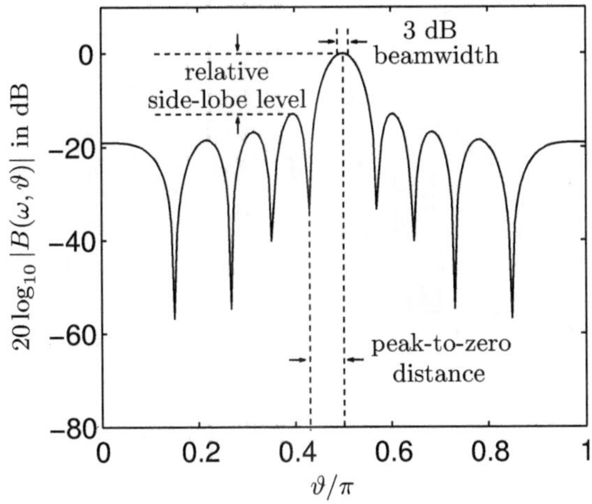

Fig. 4.4. Directivity pattern of a uniformly weighted standard ULA with $M = 9$ (adapted from [Her05])

- 3 dB beamwidth, also known as the half-power beamwidth, denoting the region of the main-lobe where the main-lobe has not decreased by more than 3 dB,
- relative side-lobe level, denoting the relative height of the first side-lobe with respect to the main-lobe, and,
- peak-to-zero distance, denoting the region from the maximum of the main-lobe to the first minimum. This quantity is also called the *Rayleigh resolution limit* since it represents the array's ability to resolve two plane-waves impinging from different directions.

All performance parameters are functions of M, ω, and the sensor geometry. Detailed mathematical derivations, along with various other less frequently used performance parameters, can be found in [Tre02].

Figure 4.5(a) depicts a top-view representation of the directivity pattern using the same parameters that were used for producing Fig. 4.4, now also including the effects of varying frequency on the beamformer response. The ordinate in Fig. 4.5 is scaled to show the microphone spacing with respect to the wavelength of the impinging plane-wave, i.e., $d_\lambda \triangleq d/\lambda$.

Figure 4.5(b) shows the result when the beamformer is steered to 'look' at direction $\vartheta_s = \pi/6$. A first obvious property of uniformly weighted ULAs is the strong dependence of the main-lobe width on frequency. With decreasing frequency, the main-lobe widens, thereby reducing spatial discrimination. The reason for this is the fact that the beamformer is unable to exploit the phase difference of the plane-wave as it travels across the array for very small values of d_λ.

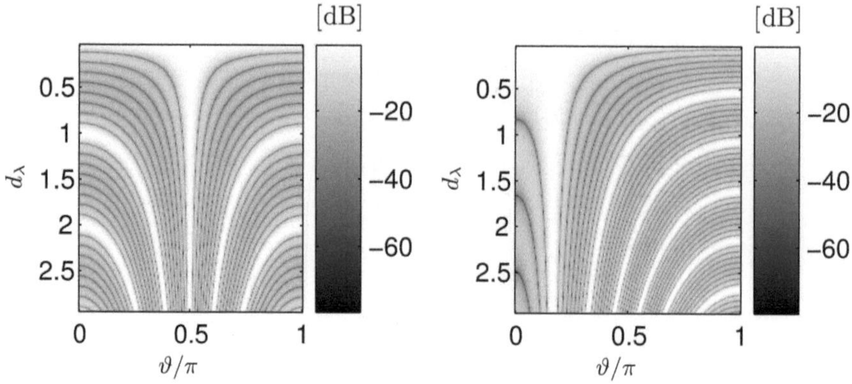

(a) $\vartheta_s = \pi/2$ (b) $\vartheta_s = \pi/6$

Fig. 4.5. Directivity pattern of a uniformly weighted ULA with $M = 9$ for varying frequency

Sampling a continuous aperture at discrete sensor positions, however, leads to phase ambiguities above a certain frequency and so-called *grating-lobes* appear in the beamformer response. This phenomenon is also denoted as spatial aliasing. The grating-lobes, which depend on the sensor spacing with respect to the wavelength of the impinging wavefield, can be found in Fig. 4.5 as the side-lobes that are just as high as the main-lobe. The properties of the grating-lobes also depend on the look-direction of the beamformer, ϑ_s. It can be shown [Tre02, Her05] that grating-lobes can be avoided if the normalized sensor spacing is chosen as,

$$d_\lambda \leq \frac{1}{1 + |\sin \vartheta_{s_{\max}}|}, \tag{4.33}$$

where $\vartheta_{s_{\max}}$ denotes the maximum steering angle to be considered for a given application. A beamformer that is steered to $\vartheta_{s_{\max}} = \pi/2$ is called a *broadside* array, and a beamformer that is steered to $\vartheta_{s_{\max}} = 0$ is called an *endfire* array. For example, $d \leq \lambda/2$ if $0 \leq \vartheta_s \leq \pi$, corresponds to an ULA that does not exhibit spatial aliasing for all wavelengths down to λ and all look-directions. These arrays are often used in applications involving speech signals. Another property of uniformly weighted ULAs is the change of main-lobe shape when the beamformer is steered to different look directions, as seen in Fig. 4.5.

It can be deduced from Eq. (4.32) that the beampattern of a uniformly weighted ULA is periodic with period π/ς, where $\varsigma = d(\cos\vartheta - \cos\vartheta_s)/c$. This is the reason for the forward-backward ambiguity inherent in all beamformers employing linear sensor arrays. This means that, for instance, a signal coming from $\vartheta = \pi/2$ cannot be discriminated from a signal coming from $\vartheta = 3\pi/2$.

Directivity

A widely used performance measure for sensor arrays is the array *directivity* factor. It is defined as [Elk04],

$$D(\omega,\vartheta,\varphi)=\frac{|B(\omega,\vartheta,\varphi)|^2}{\frac{1}{4\pi}\int_0^{2\pi}\int_0^{\pi}|B(\omega,\theta,\phi)|^2\,b(\theta,\phi)\sin\theta\,d\theta d\phi}, \tag{4.34}$$

where $B(\omega,\vartheta,\varphi)$ is given by Eq. (4.27). This quantity can be interpreted as the relationship of the output power of a sensor array due to a plane-wave impinging from (ϑ,φ) with respect to the noise power. The quantity $b(\theta,\phi)$ symbolizes the distribution of the noise power and is normalized such that,

$$\frac{1}{4\pi}\int_0^{2\pi}\int_0^{\pi} b(\theta,\phi)\sin\theta d\theta d\phi=1. \tag{4.35}$$

This type of noise is also called isotropic noise. An isotropic noise field is constituted by spatially uncorrelated plane-waves arriving with equal probability from any spatial direction [Elk04]. According to this definition, an isotropic noise field in two spatial dimensions is called a cylindrically isotropic noise field while an isotropic noise field in three spatial dimensions is denoted as spherically isotropic.

It can therefore be stated that Eq. (4.34) quantifies the array gain over isotropic noise. A second quantity in this context is the *directivity index*, which is simply defined by modifying Eq. (4.34) as,

$$D_I(\omega,\vartheta,\varphi)=10\log_{10}D(\omega,\vartheta,\varphi). \tag{4.36}$$

Specializing Eq. (4.34) to linear sensor arrays coinciding with the z-axis, cf. Fig. 4.2 and Fig. 3.23, and assuming unit gain in the direction of the incoming plane-wave as well as spherically isotropic noise – where $b(\theta,\phi)=1$ – results in,

$$D(\omega)=\left[\frac{1}{2}\int_0^{\pi}|B(\omega,\theta)|^2\sin\theta\,d\theta\right]^{-1}=\left[\frac{1}{2}\int_0^{\pi}|\boldsymbol{W}^H(\omega)\boldsymbol{V}(\omega,\theta)|^2\sin\theta d\theta\right]^{-1}. \tag{4.37}$$

In general, Eq. (4.37), needs to be evaluated numerically. However, for the special case of ULAs, Eq. (4.37) can be further simplified. It can readily be verified that for a uniformly weighted ULA, Eq. (4.37) becomes with Eq. (4.32),

$$D(d_\lambda,\vartheta_s)=M^2\left[\int_0^1\left(\frac{\sin[M\pi d_\lambda(u-\cos\vartheta_s)]}{\sin[\pi d_\lambda(u-\cos\vartheta_s)]}\right)^2du\right]^{-1}, \tag{4.38}$$

where $u \triangleq \cos\theta$. It can further be shown that for a standard ULA, i.e. $d_\lambda = 1/2$, steered to broadside, i.e. $\vartheta_s = \pi/2$, Eq. (4.38) simplifies to $D = M$ [GR65], which is the well-known result that the directivity of a standard ULA is equal to the number of sensors [Tre02]. For non-uniform weighting and $d_\lambda = 1/2$, Eq. (4.37) becomes [Tre02],

$$D(\omega) = (\mathbf{W}^H(\omega)\mathbf{W}(\omega))^{-2}. \tag{4.39}$$

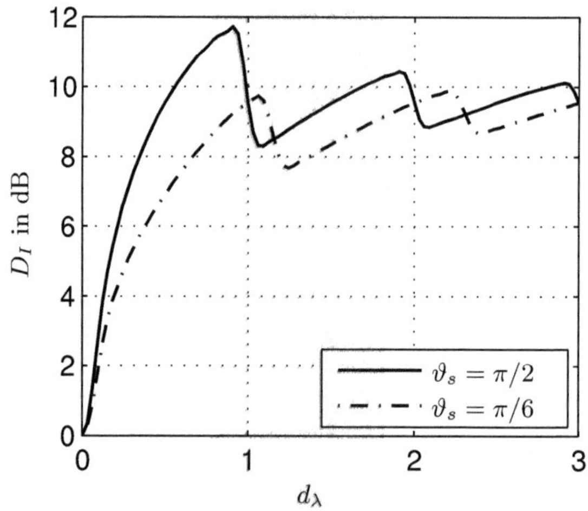

Fig. 4.6. Directivity index of a uniformly weighted ULA for $M = 9$, steered to $\vartheta_s = \pi/2$ and $\vartheta_s = \pi/6$

Continuing the example presented in Fig. 4.5, Fig. 4.6 shows the directivity index of a uniformly ULA for $M = 9$, steered to $\vartheta_s = \pi/2$ and $\vartheta_s = \pi/6$. Note that the minima present in the curves representing the directivity index can be associated with the grating-lobes from Fig. 4.5.

It can be shown [Tre02] that the uniformly weighted ULA with $d_\lambda = 1/2$ has the highest possible directivity for a given λ. Any weighting applied to the ULA will result in a decreased directivity.

Array Gain and White Noise Gain

The array gain, $A(\omega)$, is defined as the improvement in the *signal-to-interference-plus-noise ratio* (SINR) by using a sensor array compared to a single omnidirectional sensor. For WSS processes, $\text{SINR}_{\text{sensors}}$ is defined as the ratio of the average output power at the sensors to the average power of noise-plus-interference at the sensors, i.e. $\text{SINR}_{\text{sensors}}(\omega) = S_{SS}(\omega)/S_{NN}(\omega)$, where $S_{NN}(\omega)$ is the spectral density due to both noise and interference.

Omitting the spatial dependency for notational convenience, the SINR at the array output is defined as the ratio of the output power of the array due to the desired signal, i.e.,

$$S_{YY}(\omega)|_{\text{signal}} = S_{SS}(\omega)|\boldsymbol{W}^H(\omega)\boldsymbol{V}(\omega)|^2, \tag{4.40}$$

to the output power of the array due to the noise-plus-interference signal, i.e.,

$$S_{YY}(\omega)|_{\text{noise}} = \boldsymbol{W}^H(\omega)\underline{\boldsymbol{S}}_{\boldsymbol{NN}}(\omega)\boldsymbol{W}(\omega) = S_{NN}(\omega)\boldsymbol{W}^H(\omega)\underline{\tilde{\boldsymbol{S}}}_{\boldsymbol{NN}}(\omega)\boldsymbol{W}(\omega), \tag{4.41}$$

where $\underline{\tilde{\boldsymbol{S}}}_{\boldsymbol{NN}}(\omega)$ is normalized such that $\text{tr}\{\underline{\tilde{\boldsymbol{S}}}_{\boldsymbol{NN}}(\omega)\} = M$ [BS01]. Therefore, the array gain can be written as,

$$A(\omega) = \frac{S_{YY}(\omega)|_{\text{signal}}/S_{YY}(\omega)|_{\text{noise}}}{\text{SINR}_{\text{sensors}}(\omega)} = \frac{|\boldsymbol{W}^H(\omega)\boldsymbol{V}(\omega)|^2}{\boldsymbol{W}^H(\omega)\underline{\tilde{\boldsymbol{S}}}_{\boldsymbol{NN}}(\omega)\boldsymbol{W}(\omega)}. \tag{4.42}$$

For homogeneous noise fields, i.e. $\underline{\tilde{\boldsymbol{S}}}_{\boldsymbol{NN}}(\omega) = \underline{\boldsymbol{\Gamma}}_{\boldsymbol{NN}}(\omega)$, Eq. (4.42) can be expressed as,

$$A(\omega) = \frac{|\boldsymbol{W}^H(\omega)\boldsymbol{V}(\omega)|^2}{\boldsymbol{W}^H(\omega)\underline{\boldsymbol{\Gamma}}_{\boldsymbol{NN}}(\omega)\boldsymbol{W}(\omega)}, \tag{4.43}$$

where $\underline{\boldsymbol{\Gamma}}_{\boldsymbol{NN}}(\omega)$ denotes the so-called coherence matrix of size $M \times M$ whose individual entries are defined as [BS01],

$$\Gamma_{N_{\nu_1}N_{\nu_2}}(\omega) \triangleq \frac{S_{N_{\nu_1}N_{\nu_2}}(\omega)}{\sqrt{S_{N_{\nu_1}N_{\nu_1}}(\omega)S_{N_{\nu_2}N_{\nu_2}}(\omega)}}, \quad \nu_1, \nu_2 = 1(1)M. \tag{4.44}$$

For example, Eq. (4.44) becomes for spherically isotropic noise fields and equal sensor spacing d [BS01],

$$\Gamma^{(s)}_{N_{\nu_1}N_{\nu_2}}(\omega) = \text{sinc}[\omega d(\nu_1 - \nu_2)/c], \tag{4.45}$$

where $\text{sinc}\{x\} \triangleq \sin(x)/x$ and $\text{sinc}\{0\} = 1$. For cylindrically isotropic noise fields and equal sensor spacing [Elk01],

$$\Gamma^{(c)}_{N_{\nu_1}N_{\nu_2}}(\omega) = J_0[\omega d(\nu_1 - \nu_2)/c], \tag{4.46}$$

The so-called *white noise gain* quantifies the ability of an array to suppress spatially white noise – including sensor noise as well as sensor position, gain and phase errors – and is defined as [BS01],

$$A_w(\omega) \triangleq A(\omega)|_{\boldsymbol{\Gamma}=\boldsymbol{I}} = \frac{|\boldsymbol{W}^H(\omega)\boldsymbol{V}(\omega)|^2}{\boldsymbol{W}^H(\omega)\boldsymbol{W}(\omega)}. \tag{4.47}$$

For a uniformly weighted array, $\boldsymbol{W}(\omega) = \boldsymbol{V}(\omega)/M$, it follows that $A_w = M$, which is the maximum achievable white noise gain for any sensor weighting [Tre02].

A specific example, illustrating the gain and white-noise-gain measures, will be provided in Section 4.3.3, where superdirective arrays are discussed.

Sensitivity to Array Imperfections

The array sensitivity to deviations from ideal geometry and sensor characteristics is an additional performance measure which is often, rather heuristically, quantified as the inverse of the white noise gain. Deviations from ideal model assumptions mainly include errors in the sensor location, magnitude and phase deviations, as well as sensor self noise.

4.3.3 Data-Independent Waveform Estimation

In this section, some widely applied designs for data-independent beamforming are illustrated. The simplest and most common beamformer for narrowband applications is the uniformly-spaced linear array (ULA). It is then shown that for wideband applications, for instance high-quality distant speech acquisition, constant-directivity beamforming (CDB), is a more appropriate choice. Another class of beamformer are the so-called differential and superdirective beamformer, which are introduced at the end of this section.

The ULA Beamformer

The ULA is, because of its simplicity, a widely used beamformer in practice. The sensor spacing is usually chosen such that no grating-lobes appear in the frequency range of interest. For data-independent operation, the weights are often chosen as scalars, the uniformly weighted ULA being a special case, see e.g. Eq. (4.31). The design objective here is to ensure good side-lobe behavior while attaining good directivity. Two examples are shown in Fig. 4.7 utilizing Hamming weighting ($\boldsymbol{w}_H$) and Chebyshev weighting with 30 dB of side-lobe attenuation ($\boldsymbol{w}_{C30}$), compared to the uniform weighting ($\boldsymbol{w}_\mathbf{1}$) for $M = 9, d_\lambda = 1/2$. Using Eq. (4.38) it can be shown that $D_I[\boldsymbol{w}_\mathbf{1}] = D_{I_{\max}} = 9.54$ dB, $D_I[\boldsymbol{w}_H] = 5.53$ dB, and $D_I[\boldsymbol{w}_{C30}] = 6.32$ dB. Comparing these results with the respective directivity pattern in Fig. 4.7, the trade-off between directivity index and narrow main-lobe in conjunction with good side-lobe behavior becomes obvious.

The specific values for the Hamming and the Chebyshev weighting as well as a large number of other weighting functions can be found in [Tre02].

However, ULAs are not a good choice for applications where the desired source(s) and/or the interfering source(s) span a wide frequency range, since the directivity indeces for this class of beamformers are strongly frequency-dependent, see Fig. 4.6. For applications requiring a large bandwidth of operation with constant directivity using a minimum number of sensors, the sensors cannot be spaced in an equidistant configuration. Furthermore, FIR filters instead of scalar sensor weights help to reach the desired characteristics

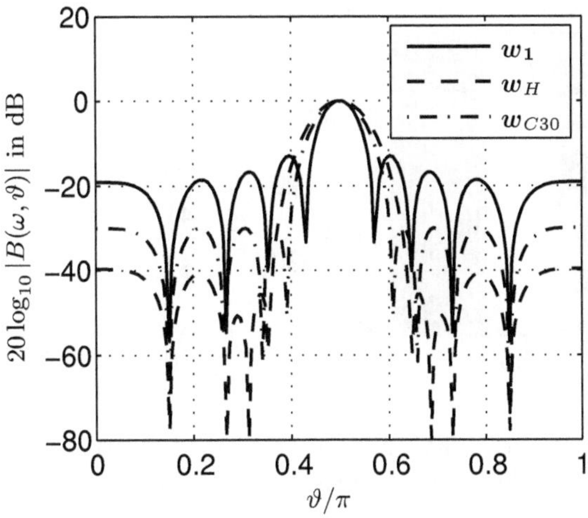

Fig. 4.7. Directivity pattern of a weighted ULA for $M = 9, d_\lambda = 1/2$, including uniform weighting ($\boldsymbol{w}_1$), Hamming weighting ($\boldsymbol{w}_H$), and Chebyshev weighting with 30 dB of side-lobe attenuation ($\boldsymbol{w}_{C30}$)

as presented in the next section. As stated earlier, a beamformer employing FIR filters are also denoted as a FSB.

The Constant-Directivity Beamformer

For certain applications, e.g. large bandwidth high-quality speech acquisition, it may be desirable that the spatial characteristics of the beamformer for varying frequency are as invariant as possible [TSH+03]. For this purpose, constant-directivity beamformer (CDB) may be employed advantageously. Several approaches to this problem are available in the literature, e.g. [DB88, GE93, Cho95, WKW95, VSd96]. The basic idea behind all CDB algorithms, however, is to jointly optimize sensor positions and sensor weights – implemented as FIR filters – such that the array aperture scales with frequency. In other words, any impinging plane-wave of arbitrary wavelength within the design frequency band 'sees' a constant length of the array aperture, l_{const}, relative to the wavelength λ. As an example, Fig. 4.8 shows the performance of a CDB design based on the algorithm presented in [VSd96]. Here, the sensor array is of length 1.72 m comprises 26 non-equally spaced sensors and FIR filters of length 128. This particular design ensures an approximately constant value of l/λ for $f = 150 \ldots 16000$ Hz. The design procedure starts with six equally spaced sensors in the center of the array, which define l_{const}. All additional sensors are then placed in a logarithmically increasing fashion. The filters are then optimized such that there are always six sensors 'visible' for each frequency of the impinging plane-wave. As can be seen from

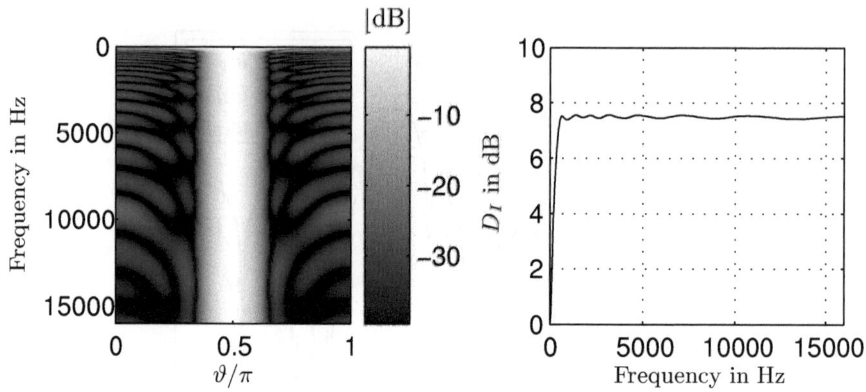

Fig. 4.8. Directivity pattern and directivity index for a 24-element CDB of length 1.72 m. Note that the lower cut-off frequency of the design is 150 Hz

the directivity pattern, the main-lobe exhibits constant spatial characteristics over the entire frequency range and the side-lobes are well suppressed (by about 20 dB). Hence, the directivity index, shown on the right-hand side of Fig. 4.8, offers fairly constant behavior. However, the directivity index is significantly lower than one would expect from a standard ULA, which is $10 \log_{10}(26) = 14.15$ dB. The reason for this is the relatively wide main-lobe of the CDB. The main-lobe width can be decreased by using more than six sensors that define l_{const}. This modification increases the total length of the sensor array, which, in turn, extends the lowest usable frequency.

Differential and Superdirective Beamformer

The beamformers discussed so far have in common that weighted and delayed sensor signals are added to yield the beamformer output. In the following, differential [Elk04] and superdirective, e.g. [CZK86, BS01], beamformers are introduced that combine the sensor outputs in a sign-alternating fashion. It will be shown that this technique yields higher directivity than conventional filter-and-sum beamformers.

Differential Beamformer

When the outputs of closely-spaced microphones are combined using sign-alternating scalars, a so-called differential microphone array (DMA) is obtained. Building upon a physical interpretation, DMAs are approximations of acoustic pressure differentials. To derive the mathematics, consider a unit-amplitude plane-wave impinging at a point on a linear aperture, see Fig. 3.23. Omitting the time-dependence, the pressure according to Eq. (3.116) is $P(k, z, \vartheta) = e^{ikz\cos\vartheta}$. Taking the n-th spatial derivative with respect to z yields,

$$\frac{\partial^n}{\partial z^n} P(k, z, \vartheta) = (ik \cos \vartheta)^n \, e^{ikz \cos \vartheta}. \tag{4.48}$$

Two observations are of importance here. Firstly, the factor k^n in Eq. (4.48) results in a $6n$ dB/octave highpass characteristic concerning the magnitude response of the n-th spatial derivative. Secondly, the factor $\cos^n \vartheta$ results in a n-th order multipole concerning the spatial characteristics, yielding n spatial nulls. Hence, these spatial derivatives are closely related to the circular/spherical harmonics derived in Chapter 3.

The paradigm of DMAs is to approximate the acoustic pressure differentials by finite differences [Elk04]. For example, a *first-order* DMA comprising two omnidirectional microphones positioned at $z_0 + d/2$ and $z_0 - d/2$ on the z-axis in Fig. 3.23 can be expressed as,

$$\begin{aligned} \frac{\Delta P(k, z, \vartheta)}{\Delta z} &= \frac{P(k, z_0 + d/2, \vartheta) - P(k, z_0 - d/2, \vartheta)}{d} \\ &= \frac{2i}{d} \sin\left(\frac{1}{2} kd \cos \vartheta\right) \, e^{ikz_0 \cos \vartheta}. \end{aligned} \tag{4.49}$$

For $kd \ll 1$, Eq. (4.49) can be approximated as,

$$\frac{\Delta P(k, z, \vartheta)}{\Delta z} \approx ik \cos \vartheta \, e^{ikz_0 \cos \vartheta}. \tag{4.50}$$

In [Elk04] it is shown that in order for the error introduced by this approximation to be less than 1 dB, the element spacing must be less than a quarter of the wavelength. Note that the minimum spacing for DSBs or FSBs, such that grating-lobes are avoided, is half of a wavelength. It is further shown in [Elk04] that, by adhering to the quarter wavelength limit, the maximum directivity in spherically isotropic and cylindrically isotropic noise is, respectively,

$$D^{(s)}_{\mathrm{DMA_{max}}} = M^2, \tag{4.51}$$

$$D^{(c)}_{\mathrm{DMA_{max}}} = 2M - 1. \tag{4.52}$$

This is also in contrast to DSBs, where the maximum directivity equals the number of microphones for uniformly weighted ULAs.

However, if a frequency-independent magnitude response is desired at the output of the DMA, the n-th order differential response has to be compensated by an n-th order integrator, which significantly boosts uncorrelated noise at low frequencies.

Returning to the first-order DMA, it can be observed that by applying a delay τ at the output of one of the microphones, *any* first-order directivity pattern can be synthesized, e.g. dipole or cardioid [Elk04]. It can be easily shown that the output of a steered first-order DMA is proportional to $\tau + d\cos(\vartheta)/c$, yielding a spatial null at $\vartheta_0 = \cos^{-1}(-c\tau/d)$. It can be further shown that an n-th-order DMA has n spatial nulls that can be controlled individually. This variability of the spatial nulls makes an n-th-order DMA a viable alternative when up to n strong directional interferers are present in a wavefield. A first- and second-order design that takes a moving interferer into account is presented in [TE01b], where the spatial null is continuously controlled by adaptively minimizing the array output.

It can be shown that $M = n+1$ omnidirectional microphones are necessary to form an n-th order DMA [Elk04]. An extension of a generic DMA of order up to $n = 3$ is shown in Fig. 4.9. Included are separate delays in each stage, i.e. τ_1, τ_2, τ_3, for synthesizing any spatial directivity pattern up to and including order $n = 3$. Assuming an equal microphone spacing of d, cumbersome but straightforward calculations show that this generic third-order structure can be re-cast into a FSB with sensor weights,

$$\boldsymbol{W}(\omega) = \frac{1}{(i\omega)^n}[W_1(\omega), W_2(\omega), \ldots, W_{n+1}(\omega)]^T, \tag{4.53}$$

where,

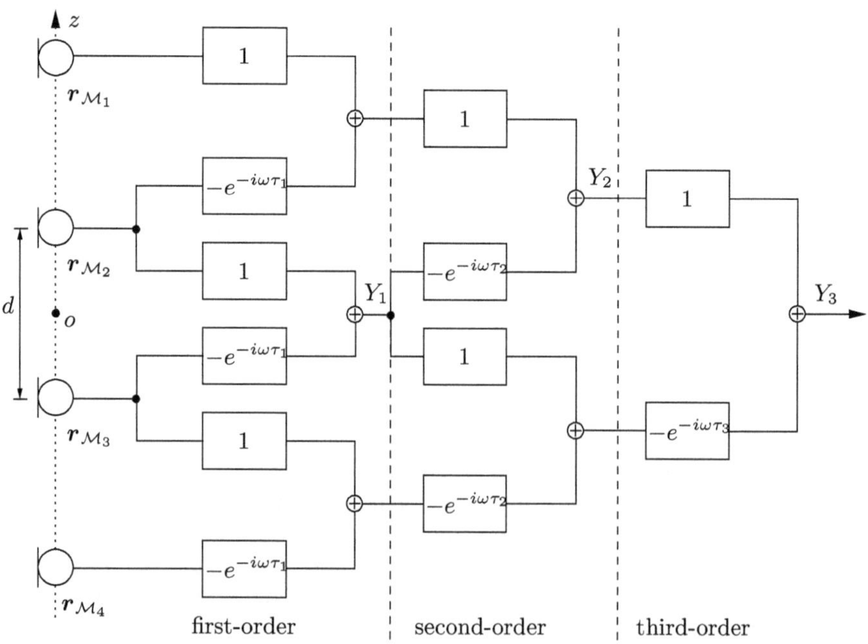

Fig. 4.9. Differential beamformer of up to order $n = 3$ including delays, cast into a FSB framework

$$W_1(\omega) = 1, \tag{4.54a}$$

$$W_2(\omega) = -\sum_{\nu_1=1}^{n} e^{-i\omega\tau_{\nu_1}}, \tag{4.54b}$$

$$W_3(\omega) = \sum_{\nu_2=0}^{n-2} e^{-i\omega\tau_{n-\nu_2}} \sum_{\nu_1=1}^{n-1-\nu_2} e^{-i\omega\tau_{\nu_1}}, \tag{4.54c}$$

$$W_4(\omega) = -e^{-i\omega \sum\limits_{\nu_1=1}^{n} \tau_{\nu_1}}. \tag{4.54d}$$

The n-th-order DMA follows with $n = 0(1)3$ in Eq. (4.53). Note that the n-th-order integrator has been included in Eq. (4.53) for obtaining a frequency-independent magnitude response.

Superdirective Beamformer

A superdirective beamformer (SDB) aims at achieving a higher directivity than the conventional delay-and-sum beamformer (DSB) does, i.e. $D_{\mathrm{DSB}_{\max}} = M$. It was stated above that differential arrays achieve $D^{(s)}_{\mathrm{DMA}_{\max}} = M^2$ by combining the microphone signals in a sign-alternating fashion using scalars. SDBs can be interpreted as a generalization of DMAs by replacing the scalars, which appear in the classical description of DSBs, by filters. SDBs are mostly used in endfire configuration, meaning that the desired signal is assumed to impinge from the prolongation of the array's axis. A widely used type of SDB is the so-called Minimum Variance Distortionless Response (MVDR) beamformer. The design objective can be verbally stated as the minimization of the output power of an array of sensors subject to a distortionless constraint for the desired signal that is assumed to impinge from the look-direction, i.e. $\vartheta_s = \vartheta$. More mathematically, this statement reads,

$$\min_{\boldsymbol{W}(\omega)} \boldsymbol{W}^H(\omega)\underline{\boldsymbol{S}}_{\boldsymbol{XX}}(\omega)\boldsymbol{W}(\omega) \quad \text{subject to} \quad \boldsymbol{W}^H(\omega)\boldsymbol{V}(\omega,\vartheta) = 1. \tag{4.55}$$

By applying the Lagrange-multiplier method [Fro72] and keeping in mind that only the noise power should be minimized, the solution of Eq. (4.55), the MVDR beamformer, can be expressed as [CZO87],

$$\boldsymbol{W}(\omega) = \frac{\underline{\boldsymbol{S}}^{-1}_{\boldsymbol{NN}}(\omega)\boldsymbol{V}(\omega)}{\boldsymbol{V}^H(\omega)\underline{\boldsymbol{S}}^{-1}_{\boldsymbol{NN}}(\omega)\boldsymbol{V}(\omega)}. \tag{4.56}$$

Note that the spatial dependency has been dropped for notational convenience. By assuming a spherically or cylindrically isotropic noise field, Eq. (4.56) can be written as,

$$\boldsymbol{W}(\omega) = \frac{\underline{\boldsymbol{\Gamma}}^{-1}_{\boldsymbol{NN}}(\omega)\boldsymbol{V}(\omega)}{\boldsymbol{V}^H(\omega)\underline{\boldsymbol{\Gamma}}^{-1}_{\boldsymbol{NN}}(\omega)\boldsymbol{V}(\omega)}, \tag{4.57}$$

where the entries of the coherence function are given by Eq. (4.45) and Eq. (4.46), respectively.

In order to deal with the spatially white noise amplification problem discussed below, a small additional scalar ϵ – the so-called white noise constraint – can be added to the diagonal entries of the coherence matrix [GM55], yielding,

$$\boldsymbol{W}_{\epsilon}(\omega) = \frac{(\boldsymbol{\Gamma}_{\boldsymbol{NN}}(\omega) + \epsilon \underline{\boldsymbol{I}})^{-1} \boldsymbol{V}(\omega)}{\boldsymbol{V}^H(\omega)(\boldsymbol{\Gamma}_{\boldsymbol{NN}}(\omega) + \epsilon \underline{\boldsymbol{I}})^{-1} \boldsymbol{V}(\omega)}. \tag{4.58}$$

Figure 4.10 shows the directivity pattern, gain and white noise gain for a 5-element SDB steered to endfire, designed according to Eq. (4.57) assuming a spherically isotropic noise field. The scalar ϵ is set to zero, corresponding to an unconstrained (superdirective) operation. Additionally, the array gain and the white noise gain for a uniformly-spaced 5-element DSB are shown for comparison, where $\epsilon = \infty$. A few interesting properties of SDBs compared to DSBs should be noted. At low frequencies, or/and small spacing, the array gain of the SDB approaches 14 dB which corresponds to M^2, while the array gain of the DSB approaches zero. Second, the WNG of the SDB approaches minus infinity while the WNG of the DSB is about 7 dB, corresponding to M as predicted in Section 4.3.2. A very small value for the WNG, or conversely a very high sensitivity, basically results in an extremely high amplification of uncorrelated white noise, making this particular design useless for practical applications at low frequencies. A third observation is that, for this particular

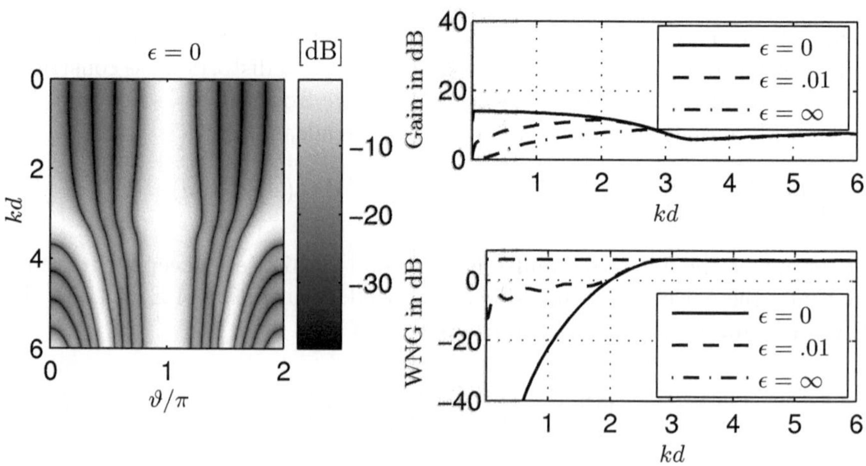

Fig. 4.10. Directivity pattern, gain and white noise gain for a uniformly-spaced 5-element SDB steered to endfire. Here, $\epsilon = 0$ corresponds to an unconstrained (superdirective) operation, while $\epsilon = \infty$ corresponds to the DSB

design, the SDB basically turns into a DSB for $kd > 3$ in terms of array gain and white noise gain.

The effect of the white noise constraint is shown on the right-hand side of Fig. 4.10. As can be seen, the array gain decreases toward the performance of the DSB. However, the WNG increases significantly for small values of kd, making the design less sensitive to random errors, like variations of the positions, gain, and phase among the sensors. Here, the trade-off between array gain and WNG becomes obvious. Specific values for ϵ depend on the application and its requirements. Hence, generic rules for its selection cannot be given.

4.3.4 Data-Dependent Waveform Estimation

Data-dependent waveform estimation algorithms adaptively control a set of filters according to the signal statistics derived from the sensor signals. A commonly used data-dependent beamformer is the so-called linearly-constrained minimum power (LCMP) beamformer [Tre02]. The design objective is here to minimize the beamformer output such that a set of linear constraints is fulfilled. Linear constraints can be imposed on the direction-of-arrival of the desired signal and/or known direction-of-arrival of strong interferers. These constraints are called *directional constraints*. Further constraints are *derivative constraints* and *eigenvector constraints* which are used to make the beamformer more robust against errors in the assumed direction-of-arrival of the desired signal and the strong interferers [Tre02]. Mathematically, this design objective can be written as,

$$\min_{\boldsymbol{W}(\omega)} \boldsymbol{W}^H(\omega)\underline{\boldsymbol{S}}_{\boldsymbol{XX}}(\omega)\boldsymbol{W}(\omega) \quad \text{subject to} \quad \boldsymbol{W}^H(\omega)\underline{\boldsymbol{C}}(\omega) = \boldsymbol{G}^H(\omega), \tag{4.59}$$

where $\underline{\boldsymbol{C}}(\omega)$ is an $M \times \tilde{C}$ constraint matrix, $\tilde{C}$ being the number of constraints used. $\boldsymbol{G}(\omega)$ is a vector containing the values of the constraints, for instance, unity for the look-direction of the beamformer toward the desired source and zero for the direction of the strong interferer(s). Note that by introducing $\tilde{C}$ constraints, there are $M - \tilde{C}$ degrees of freedom remaining in each frequency-bin for the optimization task. It was shown by Frost [Fro72] that it follows for the optimum filter coefficients that,

$$\boldsymbol{W}_{\text{opt}}(\omega) = \frac{\underline{\boldsymbol{S}}_{\boldsymbol{XX}}^{-1}(\omega)\underline{\boldsymbol{C}}(\omega)}{\underline{\boldsymbol{C}}^H(\omega)\underline{\boldsymbol{S}}_{\boldsymbol{XX}}^{-1}(\omega)\underline{\boldsymbol{C}}(\omega)}\boldsymbol{G}(\omega), \tag{4.60}$$

A so-called LCMV beamformer is obtained when replacing $\underline{\boldsymbol{S}}_{\boldsymbol{XX}}$ by $\underline{\boldsymbol{S}}_{\boldsymbol{NN}}$ in Eq. (4.60) [Tre02].

As an example, Fig. 4.11 shows the directivity pattern of a 9-sensor LCMV beamformer satisfying the distortionless constraint in the direction of the incoming desired plane-wave. A second directional constraint in the direction

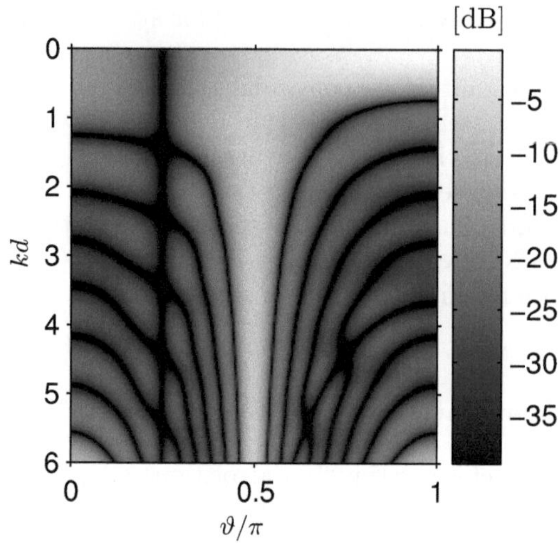

Fig. 4.11. Directivity pattern of an 9-sensor LCMV satisfying two directional constraints

of a strong interferer impinging from $\vartheta = \pi/4$ is imposed, where a deep null is formed by the beamformer.

This data-dependent beamformer can be implemented efficiently by employing the structure of the so-called *generalized side-lobe canceler* (GSC) [GJ82], as shown in Fig. 4.12, which transforms the constrained optimization problem into an unconstrained one.

The general idea is to split the optimum filter coefficients into two orthogonal structures, i.e.,

$$\boldsymbol{W}_{\text{opt}}(\omega) = \boldsymbol{W}_q(\omega) - \boldsymbol{B}(\omega)\boldsymbol{W}_a(\omega), \tag{4.61}$$

where the *quiescent* filter coefficients, $\boldsymbol{W}_q(\omega)$, form a fixed beamformer (FBF) that satisfies the constraint $\boldsymbol{W}_q^H(\omega)\underline{\boldsymbol{C}}(\omega) = \boldsymbol{G}(\omega)$. On the other hand, the *blocking matrix* (BM), $\boldsymbol{B}(\omega)$, in the orthogonal component satisfies $\underline{\boldsymbol{C}}(\omega)^H\boldsymbol{B}(\omega) = \boldsymbol{0}$. In other words, ideally, the FBF enhances the desired source while the BM blocks the desired source, only allowing the noise signals to be passed through to the output of the BM. The *interference canceler* (IC), composed of $\tilde{M} = M - \tilde{C}$ adaptive filters, driven by the respective output of the BM, then estimates the components correlated with the noise and interferers contained in $Y_{\text{FBF}}(\omega)$, see Fig. 4.12, and subtracts them from the output of the FBF, thereby further enhancing the desired signal extracted by the FBF.

The unconstrained cost function for the design of the filters comprising the IC can be written as, see Fig. 4.12,

$$\min_{\boldsymbol{W}_a(\omega)} [\boldsymbol{W}_q(\omega) - \boldsymbol{B}(\omega)\boldsymbol{W}_a(\omega)]^H \underline{\boldsymbol{S}}_{\boldsymbol{XX}}(\omega)[\boldsymbol{W}_q(\omega) - \boldsymbol{B}(\omega)\boldsymbol{W}_a(\omega)], \tag{4.62}$$

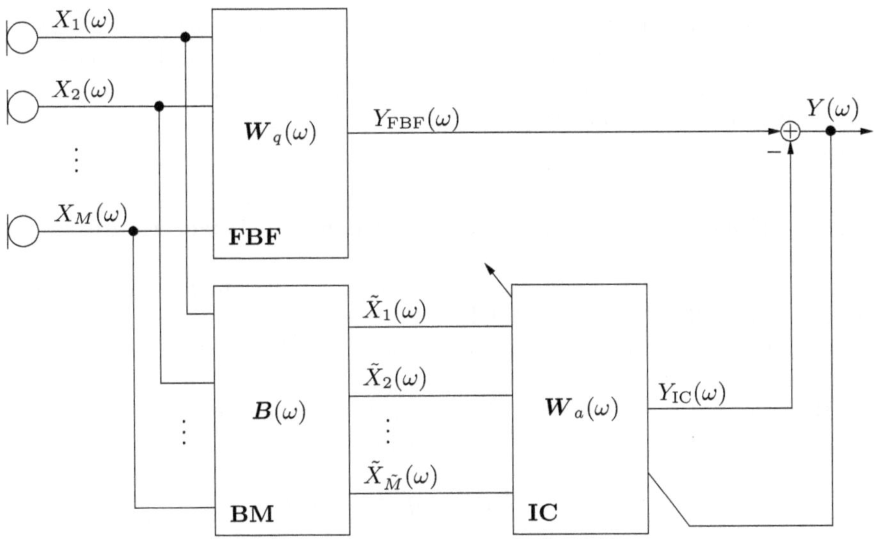

Fig. 4.12. Structure of the GSC comprising M sensors, a fixed beamformer (FBF), a blocking matrix (BM), and an interference canceler (IC)

which can be shown to result in [Tre02],

$$\boldsymbol{W}_{a,\mathrm{opt}}^H(\omega) = \boldsymbol{W}_q^H(\omega)\underline{\boldsymbol{S}}_{\boldsymbol{XX}}(\omega)\boldsymbol{B}(\omega)[\boldsymbol{B}^H(\omega)\underline{\boldsymbol{S}}_{\boldsymbol{XX}}(\omega)\boldsymbol{B}(\omega)]^{-1}. \tag{4.63}$$

Algorithms that estimate Eq. (4.63) adaptively and a plethora of subtleties related to a successful implementation of the GSC in real acoustic environments are discussed in detail in [Her05].

4.4 Parameter Estimation

In this section, parameter estimation methods are introduced. In particular, techniques that estimate the source(s) DOA are discussed. In the literature, three fundamentally different paradigms of source localization seem to exist. The first class of techniques, especially used in the localization of acoustic sources, is the paradigm of time-difference-of-arrival (TDOA). Here, two or more sensors are used to estimate the travel time of a plane-wave as it propagates across an array of sensors. Combining the TDOA with the array geometry allows the determination of the source's DOA by geometric methods. The TDOA-based techniques are sketched in Section 4.4.2. A second class of techniques are the subspace-based methods that have traditionally been used for narrowband applications. The reasons for this will become clear in the course of discussion presented in Section 4.4.3. A third class of techniques, which is not discussed in this book, are based on the methodology of a steered beamformer. The idea is simply to steer a beam in three-dimensional space and choose the direction with maximum response power as the DOA, see e.g. [DSB01] and references therein. First, however, a fundamental performance measure for parameter estimation, the Cramér-Rao lower bound, is introduced.

4.4.1 Performance Measure – The Cramér-Rao Lower Bound

The performance of parameter estimation algorithms is fundamentally bound by the Cramér-Rao lower bound (CRLB) [Tre02]. Let the vector $\boldsymbol{\eta}$ contain all parameters to be estimated from a set of data, e.g. the directions-of arrival of multiple wave-fronts, their relative spectral densities as well as the noise spectral densities. Also, let $\hat{\boldsymbol{\eta}}$ denote the result of an estimator of the parameters of interest. Then, the covariance of the estimation error, $\underline{\boldsymbol{C}}(\boldsymbol{\eta})$, is given by,

$$\underline{\boldsymbol{C}}(\boldsymbol{\eta}) \triangleq E\{(\hat{\boldsymbol{\eta}} - \boldsymbol{\eta})(\hat{\boldsymbol{\eta}} - \boldsymbol{\eta})^T\}. \tag{4.64}$$

The CRLB provides a bound on the covariance matrix of *any unbiased* estimate of $\boldsymbol{\eta}$. The covariance of the estimation error, Eq. (4.64), is therefore bounded by,

$$\|\underline{\boldsymbol{C}}(\boldsymbol{\eta})\|_F \geq \|\underline{\boldsymbol{C}}_{CRLB}(\boldsymbol{\eta})\|_F. \tag{4.65}$$

This important equation means that no estimator can perform better than the CRLB. An estimator is called *efficient* if it attains the CRLB for a long observation interval and/or high SNR.

In the following, it is assumed that only the estimation of the signals' DOA is of interest, i.e. $\boldsymbol{\eta} \equiv \boldsymbol{\Theta} = [\Theta_1, \Theta_2, \dots, \Theta_I]^T$. The model considered is the Gaussian model with unknown signal power spectral density, cf. Eq. (4.10). It is further assumed that the noise is a WSS process with covariance σ_w^2. Then, as shown in [SLG01], the CRLB is given by,

$$\begin{aligned}\underline{\boldsymbol{C}}_{CRLB}(\boldsymbol{\Theta},\omega) = &\frac{\sigma_w^2}{2K}\left\{\mathrm{Re}\left\{\underline{\boldsymbol{S}}_{SS}(\omega)\left[\left(\underline{\boldsymbol{I}} + \underline{\boldsymbol{V}}^H(\omega,\boldsymbol{\Theta})\underline{\boldsymbol{V}}(\omega,\boldsymbol{\Theta})\frac{\underline{\boldsymbol{S}}_{SS}(\omega)}{\sigma_w^2}\right)^{-1}\right.\right.\right.\\ &\left.\left.\left.\times\left(\underline{\boldsymbol{V}}^H(\omega,\boldsymbol{\Theta})\underline{\boldsymbol{V}}(\omega,\boldsymbol{\Theta})\frac{\underline{\boldsymbol{S}}_{SS}(\omega)}{\sigma_w^2}\right)\right]\odot \underline{\boldsymbol{H}}^T(\omega,\boldsymbol{\Theta})\right\}\right\}^{-1},\end{aligned} \tag{4.66}$$

where K is the number of statistically independent data snapshots and, omitting the dependencies on $\boldsymbol{\Theta}$ and ω where possible for better readability,

$$\underline{\boldsymbol{H}} = \underline{\boldsymbol{D}}^H\left[\underline{\boldsymbol{I}} - \underline{\boldsymbol{V}}(\underline{\boldsymbol{V}}^H\underline{\boldsymbol{V}})^{-1}\underline{\boldsymbol{V}}^H\right]\underline{\boldsymbol{D}}, \tag{4.67a}$$

$$\underline{\boldsymbol{D}} = \frac{\partial \underline{\boldsymbol{V}}}{\partial \boldsymbol{\Theta}} = \left[\frac{\partial \boldsymbol{V}(\omega,\Theta_1)}{\partial\Theta_1}\,\middle|\,\frac{\partial \boldsymbol{V}(\omega,\Theta_2)}{\partial\Theta_2}\,\middle|\,\dots\,\middle|\,\frac{\partial \boldsymbol{V}(\omega,\Theta_I)}{\partial\Theta_I}\right]. \tag{4.67b}$$

The symbol $\odot$ in Eq. (4.66) denotes the Hadamard product, i.e. an element-wise multiplication of the matrix elements. In general, Eq. (4.66) leads to highly intractable closed-form expressions for the CRLB. Even for the rather simple case of two desired source signals, to be considered later for subspace-based parameter estimation methods, a concise and meaningful closed-form expression seems impossible to obtain. However, for the special case of a single desired signal to be observed with a linear sensor array having sensors positioned at $z_\ell, \ell = 1(1)M$, cf. the geometric model shown in Fig. 4.2, significant simplification of Eq. (4.66) becomes possible. After applying the additional assumption of equal spectral densities at the sensors, i.e. $\underline{\boldsymbol{S}}_{\boldsymbol{SS}}(\omega) = \sigma_s^2(\omega)$, it can be shown after a series of extensive, though straightforward manipulations [MS69], that with $\Theta = \vartheta$,

$$C_{CRLB}(\vartheta, \omega) = \frac{1}{2K}\left\{\frac{\sin^2(\vartheta)}{c^2}\frac{[\sigma_s^2(\omega)/\sigma_w^2]^2}{1 + M\sigma_s^2(\omega)/\sigma_w^2}\omega^2 \sum_{\ell_1=1}^{M}\sum_{\ell_2=\ell_1+1}^{M}(z_{\ell_1} - z_{\ell_2})^2\right\}^{-1}. \tag{4.68}$$

For a standard ULA, the double sum can be expressed by $d^2M^2(M^2-1)/12$ [GR65], and therefore for a single frequency corresponding to $\lambda = 2d$,

$$C_{CRLB}(\vartheta) = \frac{6(1 + M \cdot \mathrm{SNR})}{K\pi^2 \sin^2(\vartheta) M^2(M^2 - 1) \cdot \mathrm{SNR}^2}, \tag{4.69}$$

where $\mathrm{SNR} \triangleq \sigma_s^2/\sigma_w^2$. Specific examples involving the CRLB will be shown in later sections, when the performance of subspace-based parameter estimation methods is studied. Note that the CRLB depends on the DOA of the incoming plane-wave. It can be seen that the performance of *any* standard ULA will degrade as the source moves away from broadside, which is a well-known effect that all linear arrays have in common.

A simple TDOA-estimation method for ULAs that could be used for determining the performance of time-difference-of-arrival (TDOA) methods for a single impinging plane-wave can also be derived [MS69, WW83, WWW84]. Let τ_{TDOA} denote the TDOA to be estimated. Then, for $M = 2$,

$$C_{CRLB}^{(\tau_{\mathrm{TDOA}})}(\omega) = \frac{1}{2K}\left\{\frac{[\sigma_s^2(\omega)/\sigma_w^2]^2}{1 + 2\sigma_s^2(\omega)/\sigma_w^2}\omega^2\right\}^{-1}. \tag{4.70}$$

This CRLB, however, is not too meaningful for the TDOA estimation methods presented in Section 4.4.2 since they are based on peak detection of discrete-time functions. Therefore, it can be argued, that the estimate is either absolutely correct, if the correct peak is identified (zero error), or completely unusable, if the wrong peak is identified. This is in contrast to the DOA estimate, $\hat{\boldsymbol{\Theta}}$, to be presented in Section 4.4.3, which is not limited to discrete values. As a result, the CRLB based on the TDOA, Eq. (4.70), will not be used any further.

Limitations of the CRLB

The CRLB is used extensively for evaluating the performance of any unbiased estimator. However, it must be kept in mind that the CRLB assumes that the estimation error is small. It can therefore be only compared to *local* variations of the estimate obtained by PE methods. Small estimation errors can only be obtained when the SNR is high and/or the observation interval T is sufficiently long. T is considered to be sufficiently long if $WT \gg 2\pi$, which means that the observation interval must be larger than the correlation time – i.e. inverse bandwidth – of the signal and noise [WW83]. At low SNR values and/or short observation times the probability of erroneous estimation results cannot be neglected and the performance of any estimator will degrade drastically compared to the CRLB. An overview of specific analyses regarding ambiguous estimation results for TDOA-based algorithms and subspace-based DOA algorithms will be given in Section 4.4.2 and Section 4.4.3, respectively.

However, detailed performance analysis of parameter estimation methods is beyond the scope of this book. Here, a few selected examples are presented in Section 4.4.2 and Section 4.4.3 to highlight some strengths and weaknesses of the estimation algorithms to be described below.

4.4.2 TDOA-Based Algorithms

Here, two major contributions in the area of TDOA estimation are presented. First, TDOA estimation based on the generalized cross-correlation (GCC) is introduced [KC76]. Second, the adaptive eigenvalue decomposition (AED) algorithm [Ben00] is outlined that, in contrast to the GCC method, does not rely on the free-field plane-wave propagation model but explicitly models multipath propagation by appropriate FIR filters.

Generalized Cross-Correlation-based Algorithms

It is assumed that a single desired plane-wave signal impinges on a sensor array comprising two sensors. In particular, it is assumed that the attenuation due to wave propagation can be neglected. The signal $s(t)$ and the noise signals, $n_1(t), n_2(t)$, are assumed to be jointly WSS and mutually uncorrelated processes. Then, the signal model, cf. Eq. (4.1), can be written as,

$$x_1(t) = s(t) + n_1(t), \tag{4.71}$$

$$x_2(t) = s(t - \tau_{\mathrm{TDOA}}) + n_2(t). \tag{4.72}$$

The generalized cross-correlation (GCC) method now estimates the TDOA, $\hat{\tau}_{\mathrm{TDOA}}$, by determining the value of τ that maximizes the generalized cross-correlation function given by the inverse Fourier transform of the weighted cross power spectral density [KC76]. Assuming infinite observation time T,

$$c_{x_1x_2}(\tau) = \frac{1}{2\pi} \int_{-\infty}^{\infty} A(\omega) S_{x_1x_2}(\omega) e^{-i\omega\tau} d\omega, \tag{4.73}$$

where $A(\omega)$ is a weighting function, cf. Tab. 4.1. Then, the estimate $\hat{\tau}_{\mathrm{TDOA}}$ can be obtained by,

$$\hat{\tau}_{\mathrm{TDOA}} = \arg\max_{\tau} c_{x_1x_2}(\tau). \tag{4.74}$$

Equation (4.74) therefore searches for the maximum peak in the cross-correlation (CC) function. This search becomes increasingly more difficult as the input signals decrease in bandwidth since the relative height of the maximum peak in the classical CC function, where $A(\omega) = 1$, with respect to the adjacent peaks is proportional to the signal's bandwidth [WW83]. As a remedy, high SNR values become necessary for detection as the bandwidth decreases. Another possibility would be to apply the so-called PHAT (phase transform) weighting function, see Tab. 4.1, which basically serves as a pre-whitening filter as the following discussion shows. Let $X_1(\omega)$ and $X_2(\omega)$ denote the Fourier transforms of Eq. (4.71) and Eq. (4.72), respectively. Then, by virtue of the uncorrelatedness assumption of $s(t)$ and $n_\nu(t), \nu = 1, 2$,

$$S_{X_1X_2}(\omega) = E\{X_1(\omega)X_2^*(\omega)\} = S_{SS}(\omega)e^{-i\omega\tau_{\mathrm{TDOA}}}. \tag{4.75}$$

Plugging this result and the PHAT filter function into Eq. (4.73) and carrying out the integration yields,

$$c_{x_1x_2}(\tau) = \delta(\tau + \tau_{\mathrm{TDOA}}), \tag{4.76}$$

therefore resulting in a single peak at the position of the TDOA.

A few of the commonly used weighting functions are summarized in Tab. 4.1. Further discussions and comparisons can be found in [KC76].

One of the main disadvantages of the GCC-based TDOA estimation methods applied to real-world wavefields is that the signal model only takes the direct propagation path into account. As a result, strong reflections (multipath) lead to additional peaks in the cross-correlation function which can be just as high, or even higher, than the peak corresponding to the direct propagation path [CBS96]. This effect cannot be prevented by any pre-filter, $A(\omega)$.

Algorithms based on the adaptive eigenvalue decomposition method (AED) deal with this problem by specifically including the multipath environment into the model, as shown in the following.

The Adaptive Eigenvalue Decomposition Algorithm

In contrast to the GCC signal model, the model underlying the AED algorithm explicitly incorporates the impulse responses from the source $s(t)$ to the first and second microphone, i.e. $u_1(t)$ and $u_2(t)$, respectively as,

Table 4.1. Specific weighting functions for GCC methods (after [KC76])

$A(\omega)$	Synonym
1	classical CC
$\dfrac{1}{S_{x_1x_2}(\omega)}$	ROTH filter
$\dfrac{1}{\sqrt{S_{x_1x_1}(\omega)S_{x_2x_2}(\omega)}}$	SCOT filter
$\dfrac{1}{\lvert S_{x_1x_2}(\omega)\rvert}$	PHAT filter

$$x_1(t) = u_1(t) * s(t) + n_1(t), \tag{4.77}$$
$$x_2(t) = u_2(t) * s(t) + n_2(t). \tag{4.78}$$

The TDOA is now 'coded' into the two impulse responses as the difference between the two peaks that represent the propagation due to the direct path. The AED algorithm [Ben00] now aims at adaptively identifying the two impulse responses, as indicated in Fig. 4.13. However, an adaptive algorithm is only capable of identifying a finite set of parameters. Therefore, as in Section 4.3.1, the impulse responses as well as the filters are expanded into a set of basis functions, $\chi_q(t) = \delta(t - q\Delta t)$, weighted by Q_u and Q_w expansion coefficients, respectively. In the noise-free case, it therefore follows that,

$$\begin{aligned} x_\nu(t) &= s(t) * \sum_{q=0}^{Q_u-1} u_{\nu,q}\delta(t - q\Delta t) = \sum_{q=0}^{Q_u-1} u_{\nu,q}s(t - q\Delta t) \\ &= \boldsymbol{u}_\nu^T \boldsymbol{s}(t), \end{aligned} \tag{4.79}$$

$$\begin{aligned} y_1(t) &= x_1(t) * \sum_{q=0}^{Q_w-1} w_{2,q}\delta(t - q\Delta t) = \sum_{q=0}^{Q_w-1} w_{2,q}x_1(t - q\Delta t) \\ &= \boldsymbol{x}_1(t)^T \boldsymbol{w}_2, \end{aligned} \tag{4.80}$$

and,

$$\begin{aligned} y_2(t) &= x_2(t) * \sum_{q=0}^{Q_w-1} w_{1,q}\delta(t - q\Delta t) = \sum_{q=0}^{Q_w-1} w_{1,q}x_2(t - q\Delta t) \\ &= \boldsymbol{x}_2(t)^T \boldsymbol{w}_1, \end{aligned} \tag{4.81}$$

where $\nu = 1, 2$ and,

$$\boldsymbol{s}(t) = [s(t), s(t-\Delta t), \ldots, s(t-(Q_u-1)\Delta t)]^T, \quad (4.82)$$
$$\boldsymbol{x}_\nu(t) = [x_\nu(t), x_\nu(t-\Delta t), \ldots, x_\nu(t-(Q_w-1)\Delta t)]^T, \quad (4.83)$$
$$\boldsymbol{u}_\nu = [u_\nu(0), u_\nu(1), \ldots, u_\nu(Q_u-1)]^T, \quad (4.84)$$
$$\boldsymbol{w}_\nu = [w_\nu(0), w_\nu(1), \ldots, w_\nu(Q_w-1)]^T. \quad (4.85)$$

Note that $\boldsymbol{u}_\nu$ is assumed to be time-varying if either the acoustic environment cannot be approximated as a time-invariant system, and/or if the acoustic source is moving. The task is now to identify the time-varying impulse response $\boldsymbol{u}_\nu$ by the adaptive filters $\boldsymbol{w}_\nu$.

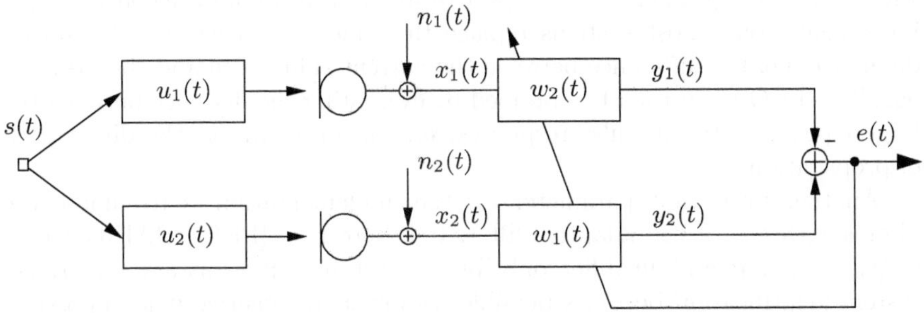

Fig. 4.13. Time-difference estimation in multipath environments using AED algorithm

Returning to the AED in the noise-free case, it can be deduced from Fig. 4.13 that,

$$e(t) = \boldsymbol{x}_1^T(t)\boldsymbol{w}_2 - \boldsymbol{x}_2^T(t)\boldsymbol{w}_1. \quad (4.86)$$

The spatio-temporal correlation matrix for the two-sensor case reads,

$$\underline{\boldsymbol{C}}_{\boldsymbol{xx}}(\tau) = \begin{bmatrix} E\{\boldsymbol{x}_1(t)\boldsymbol{x}_1^T(t-\tau)\} & E\{\boldsymbol{x}_1(t)\boldsymbol{x}_2^T(t-\tau)\} \\ E\{\boldsymbol{x}_2(t)\boldsymbol{x}_1^T(t-\tau)\} & E\{\boldsymbol{x}_2(t)\boldsymbol{x}_2^T(t-\tau)\} \end{bmatrix}, \quad (4.87)$$

Dropping the explicit time dependence, it follows with,

$$\boldsymbol{w} \triangleq [\boldsymbol{w}_2 \quad -\boldsymbol{w}_1]^T, \quad (4.88)$$
$$\boldsymbol{x} \triangleq [\boldsymbol{x}_1 \quad \boldsymbol{x}_2]^T, \quad (4.89)$$

and Eq. (4.86) satisfying $e(t) = 0$, that,

$$\underline{\boldsymbol{C}}_{\boldsymbol{xx}}\boldsymbol{w} = \boldsymbol{0}. \quad (4.90)$$

This result states that $\boldsymbol{w}$ is the eigenvector corresponding to the eigenvalue 0 of $\underline{\boldsymbol{C}}_{\boldsymbol{xx}}$. However, in the noisy case, the error $e(t)$ in Eq. (4.86) will never become zero exactly because of the additional noise components, $n_\nu(t), \nu = 1, 2,$

and because of the fact that, in general, $Q_u \neq Q_w$. Therefore, the expansion coefficients (FIR filter coefficients), $\hat{\boldsymbol{w}}$ are estimated by minimizing the squared value of the error signal, $\boldsymbol{e} = \boldsymbol{w}^T\boldsymbol{x}$, see Fig. 4.13. Then,

$$\hat{\boldsymbol{w}} = \min_{\boldsymbol{w}} E\{e^2\} = \min_{\boldsymbol{w}}\{\boldsymbol{w}^T\boldsymbol{x}\boldsymbol{x}^T\boldsymbol{w}\} = \min_{\boldsymbol{w}} \boldsymbol{w}^T \underline{\boldsymbol{C}}_{\boldsymbol{xx}}\boldsymbol{w}. \tag{4.91}$$

The search can be performed adaptively by a stochastic gradient descent algorithm, e.g. a least-mean squares (LMS) algorithm, which replaces the expected value of the squared error by its instantaneous value [SK92], i.e.,

$$\frac{d\boldsymbol{w}}{dt} = -\kappa\nabla_{\boldsymbol{w}}e^2(t) = -2\kappa e(t)\boldsymbol{x}(t), \tag{4.92}$$

where $\nabla_{\boldsymbol{w}}$ is the gradient with respect to $\boldsymbol{w}$ and κ is an update constant. For implementations, most systems replace the time derivative by a first-order difference, i.e. the difference between the current value of $\boldsymbol{w}$ and the previous one. The TDOA can then be obtained by evaluating the distance between the largest peaks in the impulse responses, $\boldsymbol{w}$, corresponding to the direct path of propagation.

Another important parameter for the implementation is to choose the filter length Q_w of the adaptive filter. It is shown in [FK97, BAK05b] that if $Q_w = Q_u$, $\boldsymbol{u}$ and $\boldsymbol{w}$ differ only by a constant but arbitrary scalar, i.e. system identification becomes possible. However, in practice it is impossible to estimate Q_u exactly. In case of $Q_w > Q_u$, the impulse responses are related through an arbitrary filter, which is undesirable. Finally, if $Q_w < Q_u$, which is often encountered in practice, the impulse response cannot be identified completely. However, for the problem of TDOA estimation, where only the difference of the peaks corresponding to the direct path of propagation in the identified impulse response is of importance, this constraint does not pose a serious restriction. A requirement for system identifiability in all three cases, however, is that the transfer functions corresponding to impulse responses do not share any common zeros [Ben00].

Performance of TDOA-Based Algorithms

A rigorous performance analysis may be obtained by comparing the estimates of the TDOA-based algorithms with the CRLB, i.e. Eq. (4.70). However, due to the fact that, for the evaluation process applied here, the estimated TDOAs are integer multiples of the sampling frequency, this performance bound is not applied here, cf. Section 4.4.1. Another reason for not using the CRLB is that at low SNR, short observation intervals, and/or strong multipath propagation, the estimates become ambiguous. This means that the probability for selecting the wrong peaks increases drastically in these scenarios and, thus, the variance of the estimation error increases well beyond the CRLB. This fact makes the CRLB far too optimistic in most scenarios, i.e. impossible to achieve in realistic environments. Weiss and Weinstein [WW83] address this problem by deriving a modified lower performance bound.

The performance measure used for the simulations presented here is the probability of correct TDOA estimation applied to the estimate returned by the GCC- and AED-based algorithms, i.e. the probability that the correct peaks are detected. As mentioned above, the TDOA estimation algorithms need to be implemented in the discrete time-domain for the purpose of performance evaluations. Therefore, the peak positions are restricted to integer multiples of the sampling interval. For taking into account rounding effects due to the algorithm, an error margin of ± 1 sample in evaluating the peak detection process was allowed. Multipath propagation was simulated by applying the widely used image method, see [AB79], with varying boundary reflection coefficients, denoted as $\beta \in [0 \ldots 1]$.

The following discussion considers a specific example for performance evaluation of the TDOA-based DOA estimation algorithms.

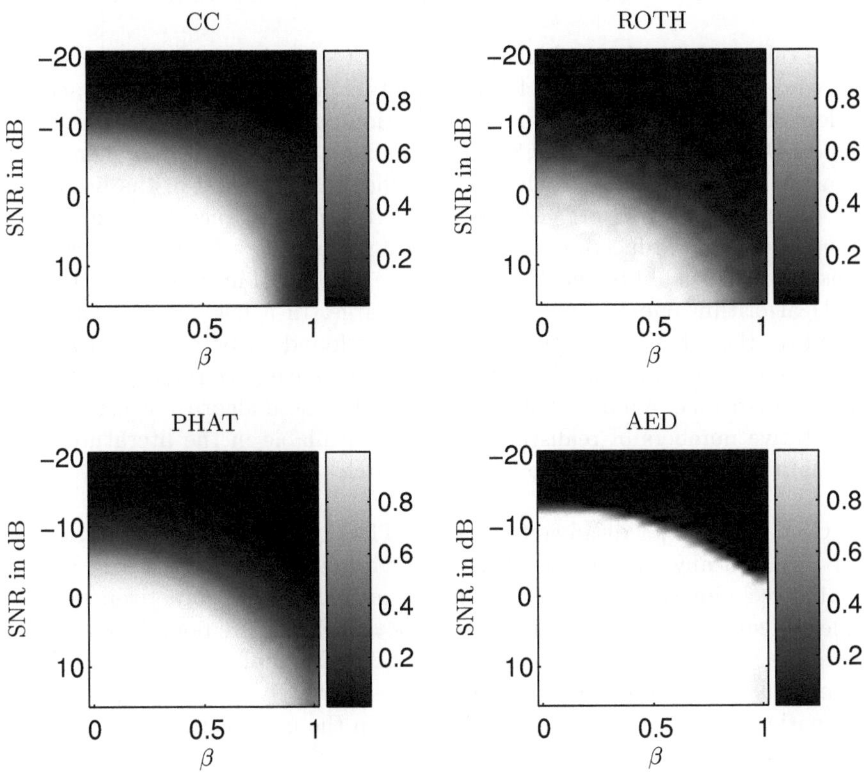

Fig. 4.14. Probability of correct TDOA estimation of GCC- and AED-based algorithms ($K = 4800$, 300 trial runs)

Figure 4.14 shows the probability of correct peak detection with respect to varying reflection coefficients on the one hand and varying power of cylindrically diffuse noise on the other hand. The two sensors, spaced 0.48 m apart, were positioned in a virtual room of dimension $10 \times 10 \times 3$ m. Spatially uncorrelated white noise of varying power was added to the sensors. The excitation signal was noise with a speech-like envelope. Results are shown for the classical cross-correlation method (CC) as well as CC weighted by the ROTH and the PHAT filters. Also shown is the performance of the AED algorithm. The adaptive filter was implemented utilizing a frequency-domain LMS algorithm [Hay02], where $Q_w = 256$ and $\kappa = 0.01$. For high SNR and small values of the wall reflection coefficients, reliable estimates of all algorithms can be expected. With increasing reflection coefficients, all algorithms begin to return unreliable estimates, the AED yielding the best performance. This result is expected since the classical CC methods do not explicitly model reverberation such as the AED algorithm. Note the distinct threshold behavior, especially of the AED algorithm, i.e. starting from a specific value for either the SNR or the reflection coefficient, the algorithms abruptly start to return erroneous estimates.

In this example, the performance of all GCC-based algorithms is very similar, while the AED algorithm clearly exhibits superior performance. Especially at low SNR and high values for the reflection coefficient β, the advantage of the AED compared to the GCC-based methods becomes obvious here. This result is in accordance with the findings that can be found in the literature, e.g. [Ben00, Hua01]. Note, however, that the performance for one specific setup only is reproduced, chosen from a large number of different scenarios considered. It should be mentioned that the amount of improvement that the AED algorithm offers is not consistently large. In a few cases, even, GCC based on the PHAT weighting function was found to be superior to AED. To the best of the author's knowledge, no exhaustive performance evaluation and comparison of a number of different TDOA-based algorithms using a representative numbers of realistic scenarios is available in the literature. This evaluation could be subject to further research.

One of the major disadvantages of all TDOA-based algorithms is the fact that, without any a-priori knowledge and without using any heuristically motivated techniques, only a single source can be active at any given point in time and space. This also means that these algorithms are bound to fail when there are strong directional interferers. The problem of estimating the DOA of multiple simultaneously active sources is addressed by the subspace-based estimation algorithms, which are presented in the next section.

4.4.3 Subspace-Based DOA Estimation Algorithms

A fundamentally different concept of DOA estimation is offered by the methodology of subspaces. One of the major advantages is the capability of resolving multiple sources. Following the introduction of the fundamental concepts of subspaces, two principal algorithms, namely the MUSIC and ESPRIT algorithms as well as some of their important variants are derived.

Preliminaries

Let I plane-waves impinge on an array of M sensors. As before, the noise is assumed to be a spatially white WSS process. The narrowband signal model is given by Eq. (4.7) with $\omega_\mu = \omega$, repeated here for convenience,

$$\boldsymbol{X}(\omega) = \underline{\boldsymbol{V}}(\omega, \boldsymbol{\Theta})\boldsymbol{S}(\omega) + \boldsymbol{N}(\omega), \tag{4.93}$$

where the array manifold matrix is,

$$\underline{\boldsymbol{V}}(\omega, \boldsymbol{\Theta}) = [\boldsymbol{V}(\omega, \Theta_1) \mid \boldsymbol{V}(\omega, \Theta_2) \mid \ldots \mid \boldsymbol{V}(\omega, \Theta_I)]. \tag{4.94}$$

Dropping the explicit dependencies on frequency and DOAs for better readability, Eq. (4.10) becomes,

$$\underline{\boldsymbol{S}}_{\boldsymbol{XX}} = \underline{\boldsymbol{V}}\underline{\boldsymbol{S}}_{\boldsymbol{SS}}\underline{\boldsymbol{V}}^H + \sigma_w^2 \underline{\boldsymbol{I}}. \tag{4.95}$$

It is further assumed that no two signals are spatially fully coherent and that the columns of $\underline{\boldsymbol{V}}$ are linearly independent. Then, the $M \times M$ spatio-spectral correlation matrix $\underline{\boldsymbol{S}}_{\boldsymbol{XX}}$ is positive definite [Tre02].

As will be shown subsequently, it is beneficial to decompose $\underline{\boldsymbol{S}}_{\boldsymbol{XX}}$ into its eigenvalues ζ_ℓ and eigenvectors $\boldsymbol{\xi}_\ell$, $\ell = 1(1)M$, as,

$$\underline{\boldsymbol{S}}_{\boldsymbol{XX}} = \sum_{\ell=1}^{M} \zeta_\ell \boldsymbol{\xi}_\ell \boldsymbol{\xi}_\ell^H. \tag{4.96}$$

Note that since correlation matrices are conjugate-symmetric (Hermitian), the eigenvectors, $\boldsymbol{\xi}_\ell$, form an orthonormal set [JD82]. These eigenvectors are now separated and grouped into so-called subspaces, the signal-plus-noise-subspace – or signal subspace for short – and the noise subspace. The grouping for positive definite matrices follows naturally from eigen-decomposition theory [Tre02]. Let the eigenvalues be arranged in decreasing order. Then, it can be shown that [Tre02],

$$\zeta_1 \geq \zeta_2 \geq \cdots \geq \zeta_I > \zeta_{I+1} = \cdots = \sigma_w^2. \tag{4.97}$$

The signal subspace is defined as a matrix of size $M \times I$ that contains the eigenvectors corresponding to the I largest eigenvalues, i.e.,

$$\underline{\boldsymbol{U}}_S \triangleq [\boldsymbol{\xi}_1 \mid \boldsymbol{\xi}_2 \mid \ldots \mid \boldsymbol{\xi}_I]. \tag{4.98}$$

Similarly, the noise subspace is defined as a matrix of size $M \times (M-I)$ that contains the eigenvectors corresponding to the $M-I$ smallest eigenvalues, i.e.,

$$\underline{\boldsymbol{U}}_N \triangleq [\boldsymbol{\xi}_{I+1} \mid \boldsymbol{\xi}_{I+2} \mid \cdots \mid \boldsymbol{\xi}_M]. \tag{4.99}$$

Note that these subspaces are orthogonal and that the noise subspace contains only components that belong to the noise component, while the signal subspace contains both desired signal components and noise components.

By virtue of grouping the eigenvectors into signal and noise subspaces, the spatio-spectral correlation can also be written as,

$$\underline{\boldsymbol{S}}_{\boldsymbol{XX}} = \underline{\boldsymbol{U}}_S \underline{\boldsymbol{\Lambda}}_S \underline{\boldsymbol{U}}_S^H + \underline{\boldsymbol{U}}_N \underline{\boldsymbol{\Lambda}}_N \underline{\boldsymbol{U}}_N^H, \tag{4.100}$$

where $\underline{\boldsymbol{\Lambda}}_S$ and $\underline{\boldsymbol{\Lambda}}_N$ are diagonal matrices containing the ordered eigenvalues corresponding to I signal eigenvectors and $M - I$ noise eigenvectors, respectively.

A fundamental observation, that is exploited in all subspace-based techniques, is that the eigenvectors that form the signal subspace are linear combinations of the array manifold vectors corresponding to the I signal sources. In mathematical terms, this observation reads,

$$\text{span}\{\underline{\boldsymbol{U}}_S\} = \text{span}\{\underline{\boldsymbol{V}}\}. \tag{4.101}$$

Of course, real systems need to obtain estimates of the spatio-spectral correlation matrix, $\hat{\underline{\boldsymbol{S}}}_{\boldsymbol{XX}}$, for instance by virtue of Eq. (4.11). Then, Eq. (4.96) becomes,

$$\hat{\underline{\boldsymbol{S}}}_{\boldsymbol{XX}} = \sum_{\ell=1}^{M} \hat{\zeta}_\ell \hat{\boldsymbol{\xi}}_\ell \hat{\boldsymbol{\xi}}_\ell^H. \tag{4.102}$$

Correspondingly, one obtains only an estimate of the signal subspace $\hat{\underline{\boldsymbol{U}}}_S$ and of the noise subspace $\hat{\underline{\boldsymbol{U}}}_N$. Indeed, for finite observation time and/or low SNR, $\hat{\underline{\boldsymbol{U}}}_S \neq \underline{\boldsymbol{U}}_S$ and $\hat{\underline{\boldsymbol{U}}}_N \neq \underline{\boldsymbol{U}}_N$. The main task for all subspace-based estimation algorithms is to retrieve the signal and noise subspaces from a finite set of noisy measurements in order to extract the desired parameters.

Eigenvalues and eigenvectors can also be estimated directly from the data matrix by a technique called the singular value decomposition (SVD) [GL89]. The advantage of this approach as opposed to estimating the eigenvalues and eigenvectors from the spatio-spectral correlation matrix, Eq. (4.102), is the fact that no squaring of the data vector, see e.g. Eq. (4.11), is necessary. This squaring operation has the problem of potentially amplifying the errors, with respect to the assumed signal model, contained in the estimated data matrix.

In the following subsections, it is assumed that the number of plane-waves I is known a-priori. If this information is not available, the techniques outlined in Section 4.4.4 may be applied.

Spatially Fully Coherent Sources

The discussion presented in this section so far has assumed that no two signals are spatially fully coherent. Fully coherent signals yield a singular signal correlation matrix, $\underline{\boldsymbol{S}}_{\boldsymbol{SS}}$, i.e.,

$$\text{rank}\{\underline{\boldsymbol{S}}_{\boldsymbol{SS}}\} < I. \tag{4.103}$$

As a result, by decomposing the observed correlation matrix, $\underline{\boldsymbol{S}}_{\boldsymbol{XX}}$, using Eqs. (4.96)–(4.100), results in the array's inability to localize the fully coherent sources [SWK85] since now,

$$\text{span}\{\underline{\boldsymbol{U}}_S\} \neq \text{span}\{\underline{\boldsymbol{V}}\}. \tag{4.104}$$

A conceptually simple pre-processing technique that is known to essentially perform signal decorrelation is *forward-backward spatial smoothing* (FBSS) of the observed spatio-spectral correlation matrix [Tre02]. Spatial smoothing is applicable to arrays exhibiting a regular structure, such as uniform linear or uniform planar arrays. The idea of spatial smoothing is to group the array into L overlapping subarrays, as shown in Fig. 4.15, resulting in data vectors $\boldsymbol{X}_l$, $l = 1(1)L$. Using these data vectors, an average of the subarray output covariance matrices is determined. For each block of data, the estimated correlation matrix, Eq. (4.11), is then replaced by [Tre02],

$$\hat{\underline{\boldsymbol{S}}}_{\boldsymbol{XX},\text{FBSS}} \triangleq \frac{1}{2KL}\sum_{l=1}^{L}\sum_{\nu=1}^{K}\boldsymbol{X}_{l,\nu}\boldsymbol{X}_{l,\nu}^H + \bar{\underline{\boldsymbol{I}}}\boldsymbol{X}_{l,\nu}^*\boldsymbol{X}_{l,\nu}^T\bar{\underline{\boldsymbol{I}}}, \tag{4.105}$$

where the $M \times M$ matrix $\bar{\underline{\boldsymbol{I}}}$, the so-called exchange matrix, is defined as,

$$\bar{\underline{\boldsymbol{I}}} \triangleq \begin{bmatrix} 0 & 0 & \dots & 1 \\ \vdots & \vdots & \iddots & \vdots \\ 0 & 1 & \dots & 0 \\ 1 & 0 & \dots & 0 \end{bmatrix}. \tag{4.106}$$

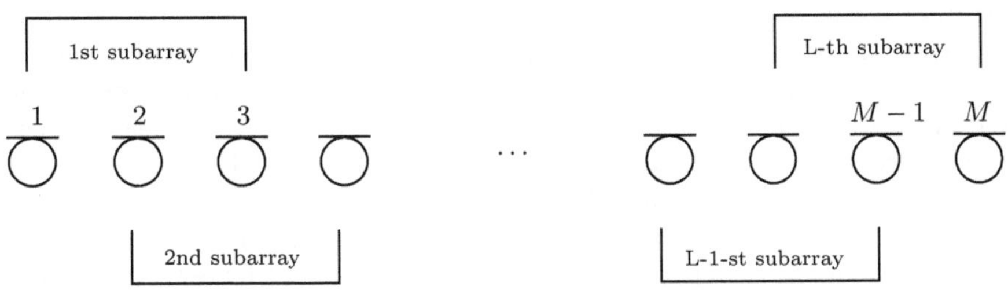

Fig. 4.15. Subarrays defined for forward-backward spatial smoothing

The first term in Eq. (4.105) is the component due to forward averaging and the second term due to backward averaging. Pillai and Kwon [PK89] have shown that for $L \geq I/2$, the spatio-temporal correlation matrix $\underline{\boldsymbol{S}}_{\boldsymbol{SS}}$ as observed by this array described by $\hat{\underline{\boldsymbol{S}}}_{\boldsymbol{XX},\mathrm{FBSS}}$, is now non-singular, irrespective of the correlation between the impinging sources

A disadvantage of this technique is, however, that FBSS effectively reduces the array's effective aperture [SWK85], thereby reducing resolution capacity, which is the capability of resolving closely-spaced signals.

Source Localization Using MUSIC-Based Algorithms

MUSIC (Multiple SIgnal Classification) was introduced by Schmidt in [Sch81, Sch86] as a method for multiple emitter location and signal parameter estimation. There are many flavors of MUSIC available in the literature, e.g. see [KV96] and references therein. The three most important ones, spectral (classical) MUSIC, root-MUSIC, and unitary root-MUSIC are outlined below.

Spectral MUSIC

Using the background provided in the previous section, the main idea of spectral MUSIC can be formulated rather simply. It searches for I vectors from the array manifold matrix which most closely fit the signal subspace, or, conversely, are 'most orthogonal' to the noise subspace. Focusing on the latter statement, the spectral MUSIC estimator computes a so-called pseudo-spectrum, $\hat{Q}(\Theta)$, by projecting the array manifold vector $\boldsymbol{V}(\Theta)$ onto the noise subspace while varying Θ, i.e. [Tre02],

$$\hat{Q}(\Theta) = \boldsymbol{V}^H(\Theta)\left(\sum_{\iota=I+1}^{M} \hat{\boldsymbol{\xi}}_\iota \hat{\boldsymbol{\xi}}_\iota^H\right)\boldsymbol{V}(\Theta) = \boldsymbol{V}^H(\Theta)\hat{\underline{\boldsymbol{U}}}_N \hat{\underline{\boldsymbol{U}}}_N^H \boldsymbol{V}(\Theta), \qquad (4.107)$$

where in the last step the relations in Eqs. (4.96), Eq. (4.98), and Eq. (4.99) have been used. The values corresponding to I minima in $\hat{Q}(\Theta)$ can then be associated with the desired $\Theta_\iota, \iota = 1(1)I$.

Note that spectral MUSIC is applicable to all sensor array geometries making it one of the most versatile subspace-based parameter estimation methods. In particular, MUSIC applies also to two- and three-dimensional array geometries. However, in order to obtain the MUSIC pseudo-spectrum, the array's manifold vector $\boldsymbol{V}(\Theta)$ has to be known, either by means of analytical expressions, or obtained via measurements. For a ULA, cf. Fig. 4.2, $\boldsymbol{\Theta} \equiv \boldsymbol{\vartheta} = [\vartheta_1, \vartheta_2, \ldots, \vartheta_I]^T$, and $\boldsymbol{V}(\vartheta) = \exp(ikz\cos\vartheta)$.

Figure 4.16 shows an example of a pseudo-spectrum of spectral MUSIC. In this simulation, two spatially uncorrelated equi-power plane-waves impinge from $\vartheta_1 = 0.2\pi$ and $\vartheta_2 = 0.4\pi$, respectively with SNR=30 dB. The sensor array used is an 10 element standard ULA, where, as usual, the term 'standard' corresponds to half-wavelength element spacing. 300 trial runs were performed

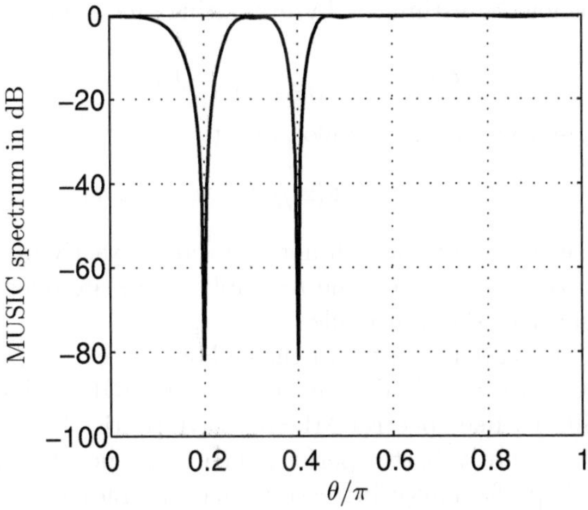

Fig. 4.16. Averaged pseudo spectrum of spectral MUSIC for two equi-power plane-waves $\vartheta_1 = 0.2\pi$ and $\vartheta_2 = 0.4\pi$, SNR=30 dB, standard ULA, $M = 10$, $K = 1000$, 300 trials

with $K = 1000$. As can be seen, deep nulls are formed in the averaged pseudo-spectrum corresponding to the directions of the impinging plane-waves.

Further performance evaluations of MUSIC and its variations comparing them with ESPRIT-based algorithms and the CRLB are presented further below.It is shown in [SN89] that spectral MUSIC is asymptotically efficient, i.e. its performance for high SNR and large K approaches the CRLB.

Root-MUSIC

Root-MUSIC [Bar83] uses a polynomial representation of the MUSIC pseudo spectrum, Eq. (4.107), applicable to arrays exhibiting an array manifold matrix of Vandermonde structure, i.e.,

$$\underline{\bar{V}} = \begin{bmatrix} z_1^0 & z_2^0 & \dots & z_I^0 \\ z_1^1 & z_2^1 & \dots & z_I^1 \\ \vdots & \vdots & \ddots & \vdots \\ z_1^{M-1} & z_2^{M-1} & \dots & z_I^{M-1} \end{bmatrix}. \tag{4.108}$$

An example for an array geometry that satisfies the Vandermonde structure is the standard ULA, where $z_\iota^\ell \triangleq \exp\left[i(\ell-1)\pi\cos\vartheta_\iota\right]$. Note that the symbol z here defines a complex number.

With $\bar{\boldsymbol{V}}(z) = [1, z, \dots, z^{M-1}]^T$, Eq. (4.107) becomes,

$$\hat{Q}(z) = \bar{\boldsymbol{V}}^T(1/z)\hat{\underline{\boldsymbol{U}}}_N\hat{\underline{\boldsymbol{U}}}_N^H\bar{\boldsymbol{V}}(z). \tag{4.109}$$

The DOAs can now be estimated by evaluating the I roots of $\hat{Q}(z)$ on the unit circle, i.e.,

$$\hat{Q}(z)\big|_{z=\exp(i\pi\cos\vartheta)} \mapsto \hat{Q}(\vartheta). \tag{4.110}$$

By denoting these roots by z_ι, it follows that,

$$\hat{\vartheta}_\iota = \cos^{-1}[\arg(z_\iota)/\pi], \quad \iota = 1(1)I. \tag{4.111}$$

In practice, the estimated roots will not be located exactly on the unit circle, i.e. $\hat{z}_\iota = |\hat{z}_\iota| \exp(i\pi \cos\hat{\vartheta})$. Hence, the I complex roots are selected that lie the closest with respect to the unit circle.

Specific examples are presented at the end of this section.An in-depth performance analysis of root-MUSIC can be found in [RH89]. There it is shown that root-MUSIC is, like spectral MUSIC, asymptotically efficient, and has the property of a slightly better performance than spectral MUSIC. An explanation for this performance increase is that the radial errors in the zeros corresponding to the pseudo-spectrum of spectral MUSIC, $|\hat{z}_\iota|$, manifest themselves in the radial component of the projection onto the unit circle [RH89]. As a result, radial errors negatively affect MUSIC's pseudo spectrum, while they do not affect the DOA estimate obtained by root-MUSIC. An additional advantage over spectral MUSIC is the fact that a computationally complex search does not need to be performed. The price that one has to pay is that the universal applicability of MUSIC to all array geometries is lost.

Unitary Root-MUSIC

Yet another widely used flavor of MUSIC is the unitary root-MUSIC algorithm. It corresponds to a particular implementation of root-MUSIC using FBSS, thereby employing, in contrast to spectral MUSIC and root-MUSIC, a real-valued eigen-decomposition [PGH00]. The first step is to obtain a real-valued estimate of the spatio-spectral correlation matrix with,

$$\underline{\hat{\boldsymbol{S}}}_{\boldsymbol{XX},\mathrm{Re}} = \underline{\boldsymbol{Q}}^H \underline{\hat{\boldsymbol{S}}}_{\boldsymbol{XX},\mathrm{FBSS}} \underline{\boldsymbol{Q}}, \tag{4.112}$$

where $\underline{\boldsymbol{Q}}$ can be any unitary, column conjugate-symmetric matrix of size $M \times M$. Note that any unitary matrix $\underline{\boldsymbol{Q}}$ satisfies the relation,

$$\underline{\boldsymbol{Q}}^H \underline{\boldsymbol{Q}} = \underline{\boldsymbol{Q}}\underline{\boldsymbol{Q}}^H = \underline{\boldsymbol{I}}. \tag{4.113}$$

In [PGH00] for M even and M odd, respectively,

$$\underline{\boldsymbol{Q}} \triangleq \frac{1}{\sqrt{2}} \begin{bmatrix} \underline{\boldsymbol{I}} & i\underline{\boldsymbol{I}} \\ \underline{\bar{\boldsymbol{I}}} & -i\underline{\bar{\boldsymbol{I}}} \end{bmatrix}, \quad \underline{\boldsymbol{Q}} \triangleq \frac{1}{\sqrt{2}} \begin{bmatrix} \underline{\boldsymbol{I}} & \underline{\boldsymbol{0}} & i\underline{\boldsymbol{I}} \\ \underline{\boldsymbol{0}}^T & \sqrt{2} & \underline{\boldsymbol{0}}^T \\ \underline{\bar{\boldsymbol{I}}} & \underline{\boldsymbol{0}} & -i\underline{\bar{\boldsymbol{I}}} \end{bmatrix}. \tag{4.114}$$

Utilizing the definition of the FBSS version of the estimated spatio-spectral correlation matrix, Eq. (4.105), it can be shown [Tre02] that Eq. (4.112) can also be expressed as,

$$\underline{\hat{S}}_{\boldsymbol{XX},\mathrm{Re}} = \mathrm{Re}\{\underline{\boldsymbol{Q}}^H \underline{\hat{S}}_{\boldsymbol{XX}} \underline{\boldsymbol{Q}}\}, \tag{4.115}$$

Defining $\tilde{\boldsymbol{V}}(z) \triangleq \underline{\boldsymbol{Q}}^H \boldsymbol{V}(z)$ and performing a real-valued eigen-decomposition of $\underline{\hat{S}}_{\boldsymbol{XX},\mathrm{Re}}$, followed by grouping of the eigenvectors into signal subspace $\underline{\hat{\boldsymbol{U}}}_{S,\mathrm{Re}}$ and noise subspace $\underline{\hat{\boldsymbol{U}}}_{N,\mathrm{Re}}$, respectively, Eq. (4.109) becomes,

$$\tilde{Q}(z) = \tilde{\boldsymbol{V}}^T(1/z)\underline{\hat{\boldsymbol{U}}}_{N,\mathrm{Re}}\underline{\hat{\boldsymbol{U}}}_{N,\mathrm{Re}}^H \tilde{\boldsymbol{V}}(z) \tag{4.116}$$

The task now is to find the I roots, $\hat{z}_\iota, \iota = 1(1)I$, that lie the closest to the unit circle. The corresponding DOAs, ϑ_ι, are related to $\hat{z}_\iota$ through Eq. (4.111). The advantage of unitary root-MUSIC is that it exhibits the same performance as FBSS root-MUSIC while offering a significant reduction in computational complexity due to the real-valued eigen-decomposition. Specific examples showing the performance of unitary root-MUSIC are presented at the end of this section.

Source Localization Using ESPRIT-Based Algorithms

The standard ESPRIT (Estimation of Signal Parameters via Rotational Invariance Techniques) algorithm was introduced by Roy and Kailath in [RK89]. In the following, classical ESPRIT and unitary-ESPRIT are discussed following the derivations given in [Tre02].

ESPRIT

ESPRIT methods can only be applied to arrays that exhibit rotational invariance. Rotational invariance means that the output of two identical sensors, or identical groups of sensors, can be related to each other via a rotation matrix. The algorithm described here is developed in the context of a ULA, as in [Tre02].

In a first step of the algorithm, two identical subarrays comprising M_{sub} elements with relative inter-subarray sensor shift by d_{sub} sensors are chosen. Two examples for subarray grouping with $M = 5$ are shown in Fig. 4.17.

The two subarrays yield two separate subarray manifold matrices, $\underline{\boldsymbol{V}}_1$ and $\underline{\boldsymbol{V}}_2$. They can be determined by using selection matrices $\underline{\boldsymbol{K}}^a$ and $\underline{\boldsymbol{K}}^b$ of size $M_{\mathrm{sub}} \times I$ [Tre02],

Fig. 4.17. Choices for subarray grouping (left: $d_{\mathrm{sub}} = 1$, $M_{\mathrm{sub}} = 4$; right: $d_{\mathrm{sub}} = 2$, $M_{\mathrm{sub}} = 3$)

$$\underline{\boldsymbol{K}}^a \triangleq [\underline{\tilde{\boldsymbol{K}}}_s \ \underline{\boldsymbol{0}}], \tag{4.117}$$
$$\underline{\boldsymbol{K}}^b \triangleq [\underline{\boldsymbol{0}} \ \underline{\tilde{\boldsymbol{K}}}_s], \tag{4.118}$$

where $\underline{\tilde{\boldsymbol{K}}}_s$ is a $M_{\text{sub}} \times M_{\text{sub}}$ identity matrix and $\underline{\boldsymbol{0}}$ is a $M_{\text{sub}} \times d_{\text{sub}}$ zero-matrix. Then,

$$\underline{\boldsymbol{V}}_1 = \underline{\boldsymbol{K}}^a \underline{\boldsymbol{V}}, \tag{4.119}$$
$$\underline{\boldsymbol{V}}_2 = \underline{\boldsymbol{K}}^b \underline{\boldsymbol{V}}. \tag{4.120}$$

Now, the ESPRIT algorithm exploits the following invariance relation,

$$\underline{\boldsymbol{V}}_2 = \underline{\boldsymbol{V}}_1 \underline{\boldsymbol{\Phi}}, \tag{4.121}$$

where,

$$\underline{\boldsymbol{\Phi}} = \text{diag}\{e^{id_{\text{sub}}\psi_1}, e^{id_{\text{sub}}\psi_2}, \ldots, e^{id_{\text{sub}}\psi_I}\}, \tag{4.122}$$

and,

$$\psi_\iota \triangleq kd \cos\vartheta_\iota, \quad \iota = 1(1)I. \tag{4.123}$$

Note that $\underline{\boldsymbol{\Phi}}$ is a diagonal matrix of the phase delays between the two subarrays, i.e. the operator $\underline{\boldsymbol{\Phi}}$ relates the output of subarray 1 to the output of subarray 2. In the following derivation, only standard ULAs with $kd = \pi$ are considered, therefore, $\psi_\iota = \pi \cos\vartheta_\iota$.

The parameter ψ_ι, or, equivalently, the DOAs, ϑ_ι, can therefore be found by estimating $\underline{\boldsymbol{\Phi}}$. It will now be shown that the subarray's rotational invariance translates into rotational invariance of its subspaces and, as a consequence, that the particular structure of the array's manifold matrix is irrelevant for the estimation procedure offered by ESPRIT.

Since the columns of $\underline{\boldsymbol{V}}$ span the signal subspace, Eq. (4.101), the following relation holds,

$$\underline{\boldsymbol{U}}_S = \underline{\boldsymbol{V}}\underline{\boldsymbol{T}}, \tag{4.124}$$

where $\underline{\boldsymbol{T}}$ may be any non-singular matrix. Now, the signal subspaces of the subarrays can be defined by,

$$\underline{\boldsymbol{U}}_{S_1} \triangleq \underline{\boldsymbol{K}}^a \underline{\boldsymbol{U}}_S = \underline{\boldsymbol{K}}^a \underline{\boldsymbol{V}}\underline{\boldsymbol{T}} = \underline{\boldsymbol{V}}_1 \underline{\boldsymbol{T}}, \tag{4.125}$$
$$\underline{\boldsymbol{U}}_{S_2} \triangleq \underline{\boldsymbol{K}}^b \underline{\boldsymbol{U}}_S = \underline{\boldsymbol{K}}^b \underline{\boldsymbol{V}}\underline{\boldsymbol{T}} = \underline{\boldsymbol{V}}_2 \underline{\boldsymbol{T}} = \underline{\boldsymbol{V}}_1 \underline{\boldsymbol{\Phi}}\underline{\boldsymbol{T}}. \tag{4.126}$$

It follows from Eq. (4.125) that,

$$\underline{\boldsymbol{V}}_1 = \underline{\boldsymbol{U}}_{S_1} \underline{\boldsymbol{T}}^{-1}, \tag{4.127}$$

and therefore with Eq. (4.126),

$$\underline{\boldsymbol{U}}_{S_2} = \underline{\boldsymbol{U}}_{S_1} \underline{\boldsymbol{T}}^{-1} \underline{\boldsymbol{\Phi}}\underline{\boldsymbol{T}}. \tag{4.128}$$

By defining,

$$\underline{\boldsymbol{\Psi}} \triangleq \underline{\boldsymbol{T}}^{-1} \underline{\boldsymbol{\Phi}}\underline{\boldsymbol{T}}, \tag{4.129}$$

it follows that,

$$\underline{\boldsymbol{U}}_{S_2} = \underline{\boldsymbol{U}}_{S_1}\underline{\boldsymbol{\Psi}}. \tag{4.130}$$

This means that $\underline{\boldsymbol{U}}_{S_1}$ and $\underline{\boldsymbol{U}}_{S_2}$ have equivalent range spaces with $\underline{\boldsymbol{\Psi}}$ being the operator that maps $\underline{\boldsymbol{U}}_{S_1}$ onto $\underline{\boldsymbol{U}}_{S_2}$. By recognizing that the eigenvalues of $\underline{\boldsymbol{\Psi}}$ are the diagonal elements of $\underline{\boldsymbol{\Phi}}$, the unknown ψ_ι can be derived. In a real system, however, $\underline{\boldsymbol{U}}_{S_1}$ and $\underline{\boldsymbol{U}}_{S_2}$ are not available but only estimations thereof, i.e. $\hat{\underline{\boldsymbol{U}}}_{S_1}$ and $\hat{\underline{\boldsymbol{U}}}_{S_2}$, respectively. Therefore, one obtains,

$$\hat{\underline{\boldsymbol{U}}}_{S_2} = \hat{\underline{\boldsymbol{U}}}_{S_1}\underline{\boldsymbol{\Psi}}. \tag{4.131}$$

There are two widely used solutions of this equations, namely the LS (least-squares) and the TLS (total least-squares) estimators [GL89]. The LS solutions can be written in terms of the following minimization problem,

$$\begin{aligned}\hat{\underline{\boldsymbol{\Psi}}}_{LS} &= \arg\min_{\underline{\boldsymbol{\Psi}}}\{||\hat{\underline{\boldsymbol{U}}}_{S_2} - \hat{\underline{\boldsymbol{U}}}_{S_1}\underline{\boldsymbol{\Psi}}||_F\} \\ &= \arg\min_{\underline{\boldsymbol{\Psi}}}\{\mathrm{tr}\{[\hat{\underline{\boldsymbol{U}}}_{S_2} - \hat{\underline{\boldsymbol{U}}}_{S_1}\underline{\boldsymbol{\Psi}}]^H[\hat{\underline{\boldsymbol{U}}}_{S_2} - \hat{\underline{\boldsymbol{U}}}_{S_1}\underline{\boldsymbol{\Psi}}]\}\},\end{aligned} \tag{4.132}$$

where the subscript F denotes the Frobenius norm and $\mathrm{tr}\{\cdot\}$ denotes the trace of the argument. The solution is known to be,

$$\hat{\underline{\boldsymbol{\Psi}}}_{LS} = \hat{\underline{\boldsymbol{U}}}_{S_1}^{\dagger} \cdot \hat{\underline{\boldsymbol{U}}}_{S_2} = [\hat{\underline{\boldsymbol{U}}}_{S_1}^H\hat{\underline{\boldsymbol{U}}}_{S_1}]^{-1}\hat{\underline{\boldsymbol{U}}}_{S_1}^H \cdot \hat{\underline{\boldsymbol{U}}}_{S_2}, \tag{4.133}$$

where † denotes the pseudo-inverse of a matrix [Tre02].

An estimate for the DOAs can then be simply obtained by calculating the argument of the eigenvalues of $\hat{\underline{\boldsymbol{\Psi}}}_{LS}$, denoted as $\hat{\zeta}_\iota$. For a standard ULA, this results in,

$$\hat{\vartheta}_\iota = \cos^{-1}[\arg(\hat{\zeta}_\iota)/(d_{\mathrm{sub}}\pi)], \quad \iota = 1(1)I. \tag{4.134}$$

Note that for a ULA with arbitrary spacing, $kd \neq \pi$,

$$\hat{\vartheta}_\iota = \cos^{-1}[\arg(\hat{\zeta}_\iota)/(d_{\mathrm{sub}}kd)], \quad \iota = 1(1)I, \tag{4.135}$$

yielding a frequency-dependent estimate. This result will be of importance later.

The solution in Eq. (4.134) is known to be biased since the LS solution assumes only the estimates of $\hat{\underline{\boldsymbol{U}}}_{S_1}$ to contain measurement errors. Since it is easy to argue that both $\hat{\underline{\boldsymbol{U}}}_{S_1}$ and $\hat{\underline{\boldsymbol{U}}}_{S_2}$ contain measurement errors, $\underline{\boldsymbol{\mathcal{E}}}_1$ and $\underline{\boldsymbol{\mathcal{E}}}_2$, it may be more appropriate to consider an TLS-based approach [GL89], modifying Eq. (4.131) as,

$$[\hat{\underline{\boldsymbol{U}}}_{S_1} + \underline{\boldsymbol{\mathcal{E}}}_1]\hat{\underline{\boldsymbol{\Psi}}} = \hat{\underline{\boldsymbol{U}}}_{S_2} + \underline{\boldsymbol{\mathcal{E}}}_2. \tag{4.136}$$

Its solution is given by,

$$\hat{\underline{\boldsymbol{\Psi}}}_{TLS} = -\underline{\boldsymbol{E}}_{12}\underline{\boldsymbol{E}}_{22}^{-1}, \tag{4.137}$$

where $\underline{\boldsymbol{E}}_{12}$ and $\underline{\boldsymbol{E}}_{22}$ are matrices of size $I \times I$ and are defined implicitly by the eigen-decomposition of the $2I \times 2I$ matrix $\tilde{\underline{\boldsymbol{U}}}$,

$$\tilde{\underline{\boldsymbol{U}}} \triangleq \begin{bmatrix} \hat{\underline{\boldsymbol{U}}}_{S1}^H \\ \hat{\underline{\boldsymbol{U}}}_{S2}^H \end{bmatrix} \begin{bmatrix} \hat{\underline{\boldsymbol{U}}}_{S1} & \hat{\underline{\boldsymbol{U}}}_{S2} \end{bmatrix} = \begin{bmatrix} \underline{\boldsymbol{E}}_{11} & \underline{\boldsymbol{E}}_{12} \\ \underline{\boldsymbol{E}}_{21} & \underline{\boldsymbol{E}}_{22} \end{bmatrix} \underline{\boldsymbol{\Lambda}}_{\tilde{U}} \begin{bmatrix} \underline{\boldsymbol{E}}_{11}^H & \underline{\boldsymbol{E}}_{21}^H \\ \underline{\boldsymbol{E}}_{12}^H & \underline{\boldsymbol{E}}_{22}^H \end{bmatrix}, \tag{4.138}$$

where $\underline{\boldsymbol{\Lambda}}_{\tilde{U}} = \mathrm{diag}\{\zeta_1, \zeta_2, \ldots, \zeta_{2I}\}$, and $\zeta_\nu, \nu = 1(2)2I$ denote the ordered eigenvalues of the matrix $\tilde{\underline{\boldsymbol{U}}}$.

Analogously to the LS-based algorithm, the DOAs can then be obtained by evaluating the argument of the eigenvalues of $\hat{\underline{\boldsymbol{\Psi}}}_{TLS}$. An in-depth performance analysis of TLS-ESPRIT can be found in [OVK91].

Unitary-ESPRIT

From a computational point of view, ESPRIT has one disadvantage which is the necessity to perform three complex-valued eigen-decompositions. An algorithm that results in purely real-valued eigen-decomposition is the so-called unitary-ESPRIT algorithm [HN95]. Unitary-ESPRIT is applicable to centro-symmetric arrays. Conjugate centro-symmetry can be mathematically formulated in terms of the array manifold vector $\boldsymbol{V}(\psi)$ as,

$$\bar{\underline{\boldsymbol{I}}}_M \boldsymbol{V}(\psi) \triangleq \boldsymbol{V}^*(\psi), \tag{4.139}$$

where $\bar{\underline{\boldsymbol{I}}}$ is given by Eq. (4.106) and here and in the following the subscripts 'M' and '$M-1$' indicate the dimension of the associated quadratic matrix. Pre-multiplying a matrix with $\bar{\underline{\boldsymbol{I}}}_M$ results in swapping of its rows while post-multiplication results in the columns of the matrix being swapped, in particular $\bar{\underline{\boldsymbol{I}}}_M \bar{\underline{\boldsymbol{I}}}_M = \underline{\boldsymbol{I}}_M$.

It can be easily shown that the inner product between any two conjugate centro-symmetric vectors is real. This means that any matrix whose rows are individually conjugate centro-symmetric can be used to transform the complex-valued manifold vector, $\boldsymbol{V}(\psi)$, into a real-valued one.

Now, using a unitary matrix $\underline{\boldsymbol{Q}}_M^H$, such as the one defined in Eq. (4.114), a complex-valued centro-symmetric array manifold matrix $\underline{\boldsymbol{V}}(\psi) \in \mathbb{C}^{M \times I}$ can be transformed into a real-valued one, $\underline{\boldsymbol{B}}(\psi) \in \mathbb{R}^{M \times I}$, by,

$$\underline{\boldsymbol{B}}(\psi) = \underline{\boldsymbol{Q}}_M^H \underline{\boldsymbol{V}}(\psi). \tag{4.140}$$

Since the transformed sensor-space array manifold is now real-valued, the signal eigenvectors necessary to form an estimate of the real-valued signal subspace $\underline{\boldsymbol{U}}_{S,\mathrm{Re}}$ can be obtained by performing an eigen-decomposition of the real-valued matrix given in Eq. (4.115), where $\underline{\boldsymbol{Q}}$ has been replaced by $\underline{\boldsymbol{Q}}_M$. Note that all complex-valued centro-Hermitian covariance matrices can be thus transformed to become matrices having real entries only. Note also an additional advantage of unitary-ESPRIT over the conventional ESPRIT algorithm which is due to the forward-backward averaging effect of this transformation. This manipulation effectively results in signal decorrelation which can be exploited advantageously in scenarios of spatially correlated and/or fully coherent signals, as in the unitary root-MUSIC algorithm.

Using this mathematical background and considering the special case of maximally overlapping subarrays (see Fig. 4.17) and a single impinging plane-wave signal, i.e. $d_{\text{sub}} = 1$, $M_{\text{sub}} = M - 1$, and $I = 1$, it can be easily verified using Eq. (4.139) and Eq. (4.140) that the ESPRIT invariance relation Eq. (4.121) can be written as,

$$e^{i\psi}\underline{\boldsymbol{K}}^{a}\boldsymbol{V}(\psi) = \underline{\boldsymbol{K}}^{b}\boldsymbol{V}(\psi), \tag{4.141}$$

where the matrices $\underline{\boldsymbol{K}}^a$ and $\underline{\boldsymbol{K}}^b$ that select the first and last $M - 1$ entries of $\underline{\boldsymbol{V}}(\psi)$, respectively, are,

$$\underline{\boldsymbol{K}}^{a} = \begin{bmatrix} 1\,0 \ldots 0\,0 \\ 0\,1 \ldots 0\,0 \\ \vdots\;\vdots\;\ddots\;\vdots\;\vdots \\ 0\,0 \ldots 1\,0 \end{bmatrix}, \qquad \underline{\boldsymbol{K}}^{b} = \begin{bmatrix} 0\,1\,0 \ldots 0 \\ 0\,0\,1 \ldots 0 \\ \vdots\;\vdots\;\vdots\;\ddots\;\vdots \\ 0\,0\,0 \ldots 1 \end{bmatrix}. \tag{4.142}$$

These selection matrices have the property that,

$$\underline{\bar{\boldsymbol{I}}}_{M-1}\underline{\boldsymbol{K}}^{b}\underline{\bar{\boldsymbol{I}}}_{M} = \underline{\boldsymbol{K}}^{a}. \tag{4.143}$$

Now, utilizing unitarity of $\underline{\boldsymbol{Q}}_M$ and Eq. (4.140), Eq. (4.141) becomes,

$$e^{i\psi}\underline{\boldsymbol{K}}^{a}\underline{\boldsymbol{Q}}_{M}\boldsymbol{B}(\psi) = \underline{\boldsymbol{K}}^{b}\underline{\boldsymbol{Q}}_{M}\boldsymbol{B}(\psi). \tag{4.144}$$

This invariance relationship can modified by a pre-multiplication with $\underline{\boldsymbol{Q}}_{M-1}^{H}$ to yield,

$$e^{i\psi}\underline{\boldsymbol{Q}}_{M-1}^{H}\underline{\boldsymbol{K}}^{a}\underline{\boldsymbol{Q}}_{M}\boldsymbol{B}(\psi) = \underline{\boldsymbol{Q}}_{M-1}^{H}\underline{\boldsymbol{K}}^{b}\underline{\boldsymbol{Q}}_{M}\boldsymbol{B}(\psi). \tag{4.145}$$

With Eq. (4.143) and $\underline{\bar{\boldsymbol{I}}}_{M}\underline{\boldsymbol{Q}}_{M} = \underline{\boldsymbol{Q}}_{M}^{*}$ one obtains,

$$\underline{\boldsymbol{Q}}_{M-1}^{H}\underline{\boldsymbol{K}}^{b}\underline{\boldsymbol{Q}}_{M} = (\underline{\boldsymbol{Q}}_{M-1}^{H}\underline{\boldsymbol{K}}^{a}\underline{\boldsymbol{Q}}_{M})^{*}, \tag{4.146}$$

and therefore for the invariance relationship Eq. (4.145),

$$e^{i\frac{\psi}{2}}(\underline{\boldsymbol{G}}_1 - i\underline{\boldsymbol{G}}_2)\boldsymbol{B}(\psi) = e^{-i\frac{\psi}{2}}(\underline{\boldsymbol{G}}_1 + i\underline{\boldsymbol{G}}_2)\boldsymbol{B}(\psi), \tag{4.147}$$

where,

$$\underline{\boldsymbol{G}}_1 \triangleq \operatorname{Re}\{\underline{\boldsymbol{Q}}_{M-1}^{H}\underline{\boldsymbol{K}}^{b}\underline{\boldsymbol{Q}}_{M}\}, \tag{4.148a}$$

$$\underline{\boldsymbol{G}}_2 \triangleq \operatorname{Im}\{\underline{\boldsymbol{Q}}_{M-1}^{H}\underline{\boldsymbol{K}}^{b}\underline{\boldsymbol{Q}}_{M}\}. \tag{4.148b}$$

Rearranging terms and using,

$$\tan\alpha = \frac{e^{i\alpha} - e^{-i\alpha}}{i(e^{i\alpha} + e^{-i\alpha})}, \tag{4.149}$$

one finally arrives at,

$$\tan\left(\frac{\psi}{2}\right)\underline{\boldsymbol{G}}_1\boldsymbol{B}(\psi) = \underline{\boldsymbol{G}}_2\boldsymbol{B}(\psi). \tag{4.150}$$

When I plane-wave signals are impinging on the array the now *real-valued* array manifold matrix becomes $\underline{\boldsymbol{V}}_{\mathrm{Re}} \triangleq [\boldsymbol{B}(\psi_1) \mid \ldots \mid \boldsymbol{B}(\psi_I)]$ and therefore the invariance relationship equivalent to Eq. (4.121) is now,

$$\underline{\boldsymbol{G}}_2\underline{\boldsymbol{V}}_{\mathrm{Re}} = \underline{\boldsymbol{G}}_1\underline{\boldsymbol{V}}_{\mathrm{Re}}\underline{\boldsymbol{\Phi}}, \tag{4.151}$$

where,

$$\underline{\boldsymbol{\Phi}} = \mathrm{diag}\left\{\tan\left(\frac{\psi_1}{2}\right), \tan\left(\frac{\psi_2}{2}\right), \ldots, \tan\left(\frac{\psi_I}{2}\right)\right\}. \tag{4.152}$$

Following the same discussion as for the ESPRIT algorithm, though employing only real-valued matrix algebra, it holds that,

$$\underline{\boldsymbol{U}}_{S,\mathrm{Re}} = \underline{\boldsymbol{V}}_{\mathrm{Re}}\underline{\boldsymbol{T}}, \tag{4.153}$$

and therefore,

$$\underline{\boldsymbol{V}}_{\mathrm{Re}} = \underline{\boldsymbol{U}}_{S,\mathrm{Re}}\underline{\boldsymbol{T}}^{-1}. \tag{4.154}$$

Plugging this equation into Eq. (4.151), it follows that,

$$\underline{\boldsymbol{G}}_1\underline{\boldsymbol{U}}_{S,\mathrm{Re}}\underline{\boldsymbol{\Psi}} = \underline{\boldsymbol{G}}_2\underline{\boldsymbol{U}}_{S,\mathrm{Re}}, \tag{4.155}$$

where,

$$\underline{\boldsymbol{\Psi}} = \underline{\boldsymbol{T}}^{-1}\underline{\boldsymbol{\Phi}}\underline{\boldsymbol{T}}. \tag{4.156}$$

As before, the unknown parameters $\psi_\iota, \iota = 1(1)I$ can be obtained by computing the LS- or TLS-based solution for Eq. (4.155) and determining the eigenvalues of this result.

The following list summarizes the steps required for estimating the DOAs of I plane-waves using TLS unitary ESPRIT applied to a standard ULA with maximally overlapping subarrays.

1. Estimate the spatio-spectral covariance matrix, $\hat{\underline{\boldsymbol{S}}}_{\boldsymbol{XX}}$, using Eq. (4.11).
2. Transform the complex spatio-spectral covariance matrix into a real-valued one by computing, $\hat{\underline{\boldsymbol{S}}}_{\boldsymbol{XX},\mathrm{Re}} \triangleq \underline{\boldsymbol{Q}}_M^H\hat{\underline{\boldsymbol{S}}}_{\boldsymbol{XX}}\underline{\boldsymbol{Q}}_M$, cf. Eq. (4.112).
3. Perform a real-valued eigen-decomposition of $\hat{\underline{\boldsymbol{S}}}_{\boldsymbol{XX},\mathrm{Re}}$ and estimate the resulting signal subspace, $\hat{\underline{\boldsymbol{U}}}_{S,\mathrm{Re}}$, cf. Eq. (4.98).

4. Replace $\underline{\hat{U}}_{S_1}$ and $\underline{\hat{U}}_{S_2}$ in Eq. (4.138) by $\underline{G}_1\underline{\hat{U}}_{S,\mathrm{Re}}$ and $\underline{G}_2\underline{\hat{U}}_{S,\mathrm{Re}}$, respectively, cf. Eqs. (4.148) and (4.155).
5. Compute a real-valued eigen-decomposition of the thus modified Matrix in Eq. (4.138) to obtain a TLS solution of Eq. (4.137).
6. Denoting the eigenvalues of Eq. (4.137) by $\hat{\zeta}_\iota$, $\iota = 1(1)I$ the DOAs are given by, $\hat{\vartheta}_\iota = 2\tan^{-1}\{\hat{\zeta}_\iota\}$.

Performance of Subspace-Based DOA Estimation

In the following, the performance of subspace-based DOA estimation is examined. Instead of considering an in-depth performance evaluation, which would by far exceed the scope of this book, only a few representative experiments by means of simulations are presented. The simulations were designed such that meaningful conclusions can be drawn regarding the capabilities of the various algorithms.

The first set of experiments tries to evaluate the estimation error covariance – $C(\vartheta)$, cf. Eq. (4.69) – of subspace-based DOA estimation of a single desired signal in spatially uncorrelated white noise. Figure 4.18 shows the performance of unitary root-MUSIC (denoted as 'urMUSIC') and unitary ESPRIT (denoted as 'uESPRIT') with respect to the CRLB for a single

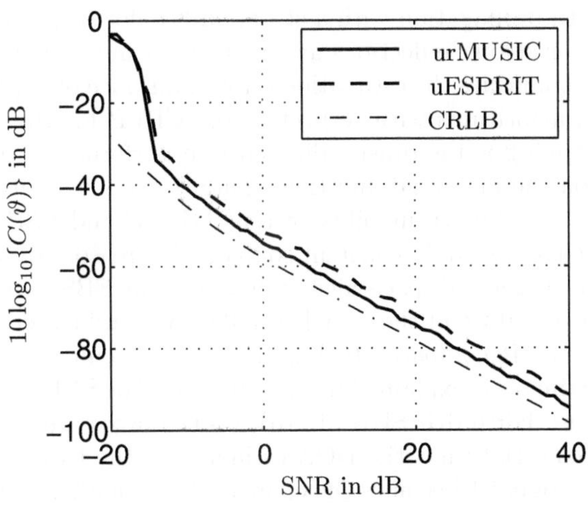

Fig. 4.18. Estimation error covariance of unitary root-MUSIC and unitary ESPRIT w.r.t. the CRLB for a single plane-wave ($\vartheta = 2\pi/9$), standard ULA, $M = 10$, $K = 1000$, 300 trial runs

narrowband plane-wave impinging from $\vartheta = 2\pi/9$ on a standard ULA comprising $M = 10$ sensors. 300 trial runs were performed and $K = 1000$. Note that all curves depend directly on the absolute angle. It can be seen that for SNR>-15 dB, both estimators closely follow the CRLB. Both estimators have been found to be efficient, for they asymptotically approach the CRLB in the limits of infinite SNR and infinite K [PGH00], [HN95]. However, the unitary ESPRIT algorithm always has a slightly higher variance than the unitary root-MUSIC algorithm. This fact is well known in the literature, e.g. [MZ94]. For SNR<-15 dB the variances of both estimators are almost identical and well above the CRLB. The value for which the variance starts to deviate from the CRLB is denoted as 'threshold'. The reason for this is that the CRLB evaluates only local variances, i.e. the CRLB assumes small estimation errors. This assumption is violated in low SNR scenarios.

Qualitatively, the situation will be similar if any of the simulation parameters is varied. In general, it can be stated, that performance will improve proportionally with respect to M and K, which can be gleaned from Eq. (4.69). The estimation variances of classical ESPRIT and root-MUSIC are not included in Fig. 4.18 since in this example, for a single source, their performance is almost identical with unitary ESPRIT and unitary root-MUSIC, respectively.

Drastic deviations in performance become obvious in the case of two fully coherent equi-power narrowband plane-waves impinging from $\boldsymbol{\vartheta} = [\pi/6, 2\pi/9]^T$ on the same array as considered before. The choice of the two sources being fully coherent can also be interpreted as a single source experiencing multipath propagation with a strong first reflection. Note that desired sources as well as strong reflections are modeled as individual sources.

Figure 4.19 quantifies the estimation error covariance of MUSIC-based and ESPRIT-based algorithms with respect to the CRLB for the second source. Immediately striking is the drastic difference in performance of root-MUSIC and ESPRIT on the one hand and their unitary variants on the other hand, resulting in the well-known inability of root-MUSIC and ESPRIT to handle spatially correlated and fully coherent sources. The reason for this is that the unitary version of the subspace algorithms perform FBSS which effectively results in a decorrelation of the signals [Tre02]. As can be seen, the threshold starts at higher SNR compared to Fig. 4.18. The peculiar characteristics of the ESPRIT estimate is explained in the following. For SNR>10 dB, ESPRIT finds two sources. For 0 dB<SNR<10 dB, both sources are still found. However, during some trial runs, the DOA estimates of the two signals swap, i.e. the estimate for signal 1 becomes the estimate for signal 2 and vice versa. For -10 dB<SNR<0 dB, one DOA is found that corresponds to a signal if it were positioned between the actual sources in space. The other estimate still swaps between the two true DOAs with some severe outliers. For SNR<-10 dB, the estimate corresponding to the virtual source in between the two true DOAs now has even less variance than before. The reason for this is that the second

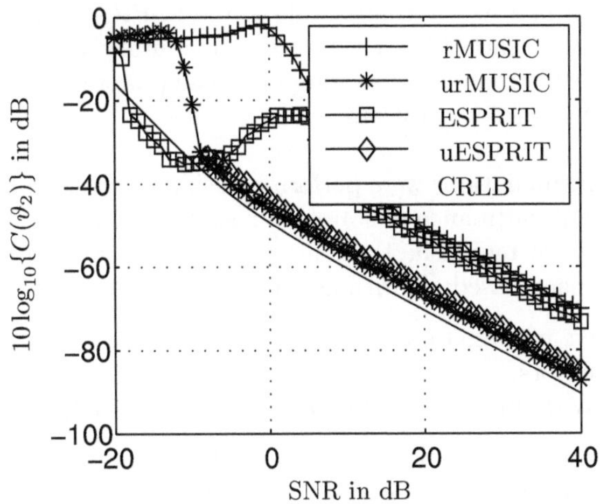

Fig. 4.19. Estimation error covariance of MUSIC-based and ESPRIT-based algorithms w.r.t. the CRLB for two fully coherent plane-waves ($\boldsymbol{\vartheta} = [\pi/6, 2\pi/9]^T$, second source shown), standard ULA, $M = 10$, $K = 1000$, 300 trial runs

estimate now returns completely random values, and that the algorithm can therefore 'concentrate' on the other estimate.

The absence of an estimate for unitary ESPRIT where SNR<-10 dB is due to an implementation detail realizing a reliability test described in [HN95]. This test verifies whether the eigenvalues obtained in step 6 of the TLS unitary ESPRIT algorithm are all real, as anticipated due to purely real-valued operations, and does not return an estimate if this test fails.

The performance for spatially uncorrelated and correlated but not fully coherent sources is not explicitly shown here by examples. Qualitatively, it can be stated, however, that the performance of all algorithms for spatially uncorrelated sources are comparable and that all algorithms are asymptotically efficient, see [Tre02] and references therein. Introducing signal correlation introduces performance degradation, especially for the non-unitary version of the algorithms, where the worst-case scenario, i.e. fully coherent sources, is shown in Fig. 4.19.

A question that has not been answered is concerned with the unambiguous resolvability of two closely-spaced sources. This rather complex problem is treated in [Tre02]. General answers can only be given through simulations due to the dependency on many parameters, such as K, M, SNR, and the DOAs themselves. In the particular case of Fig. 4.19, the two sources are separated by $\Delta\vartheta = \pi/18$. By applying unitary versions of the subspace methods the two sources can be resolved for an SNR of approximately SNR>-10 dB.

In summary, the unitary version of the subspace algorithms perform much better than the algorithms not utilizing FBSS. The performance of the

MUSIC-based estimators are comparable to the ESPRIT-based methods, the latter ones having slightly higher variance. The unitary versions have the additional advantage of requiring less computational power due to the utilization of real-valued eigen-decompositions.

Note that all simulations were performed using a standard ULA. It can be verified that the performance of subspace-based algorithms deteriorate if the assumptions are not met. The three most important assumptions that most traditional subspace-based algorithms rely on are,

- free-field wave propagation,
- calibrated sensors,
- a sensor array that is designed for half-wavelength (narrowband) operation.

Therefore, a significant problem present in all but one subspace-based DOA estimation algorithms is their inability to handle wideband signals, such as speech. This can be verified, for instance, by examining Eq. (4.135), where the DOA estimate is shown to be frequency-dependent as the wavelength of the impinging plane-wave differs from $2d$. In [HK88, DW92], this problem is addressed by introducing so-called focusing matrices that relate the correlation matrices, corresponding to individual frequencies, to a reference frequency, and then applying the standard narrowband algorithms.

A subspace-based algorithm that, to some extent, allows for the analysis of wideband signals without the need for focusing matrices is classical MUSIC. However, the restriction here is that wideband signals can only be analyzed in the frequency-domain by applying the MUSIC algorithm to each frequency-bin independently followed by an averaging process, see [WK85a] and references therein. Note that this computationally very intensive procedure can neither be applied to classical MUSIC in the time-domain nor to any of the MUSIC derivations or ESPRIT algorithms.

A novel and fundamentally different approach that does not require focusing matrices and that relaxes the narrowband requirement, is derived in detail in Chapter 5. There it is shown that eigenbeam processing applied to circular and spherical sensor arrays yield inherently wideband subspace-based algorithms for WE and PE tasks.

4.4.4 Detection of the Number of Active Sources

It has been tacitly assumed that the number of sources, i.e. the rank of the signal correlation matrix, i.e,

$$I = \mathrm{rank}\{\underline{\boldsymbol{S}}_{\boldsymbol{SS}}\}, \tag{4.157}$$

is known. In many applications, this knowledge is not available a-priori, and must therefore be estimated from the sensor correlation matrix. It is assumed that the array manifold matrix is full rank and that the $M - I$ noise eigenvalues cluster around σ_w^2, see Eq. (4.97). Historically, the first approach to this problem focused on sequential hypotheses tests, whose task is to find the likelihood ratio between the hypothesis that the $(M - I)$ smallest eigenvalues are equal and the hypothesis that the $(M - I - 1)$ smallest eigenvalues are equal. See [Tre02] for a review on hypotheses tests.

Anderson [And63] showed that if $K \gg M$, a sufficient statistic is given by,

$$\mathcal{L}(\iota) = K(M - \iota) \ln \left\{ \frac{\frac{1}{M-\iota} \sum\limits_{\nu=\iota+1}^{M} \hat{\zeta}_\nu}{\left(\prod\limits_{\nu=\iota+1}^{M} \hat{\zeta}_\nu \right)^{1/(M-\iota)}} \right\}, \tag{4.158}$$

The resulting hypotheses tests have the problem that thresholds need to be selected [Tre02]. Modifying Eq. (4.158) by a penalty function and finding the minimum of the resulting function circumvents the problem of threshold selection [WK85b]. Two popular penalty functions are due to Akaike [Aka74] and Schwartz [Sch78] yielding,

$$\mathcal{L}_1(\iota) \triangleq \mathrm{AIC}(\iota) = \mathcal{L}(\iota) + \iota(2M - \iota), \tag{4.159}$$

$$\mathcal{L}_2(\iota) \triangleq \mathrm{MDL}(\iota) = \mathcal{L}(\iota) + [\iota(2M - \iota) + 1]/2 \cdot \ln K, \tag{4.160}$$

respectively. The number of sources can then be estimated by,

$$\hat{I} = \arg\min_{\iota} \mathcal{L}_{1,2}(\iota), \quad \iota = 0(1)(M-1). \tag{4.161}$$

It is shown in [WK85b] that the MDL estimator is asymptotically consistent, i.e. $\hat{I} = I$ for $K, \mathrm{SNR} \to \infty$, whereas the AIC estimator is asymptotically inconsistent exhibiting $\hat{I} > I$.

Figure 4.20 depicts an example for the probability of detection using AIC and MDL for two incoherent plane-waves impinging from $\boldsymbol{\vartheta} = [\pi/6, 2\pi/9]^T$ utilizing a ten-element standard ULA in spatially uncorrelated white noise. This representative example shows that the MDL is not capable of detecting two sources in all cases, even for very high values for the SNR. On the other hand, the performance of AIC degrades smoothly for very low values of the SNR compared to the MDL. Therefore, for high SNR the MDL approach is to be preferred while for low SNR, the AIC yields better results. Penalty functions that are able to address the problem of spatially correlated and fully coherent sources by applying the notion of FBSS, are given in [XRK94].

Note, again, that this example is based on simulations utilizing a standard ULA. For sensor geometries where the spacing does not conform to the requirement of $d = \lambda/2$, the performance degrades up to the extent where the detection estimate becomes unusable. A solution to this problem will be given in Chapter 5, by successfully employing the notion of eigenbeam processing.

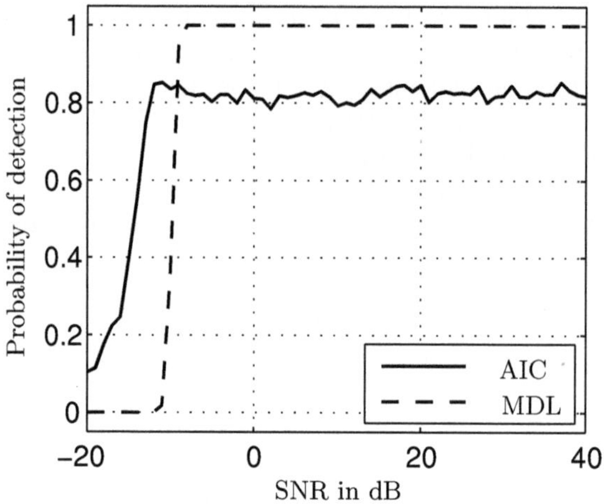

Fig. 4.20. Probability of detection using AIC and MDL for two incoherent plane-waves ($\vartheta = [\pi/6, 2\pi/9]$, standard ULA, $M = 10$, $K = 1000$, 1000 trial runs)

5

Acoustic Scene Analysis Using Modal Array Signal Processing

In this chapter, the concepts of classical acoustics and wavefield decomposition as derived in Chapters 2 and 3 on the one hand, and classical array signal processing, as discussed in Chapter 4, on the other hand, are combined to yield novel solutions for sensor array signal processing tasks. As before, the signal processing tasks considered are beamforming, i.e. waveform estimation (WE), and the localization of possibly multiple acoustic sources, i.e. parameter estimation (PE). In addition, the problem of how to determine the number of active sources in a wavefield is addressed. One of the central observations to be made here, which makes the combination of the two paradigms possible, is that the individual circular or spherical harmonics resulting from the wavefield decomposition step can be regarded as individual sensors in the classical sensor array processing framework. In the following, the circular and spherical harmonics are jointly denoted as *eigenbeams*, a concept first applied to acoustic signal processing by Elko et al. in [EKM03] and [ME04].

Building upon the signal model and terminology introduced in the previous chapters, this chapter is organized as follows. Section 5.1 discusses WE using eigenbeams and parallels the discussion presented in Section 4.3, while Section 5.2 details the concept of PE using eigenbeams, thereby paralleling the discussion of Section 4.4.

5.1 Waveform Estimation Using Eigenbeam Processing

This section introduces the notion of a beamformer that does not operate on the sensor signals directly but on eigenbeams which are obtained by decomposing a two- or three-dimensional wavefield into orthogonal eigen-solutions of the acoustic wave equation in circular/cylindrical and spherical coordinates, respectively, utilizing circular and spherical apertures.

The synthesis of a beampattern performed by the so-called *modal beamformer*, or eigenbeamformer (EBF), is detailed in Section 5.1.1 for both circular and spherical harmonics. In Section 5.1.2 optimum beampattern design utilizing eigenbeams is discussed, where the term 'optimum' is related to the notion of maximum directivity in isotropic noise fields. Finally, in Section 5.1.3, an eigenbeam GSC (EB-GSC) structure is proposed that parallels the discussion of Section 4.3.4.

5.1.1 The Modal Beamformer – Pattern Synthesis

In this section, it is shown how an *arbitrary* pattern of order $\mathcal{N}$ can be synthesized by utilizing $\mathcal{N}+1$ eigenbeams based on either circular or spherical harmonics.

Pattern Synthesis Using Circular Harmonics

Any square-integrable function on a circle can be expanded into a series of circular harmonics, $G_{d_n}(\cdot)$. In particular, any desired beampattern, $G_d(kR,\phi)$, with look direction – or rotation angle – $\varphi_s = 0\ldots 2\pi$, can be written as a Fourier series, cf. Section A.2,

$$G_d(kR,\phi,\varphi_s) = \sum_{n=-\infty}^{\infty} G_{d_n}(kR)e^{in(\phi-\varphi_s)}, \tag{5.1}$$

$$G_{d_n}(kR) = \frac{1}{2\pi}\int_0^{2\pi} G_d(kR,\phi,\varphi_s)e^{-in(\phi-\varphi_s)}. \tag{5.2}$$

Truncating the series to the $\mathcal{N}$-th harmonic, *any* pattern of order up to and including $\mathcal{N}$ can be synthesized. Examples for pattern synthesis of beamformers of order $\mathcal{N}=1$ were actually already given in Section 3.1.2 where dipole and cardioid apertures were examined.

An example for the pattern synthesis of a higher-order pattern will be given in the following where the pattern synthesis using spherical harmonics is presented.

Pattern Synthesis Using Spherical Harmonics

The fundamental concept underlying pattern synthesis using spherical harmonics is very similar to the pattern synthesis procedure shown in the previous section for circular harmonics. Here, the spherical harmonics transform pair, see Eq. (A.7) and Eq. (A.8) is used for pattern synthesis. A difference with respect to pattern synthesis using circular harmonics is the fact that the spatial beamformer response can now be specified on a sphere where two-dimensional rotation in azimuth an elevation of the pattern becomes possible.

To make the mathematics more tractable, it is first assumed that the desired pattern is rotationally symmetric about the z-axis [ME04], which results in $m = 0$ in the expressions for the spherical harmonics. Then the spherical harmonics transform pair becomes for an unrotated desired beampattern,

$$G_d(kR,\theta) = \sqrt{4\pi}\sum_{n=0}^{\infty} G_{d_n}(kR)Y_n^0(\theta,\phi) = \sum_{n=0}^{\infty} G_{d_n}(kR)\sqrt{2n+1}P_n(\cos\theta), \tag{5.3}$$

$$G_{d_n}(kR) = \frac{1}{2}\sqrt{2n+1}\int_0^{\pi} G(kR,\theta)P_n(\cos\theta)\sin\theta d\theta. \tag{5.4}$$

Now, the rotation of the pattern toward (ϑ_s, φ_s) is considered. The term $\cos\theta$ is hereby defined as the cosine angle between the look-direction (ϑ_s, φ_s) and a point on the sphere (ϑ, φ). Then, by making use of the spherical harmonics addition theorem given by Eq. (B.59), the desired pattern can be obtained by,

$$G_d(\vartheta,\varphi,\vartheta_s,\varphi_s) = 4\pi\sum_{n=0}^{\infty}\frac{1}{\sqrt{2n+1}}\sum_{m=-n}^{n} G_{d_n}Y_n^m(\vartheta,\varphi)Y_n^m(\vartheta_s,\varphi_s)^*, \tag{5.5}$$

$$G_{d_n} = \frac{\sqrt{2n+1}}{4\pi Y_n^m(\vartheta_s,\varphi_s)^*}\int_0^{2\pi}\int_0^{\pi}\tilde{G}_d(\vartheta,\varphi,\vartheta_s,\varphi_s)Y_n^m(\vartheta,\varphi)^*\sin\vartheta d\vartheta d\varphi, \tag{5.6}$$

where the dependence on kR has been dropped for notational convenience.

As was shown in the two-dimensional case, by truncating the series to the $\mathcal{N}$-th harmonic, $n \leq \mathcal{N}$, *any* pattern of order up to and including $\mathcal{N}$ can be synthesized.

At this point it is illustrative to work with a few numbers. Consider a specific application requiring a third-order – $\mathcal{N} = 3$ – hypercardioid pattern, which is defined as [Elk04],

$$G_d(\theta) = \prod_{\nu=1}^{3}\varepsilon_\nu + (1-\varepsilon_\nu)\cos\theta, \tag{5.7}$$

where $\varepsilon_1 \approx 0.45, \varepsilon_2 \approx 0.15$ and $\varepsilon_3 \approx -1.35$. Then, after evaluating Eq. (5.4), the expansion coefficients for the unrotated pattern are obtained as $G_{d_0} \approx 0.0625, G_{d_1} \approx 0.1082, G_{d_2} \approx 0.1397, G_{d_3} \approx 0.1654$, and $G_{d_n} = 0,\ n > 3$. Figure 5.1(a) shows a two-dimensional view of the polar pattern of the unrotated synthesized hypercardioid using Eq. (5.3). The pattern can be easily rotated in three-dimensional space by plugging the expansion coefficients into Eq. (5.5). Figure 5.1(b) depicts the result of this procedure for $(\vartheta_s, \varphi_s) = (\pi/4, -\pi/4)$.

Note that synthesizing a hypercardioid pattern using circular harmonics involves the same steps using Eq. (5.1) and Eq. (5.2). The only difference is that the pattern can only be rotated in azimuth.

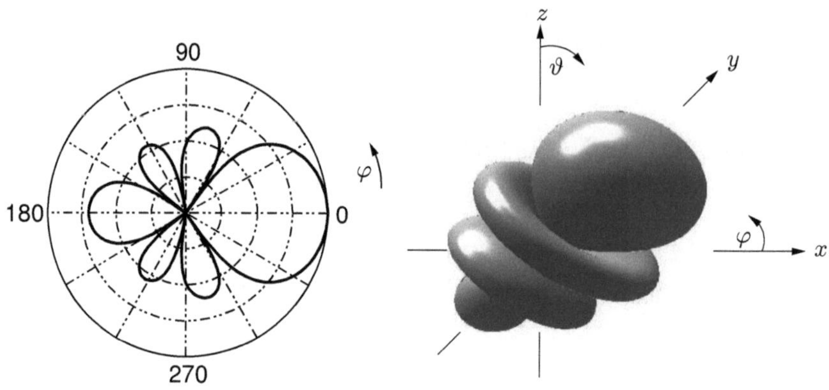

(a) 2D representation; radial scale: 10dB/div.

(b) 3D representation; pattern steered to $(\vartheta_s, \varphi_s) = (\pi/4, -\pi/4)$

Fig. 5.1. Pattern synthesis of a third-order hypercardioid using spherical harmonics

5.1.2 Optimum Beampattern Design Using Eigenbeams

This section details a procedure on how to perform optimum beampattern design, i.e. patterns exhibiting maximum directivity in isotropic noise fields, utilizing the circular or spherical eigenbeams derived in Chapter 3.

Suppressing the dependence on frequency, the array directivity given in Eq. (4.34) using vector notation can be written as,

$$D = \frac{\boldsymbol{W}^H \boldsymbol{V}_{EB}(\Theta_s)\boldsymbol{V}_{EB}^H(\Theta_s)\boldsymbol{W}}{\boldsymbol{W}^H \left(\frac{1}{\gamma}\int_{\Omega} \boldsymbol{V}_{EB}(\Theta)\boldsymbol{V}_{EB}^H(\Theta) d\Omega\right)\boldsymbol{W}} \triangleq \frac{\boldsymbol{W}^H \underline{\boldsymbol{S}}_{SS}(\Theta_s)\boldsymbol{W}}{\boldsymbol{W}^H \underline{\boldsymbol{S}}_{NN}\boldsymbol{W}}, \tag{5.8}$$

where Θ_s denotes the look direction which is assumed to coincide with the direction of incidence of an impinging plane-wave. The integration is performed either over the unit circle for circular eigenbeams, where $\gamma = 2\pi$, or over the unit sphere for spherical eigenbeams, where $\gamma = 4\pi$.

One of the central statements in this chapter follows from the observation that the array manifold vector, defined in Eq. (4.23) for classical array processing, translates into a *modal array manifold vector* where the vector entries are defined as,

$$[\boldsymbol{V}_{EB}]_n = V_{n,EB}, \tag{5.9}$$

where, for circular apertures,

$$V_{n,EB} = i^n B_n(kR) e^{-in\varphi}, \quad n = -\mathcal{N}(1)\mathcal{N}, \tag{5.10}$$

$$B_n(kR) \triangleq \begin{cases} J_n(kR) & \text{unbaffled aperture} \\ J_n(kR) - \dfrac{J_n'(kR)}{H_n'(kR)} H_n(kR) & \text{baffled aperture} \end{cases}, \tag{5.11}$$

and for spherical apertures,

$$V_{n,EB} = \sqrt{4\pi} i^n b_n(kR) Y_n^0(\vartheta, \varphi)^*, \quad n = 0(1)\mathcal{N}, \tag{5.12}$$

$$b_n(kR) \triangleq \begin{cases} j_n(kR) & \text{unbaffled aperture} \\ j_n(kR) - \dfrac{j_n'(kR)}{h_n'(kR)} h_n(kR) & \text{baffled aperture} \end{cases}. \tag{5.13}$$

Note that, again, only spherical eigenbeams of degree $m = 0$ are considered here. The resulting pattern can be easily rotated by the techniques developed in the previous section.

For continuous circular apertures, the integral in Eq. (5.8) can be evaluated in closed-form to yield,

$$\underline{\boldsymbol{S}}_{\boldsymbol{NN}} = \operatorname{diag}\{|B_{-\mathcal{N}}|^2, |B_{-\mathcal{N}+1}|^2, \ldots, |B_0|^2, \ldots, |B_{\mathcal{N}}|^2\}, \tag{5.14}$$

and for continuous spherical apertures,

$$\underline{\boldsymbol{S}}_{\boldsymbol{NN}} = \operatorname{diag}\{|b_0|^2, |b_1|^2, \ldots, |b_{\mathcal{N}}|^2\}, \tag{5.15}$$

by virtue of the orthogonality of the exponential function and Legendre polynomials, cf. Section B.3, respectively. This integral, however, needs to be evaluated numerically when sampled continuous apertures are considered.

The task of optimum beampattern design using eigenbeams is to maximize the directivity D, given by Eq. (5.8). However, as can be deduced from the frequency-dependent characteristics of the eigenbeams, cf. Figs. 3.2 and 3.5 for circular apertures and Figs. 3.16 and 3.18 for spherical apertures, an unconstrained optimization of the directivity, which tries to compensate for these characteristics, would lead to an undesired high amplification of noise components, especially at low frequencies. As was shown in Section 4.3.2, this amplification can be quantified by the white noise gain as,

$$A_w = \frac{\boldsymbol{W}^H \underline{\boldsymbol{S}}_{\boldsymbol{SS}}(\Theta_s) \boldsymbol{W}}{\boldsymbol{W}^H \boldsymbol{W}}. \tag{5.16}$$

Therefore, a practical optimization procedure optimizes the directivity subject to a white noise constraint [CZO87, ME04]. Cox et al. have shown [CZO87] that in order to solve this optimization problem, a weight vector $\boldsymbol{W}_{\text{opt}}$ needs to be found that minimizes,

$$\min_{\boldsymbol{W}} \left(\frac{1}{D} + \epsilon \frac{1}{A_w} \right) = \min_{\boldsymbol{W}} \frac{\boldsymbol{W}^H (\underline{\boldsymbol{S}}_{\boldsymbol{NN}} + \epsilon \underline{\boldsymbol{I}}) \boldsymbol{W}}{\boldsymbol{W}^H \underline{\boldsymbol{S}}_{\boldsymbol{SS}}(\Theta_s) \boldsymbol{W}}, \tag{5.17}$$

where the positive scalar ϵ quantifies the amount of deviation from the optimum design. The optimum constrained filters have been shown to equal [CZO87],

$$\boldsymbol{W}_{\text{opt}} = [\underline{\boldsymbol{S}}_{\boldsymbol{NN}} + \epsilon \underline{\boldsymbol{I}}]^{-1} \boldsymbol{V}_{EB}(\Theta_s). \tag{5.18}$$

Note the similarities between this result and the design procedure used for designing superdirective arrays as outlined in Section 4.3.3. One significant difference, however, is that the filter coefficients are applied to the individual eigenbeams and not to the infinitesimal segments of a continuous aperture or to the individual microphones.

As an example, Fig. 5.2 shows the result of this design procedure using continuous spherical eigenbeams, for both the unconstrained beampattern, i.e. $\epsilon = 0$, as well as for a constrained beampattern, here $\epsilon = 10^{-3}$. There are a number of interesting observations that can be made. First, the directivity index for an unconstrained third-order optimum beamformer design is equivalent to the maximum directivity index that can be obtained with a DMA/SDB in spherically isotropic noise, cf. Section 4.3.3. The only difference is that the DMA/SDB is based on individual microphones, while the EBF is based on individual eigenbeams. Therefore, in accordance with Eq. (4.51), the maximum directivity index due to spherically isotropic noise is, $D^{(s)}_{I_{\max}} = 10\log_{10}\{\mathcal{N}+1\}^2$. It is well known that the $\mathcal{N}$-th order beamformer, which yields the highest directivity, corresponds to an $\mathcal{N}$-th order hypercardioid [Elk04]. The second observation that can be made is the fact that the WNG for a baffled aperture is higher than for an unbaffled aperture. An explanation for this has been given in Section 3.2.1 where it was shown that a baffled spherical aperture yields a virtual modal aperture increase, cf. Fig. 3.19. A third observation is that the WNG of the unconstrained optimum EBF is of similar shape as the modal magnitude response of order three, see Fig. 3.16 and Fig. 3.18. The reason for this is that the optimization algorithm tries to equalize the highpass characteristic of the third-order modal magnitude response which, hence, results in a higher sensitivity at low frequencies.

It is important to note that the WNG in Fig. 5.2 is related to the eigenbeams and not to individual sensors. The resulting WNG for a *sampled* continuous spherical aperture using M sampling points is $10\log_{10}\{M\}$ *higher* than the one shown in Fig. 5.2. Note also that the evaluation of $\underline{\boldsymbol{S}}_{\boldsymbol{SS}}$ and $\underline{\boldsymbol{S}}_{\boldsymbol{NN}}$ depend on the sampling scheme used, and need to be obtained numerically. For the sake of brevity of the presentation, specific examples including various sampling schemes showing the effects of spatial aliasing, such as the ones presented in Section 3.2.3 are omitted here.

It is clear that the WNG obtained by an unconstrained design is not useful in practice. The right-hand side of Fig. 5.2 exemplifies a design constraining the WNG to a minimum value of -20 dB, corresponding to $\epsilon = 10^{-3}$. The trade-off between WNG and directivity becomes obvious.

The optimum beampattern design using circular eigenbeams is almost identical to the one illustrated for spherical eigenbeams. A significant dif-

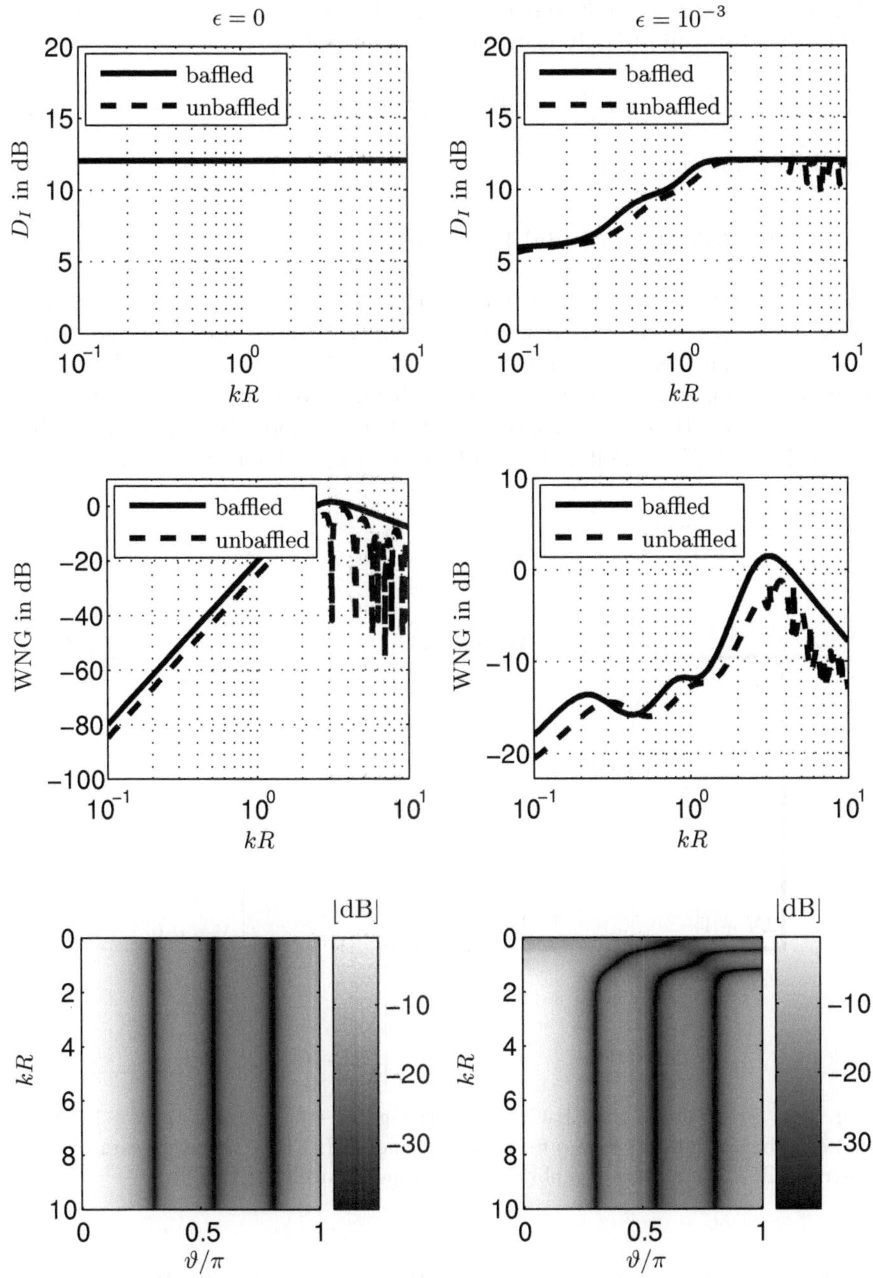

Fig. 5.2. Unconstrained (left column) and constrained (right column) optimum beampattern design of order $\mathcal{N} = 3$ w.r.t. D_I, WNG and directivity pattern using continuous spherical eigenbeams

ference, however, is that the maximum directivity now corresponds to a DMA/SDB in cylindrically isotropic noise, i.e. where,

$$D^{(c)}_{I_{\max}} = 10 \log_{10}\{2(\mathcal{N}+1) - 1\}, \tag{5.19}$$

see [Elk04] and Eq. (4.52).

5.1.3 The Adaptive Modal Beamformer

In the previous sections, data-independent pattern synthesis methods using eigenbeam processing were discussed. In this section, a simple system is proposed that parallels the discussion of Section 4.3.4, where a GSC structure is used to implement a data-dependent beamformer. Here, a modified GSC structure, the EB-GSC, is proposed that employs modal beamformers (EBFs) for both the FBF as well as for the BM, see Fig. 5.3 and cf. Fig. 4.12.

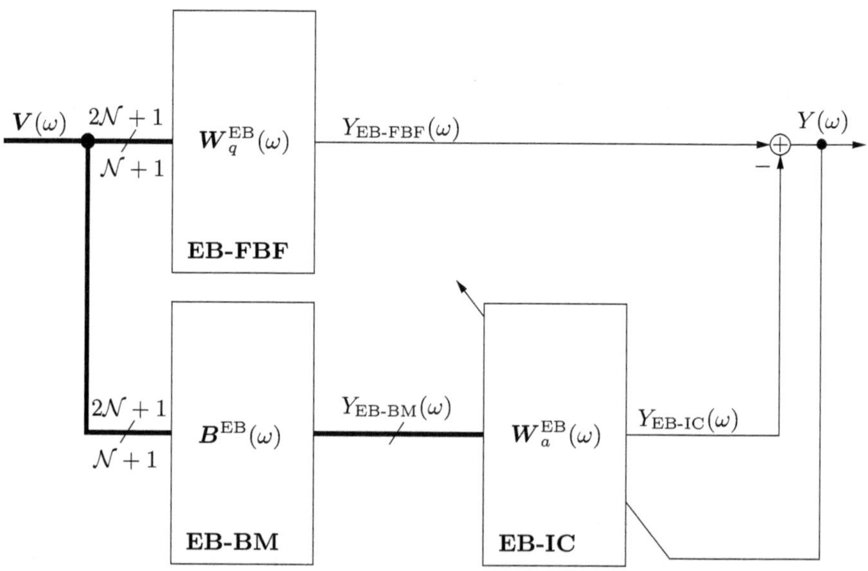

Fig. 5.3. Structure of the EB-GSC comprising $2\mathcal{N}+1$ (circular), or $\mathcal{N}+1$ (spherical) eigenbeams, an eigenbeam fixed beamformer (EB-FBF), an eigenbeam blocking matrix (EB-BM), and an eigenbeam interference canceler (EB-IC)

The main idea can be simply stated. Both the FBF and the BM are formed by the eigenbeams, either circular or spherical, by using, for example, the pattern synthesis methods discussed in the previous section. The resulting EB-FBF is then steered such that its highest spatial selectivity is aiming at the desired source. In contrast, the EB-BM is steered such that its lowest spatial selectivity is aiming at the desired source. Therefore, the EB-BM can be regarded as a null-beamformer. The EB-IC can then be adaptively determined by appropriately modifying Eq. (4.63), the estimate of the IC used by the classical GSC structure. A potential advantage of the EB-GSC over the classical GSC is its ability to handle multiple desired sources using the techniques to be detailed in Section 5.2, where the localization of multiple simultaneously active acoustic sources is discussed.

An in-depth performance analysis of the EB-GSC is beyond the scope of this book and will not be discussed any further.

5.2 Parameter Estimation Using Eigenbeam Processing

This section details parameter estimation using eigenbeam processing. The discussion parallels the discussion presented in Section 4.4. However, the algorithms to be derived here are based on an inherent wideband formulation by employing the paradigm of eigenbeam processing. This fact makes eigenbeam processing a prime candidate for wideband processing, especially in the realm of multiple wideband acoustic source localization. Section 5.2.1 introduces the eigenbeam array manifold vectors the algorithms are based upon. Section 5.2.2 presents an eigenbeam signal model as well as the notion of a modal signal subspace. Section 5.2.3 discusses a CRLB that may, under certain conditions, be used for the performance analysis of eigenbeam-based DOA estimation algorithms. Section 5.2.4 derives ESPRIT-based algorithms for single-source and multi-source DOA estimation, denoted as EB-ESPRIT [TK05a], using circular harmonics. Note that, although not further detailed, an EB-MUSIC algorithm can be straightforwardly derived. Eigenbeam processing using spherical harmonics is treated in Section 5.2.5. This chapter concludes with a discussion that parallels Section 4.4.4, i.e. the estimation of the number of active sources. Here, however, the estimation is not based on information-theoretic criteria applied to sensor array correlation matrices but on a re-synthesis of a plane-wave by superposition of the individual circular/spherical harmonics [TK05c].

5.2.1 Eigenbeam Array Manifold Vectors

Eigenbeam array manifold vectors useful for designing optimum beampatterns are given in Section 5.1.2, see Eq. (5.9). For several parameter estimation problems these manifold vectors have to be modified slightly to deal with this different signal processing task. For circular apertures, Eq. (5.10) is modified as,

$$[\breve{\boldsymbol{V}}_{EB}]_n = \breve{V}_{n,EB}, \quad n = -\mathcal{N}(1)\mathcal{N}, \tag{5.20}$$

where,

$$\breve{V}_{n,EB} = \frac{V_{n,EB}}{i^n B_n(kR)} = e^{-in\varphi}. \tag{5.21}$$

I.e. a filter is applied to each harmonic in order to yield a frequency-independent modal response, $e^{-in\varphi}$. Due to the zeros present in the modal magnitude response corresponding to unbaffled apertures, see Fig. 3.2, it is assumed here and in the following that, whenever a filter is applied to compensate for the modal magnitude response, the considered frequency range of operation does not include any zeros. Note, however, that due to WNG restrictions the necessary compensation filter cannot be used at arbitrarily low frequencies in practice, especially for eigenbeams of high order.

For spherical apertures, cf. Eq. (5.12), the order of each individual harmonic is kept fixed to $n = \mathcal{N}$, and the degree is varied as,

$$\check{\boldsymbol{V}}_{EB}(\vartheta,\varphi) = [\check{V}^{-\mathcal{N}}_{\mathcal{N},EB}(\vartheta,\varphi), \check{V}^{-\mathcal{N}+1}_{\mathcal{N},EB}(\vartheta,\varphi), \ldots, \check{V}^{0}_{\mathcal{N},EB}(\vartheta,\varphi), \ldots, \check{V}^{\mathcal{N}}_{\mathcal{N},EB}(\vartheta,\varphi)]^T, \tag{5.22}$$

where with Eq. (5.13),

$$\begin{aligned}\check{V}^{m}_{\mathcal{N},EB}(\vartheta,\varphi) &= \sqrt{4\pi} i^{\mathcal{N}} b_{\mathcal{N}}(kR) Y^{m}_{\mathcal{N}}(\vartheta,\varphi)^* \\ &= \sqrt{4\pi} i^{\mathcal{N}} b_{\mathcal{N}}(kR) \sqrt{\frac{2\mathcal{N}+1}{4\pi}\frac{(\mathcal{N}-m)!}{(\mathcal{N}+m)!}} P^{m}_{\mathcal{N}}(\cos\vartheta) e^{-im\varphi},\end{aligned} \tag{5.23}$$

where $m = -\mathcal{N}(1)\mathcal{N}$.

The reason for expressing $Y^{m}_{\mathcal{N}}(\vartheta,\varphi)$ by the associated Legendre function, $P^{m}_{\mathcal{N}}(\cos\vartheta)$, will become clear in Section 5.2.5.

5.2.2 Eigenbeam Signal Model and Modal Signal Subspaces

In this section, the frequency-domain signal model used for the further discussion on eigenbeam processing is stated. Since the idea underlying the paradigm of eigenbeam processing is that eigenbeams correspond to sensors, the signal model given in Eq. (4.7) and Eq. (4.10) remains essentially unchanged. In particular, the eigenbeam – or modal – spectral matrix is,

$$\underline{\boldsymbol{S}}^{EB}_{\boldsymbol{XX}} = \underline{\boldsymbol{V}}_{EB} \underline{\boldsymbol{S}}_{\boldsymbol{SS}} \underline{\boldsymbol{V}}^{H}_{EB} + \underline{\boldsymbol{S}}^{EB}_{\boldsymbol{NN}}, \tag{5.24}$$

where $\underline{\boldsymbol{S}}^{EB}_{\boldsymbol{NN}}$ is the matrix containing the modal noise spectral densities. In the following, two special cases for $\underline{\boldsymbol{S}}^{EB}_{\boldsymbol{NN}}$ are considered, i.e. spatially white noise, and spatially isotropic noise, where $\underline{\boldsymbol{S}}^{EB}_{\boldsymbol{NN}}$ is given by Eq. (5.14) and Eq. (5.15) for cylindrically and spherically isotropic noise fields, respectively. Isotropic noise is a reasonable model for ambient noise that cannot be classified as directional. Note that the spatially white noise model, that can be applied to many real-world noise fields, satisfy the 'whiteness' property *before* the decomposition process only. *After* the decomposition process, the white noise model cannot be, in general, regarded as spatially white with respect to the individual eigenbeams.

The modal array manifold matrix, $\underline{\boldsymbol{V}}_{EB}$, in Eq. (5.24) is either defined by stacking the modal array manifold vector given by Eq. (5.9) or by the definitions given in Section 5.2.1, see Eq. (5.28). Note that the matrix containing the signal spectral densities, $\underline{\boldsymbol{S}}_{\boldsymbol{SS}}$, is defined by both the desired sources as well as by strong directional interferer.

The idea of decomposing the spatio-spectral matrix into its eigenvalues and eigenvectors, as given in Section 4.4.3, is directly applied to eigenbeam-based processing as,

$$\underline{\boldsymbol{S}}_{\boldsymbol{XX}}^{EB} = \sum_{\ell=1}^{\mathcal{L}} \zeta_{\ell,EB} \boldsymbol{\xi}_{\ell,EB} \boldsymbol{\xi}_{\ell,EB}^{H}, \tag{5.25}$$

where $\zeta_{\ell,EB}$ and $\boldsymbol{\xi}_{\ell,EB}$ are the ℓ-th ordered eigenbeam eigenvalues and eigenbeam eigenvectors, respectively and,

$$\mathcal{L} = \begin{cases} 2\mathcal{N}+1 & \text{for Eq. (5.10) and Eq. (5.23)} \\ \mathcal{N}+1 & \text{for Eq. (5.12)} \end{cases}. \tag{5.26}$$

The modal signal subspace is then defined as,

$$\underline{\boldsymbol{U}}_{S,EB} \triangleq [\boldsymbol{\xi}_{1,EB} \mid \boldsymbol{\xi}_{2,EB} \mid \cdots \mid \boldsymbol{\xi}_{I,EB}]. \tag{5.27}$$

All statements and implications given in Section 4.4.3 remain qualitatively valid and are not reproduced here.

5.2.3 Eigenbeam Processing and the CRLB

The CRLB for any array geometry in its most general form is given by Eq. (4.66). The CRLB for eigenbeam-based parameter estimation using circular apertures can be derived from this equation by defining the modal array manifold matrix utilizing Eq. (5.20) as,

$$\underline{\check{\boldsymbol{V}}}_{EB}(\boldsymbol{\varphi}) \triangleq [\check{\boldsymbol{V}}_{EB}(\varphi_1) \mid \check{\boldsymbol{V}}_{EB}(\varphi_2) \mid \ldots \mid \check{\boldsymbol{V}}_{EB}(\varphi_I)]. \tag{5.28}$$

In accordance with classical array signal processing, closed-form expressions for the eigenbeam CRLB (EB-CRLB) are very difficult to obtain. Assuming equal spectral densities for the signal components at each eigenbeam, an expression similar in nature to Eq. (4.69) for a single plane-wave impinging on a circular aperture can be obtained after a several straightforward algebraic manipulations, see Appendix D, as,

$$C_{\text{EB-CRLB}}(\omega) = 3\left[2K\mathcal{N}(\mathcal{N}+1)(2\mathcal{N}+1)^2 \frac{\text{SNR}^2(\omega)}{1+(2\mathcal{N}+1)\text{SNR}(\omega)}\right]^{-1}, \tag{5.29}$$

where 'SNR' denotes the signal-to-noise ratio at the eigenbeams. A few comments on the validity of Eq. (5.29) are in order. This result assumes that the noise spectral density is constant over all eigenbeams. While this may be a valid assumption considering wideband spatially white noise, it may be questionable for wideband isotropic noise, cf. Eq. (5.14) and Eq. (5.15). However, for reasonably large values of kR the modal magnitude response of the eigenbeams are almost identical for baffled apertures, see Fig. 3.5 and Fig. 3.18, and the assumption becomes reasonable. Note that the magnitude response of the eigenbeams can be 'forced' to be identical by applying the filtering operation introduced in Eq. (5.21). The result expressed by Eq. (5.29) also assumes that the signal spectral density is constant over all eigenbeams. It should be noted that this assumption is, strictly speaking, only valid for $\varphi = \varphi_1 \approx 0$ for unrotated eigenbeams.

It is important to note that one significant difference between Eq. (5.29) and Eq. (4.69) is that the EB-CRLB does not depend on the DOA of the incoming plane-wave, which is a result of the symmetry of circular apertures.

In analogy to Section 4.4.1, a compact closed-form expression for the EB-CRLB dealing with multiple plane-wave signals does not seem to be available. However, simulations can be performed by numerically evaluating Eq. (4.66) applied to eigenbeams.

Note that Eq. (4.66) also applies to spherical apertures. Hereby, the array manifold matrix has to be replaced by the eigenbeam array manifold matrix given by Eq. (5.59). Since this matrix is a function of both azimuth and elevation, two different EB-CRLB are obtained as a result of evaluating the differentiation operation in Eq. (4.67) with respect to azimuth and elevation. For the sake of brevity, explicit mathematical expression are omitted here.

5.2.4 Eigenbeam-Based DOA Estimation Using Circular Apertures

The developments in this section stem from the observations resulting from Eq. (5.9) for circular apertures, that an eigenbeam array manifold vector can be defined in a similar way as for an array of individual sensors. Hence, many signal processing algorithms based on traditional sensor array processing techniques can also be formulated by using the paradigm of eigenbeams.

As a representative example, this section derives a DOA estimation algorithm based on the unitary ESPRIT algorithm, which has been outlined in Section 4.4.3 for classical sensor arrays. By applying the ideas presented in this section, any other variant of the subspace-based DOA estimation algorithms, such as MUSIC or MODE (Method Of Direction Estimation)[SS90], can be straightforwardly extended to eigenbeam-based DOA estimation.

Figure 5.4 summarizes the main steps required for making the transition from sensor-space, i.e. classical array processing, to eigen-space, i.e. modal array processing for $\mathcal{N} = 2$. The $M = 5$ sensors on the left-hand side of Fig. 5.4 are numbered from -2 to 2. By performing the transition, these individual sensors are replaced by the respective circular harmonics. This means that the microphone denoted as -2 is replaced by the circular harmonic of order n=-2, and so on. Then, as in the traditional approach, two subarrays are chosen,

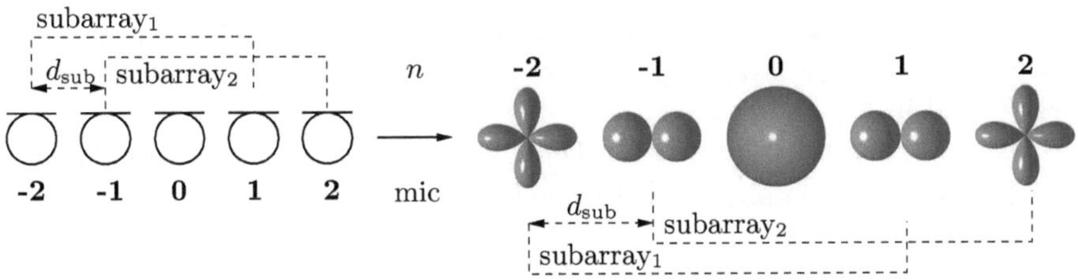

Fig. 5.4. Transition from sensor-space to eigen-space for $\mathcal{N} = 2$, $M = 5$, and $d_{\text{sub}} = 1$

here denoted as *sub-modal* arrays. As indicated in the figure, it is assumed that the first element in the original sensor-space (eigen-space) array is the first element in the first subarray (sub-modal array) and that the $(d_{\text{sub}} + 1)$th element in the original array is the first element in the second subarray. Note that d_{sub}, therefore, does *not* correspond to a physical shift in either array configuration. The motivation for this transition can be justified by realizing that the modal array manifold matrix $\underline{\boldsymbol{V}}_{EB}$ in Eq. (5.24) is Vandermonde, just like the array manifold matrix of a ULA, see Section 4.4.3.

ESPRIT-Based Algorithm for Single-Source Localization Tasks

In the following, an algorithm is presented that is able to estimate the DOA of a single impinging plane-wave with unity amplitude. A slightly modified algorithm is presented in the next section with which DOA estimation of multiple impinging plane-waves becomes possible.

It can be straightforwardly verified that for eigenbeam-based DOA estimation, the invariance relation for unitary ESPRIT given by Eq. (4.141) becomes,

$$e^{-i\varphi}\underline{\boldsymbol{O}}_{2\mathcal{N}}\underline{\boldsymbol{K}}^{a}\boldsymbol{V}_{EB}(\varphi) = \underline{\boldsymbol{K}}^{b}\boldsymbol{V}_{EB}(\varphi), \tag{5.30}$$

where $\boldsymbol{V}_{EB}(\varphi)$ is the eigenbeam array manifold vector defined in Eq. (5.9), and,

$$\underline{\boldsymbol{O}}_{2\mathcal{N}} = \text{diag}\left\{\frac{\bar{V}_{\mathcal{N}-1}}{\bar{V}_{\mathcal{N}}}, \frac{\bar{V}_{\mathcal{N}-2}}{\bar{V}_{\mathcal{N}-1}}, \ldots, \frac{\bar{V}_0}{\bar{V}_1}, \frac{\bar{V}_1}{\bar{V}_0}, \ldots, \frac{\bar{V}_{\mathcal{N}}}{\bar{V}_{\mathcal{N}-1}}\right\}, \tag{5.31}$$

where $\bar{V}_{\mathcal{N}} \triangleq i^{\mathcal{N}} B_{\mathcal{N}}(kR)$. Note that it is assumed that, for unbaffled apertures, $\bar{V}_{\mathcal{N}} \neq 0$. Here and in the following, a subscripted matrix, where the subscript contains the symbol '$\mathcal{N}$', defines its dimension.

It can be readily verified that a few important properties of the special structure of $\underline{\boldsymbol{O}}_{2\mathcal{N}}$ are,

$$\underline{\bar{\boldsymbol{I}}}_{2\mathcal{N}} = \underline{\boldsymbol{O}}_{2\mathcal{N}}\underline{\bar{\boldsymbol{I}}}_{2\mathcal{N}}\underline{\boldsymbol{O}}_{2\mathcal{N}}, \tag{5.32}$$

$$\underline{\boldsymbol{O}}_{2\mathcal{N}} = \underline{\bar{\boldsymbol{I}}}_{2\mathcal{N}}\underline{\boldsymbol{O}}_{2\mathcal{N}}^{-1}\underline{\bar{\boldsymbol{I}}}_{2\mathcal{N}}, \tag{5.33}$$

$$\underline{\boldsymbol{O}}_{2\mathcal{N}}^{-1} = \underline{\bar{\boldsymbol{I}}}_{2\mathcal{N}}\underline{\boldsymbol{O}}_{2\mathcal{N}}\underline{\bar{\boldsymbol{I}}}_{2\mathcal{N}}. \tag{5.34}$$

The $2\mathcal{N} \times (2\mathcal{N}+1)$ selection matrices $\underline{\boldsymbol{K}}^a$ and $\underline{\boldsymbol{K}}^b$ in Eq. (5.30), that select the $2\mathcal{N}$ first and $2\mathcal{N}$ last components of the eigenbeam array manifold vector, are defined in Eq. (4.142).

By applying the unitarity of $\underline{\boldsymbol{Q}}_{2\mathcal{N}+1}$ to the eigenbeam manifold vector $\boldsymbol{V}_{EB}(\varphi)$ one obtains a real-valued eigenbeam manifold vector as, cf. Eq. (4.140),

$$\boldsymbol{B}_{EB}(\varphi) = \underline{\boldsymbol{Q}}_{2\mathcal{N}+1}^H \boldsymbol{V}_{EB}(\varphi). \tag{5.35}$$

Therefore, Eq. (5.30) becomes,

$$e^{-i\varphi} \underline{\boldsymbol{O}}_{2\mathcal{N}} \underline{\boldsymbol{K}}^a \underline{\boldsymbol{Q}}_{2\mathcal{N}+1} \boldsymbol{B}_{EB}(\varphi) = \underline{\boldsymbol{K}}^b \underline{\boldsymbol{Q}}_{2\mathcal{N}+1} \boldsymbol{B}_{EB}(\varphi). \tag{5.36}$$

Pre-multiplication with $\underline{\boldsymbol{O}}_{2\mathcal{N}}^{-1/2}$ yields,

$$e^{-i\varphi} \underline{\boldsymbol{O}}_{2\mathcal{N}}^{1/2} \underline{\boldsymbol{K}}^a \underline{\boldsymbol{Q}}_{2\mathcal{N}+1} \boldsymbol{B}_{EB}(\varphi) = \underline{\boldsymbol{O}}_{2\mathcal{N}}^{-1/2} \underline{\boldsymbol{K}}^b \underline{\boldsymbol{Q}}_{2\mathcal{N}+1} \boldsymbol{B}_{EB}(\varphi), \tag{5.37}$$

which, by using Eq. (5.34) can be expressed as,

$$e^{-i\varphi} \underline{\boldsymbol{O}}_{2\mathcal{N}}^{1/2} \underline{\boldsymbol{K}}^a \underline{\boldsymbol{Q}}_{2\mathcal{N}+1} \boldsymbol{B}_{EB}(\varphi) = \underline{\bar{\boldsymbol{I}}}_{2\mathcal{N}} \underline{\boldsymbol{O}}_{2\mathcal{N}}^{1/2} \underline{\bar{\boldsymbol{I}}}_{2\mathcal{N}} \underline{\boldsymbol{K}}^b \underline{\boldsymbol{Q}}_{2\mathcal{N}+1} \boldsymbol{B}_{EB}(\varphi). \tag{5.38}$$

This result can be further modified by utilizing $\underline{\bar{\boldsymbol{I}}}_{2\mathcal{N}+1} \underline{\bar{\boldsymbol{I}}}_{2\mathcal{N}+1} = \underline{\boldsymbol{I}}_{2\mathcal{N}+1}$ and Eq. (4.143) to arrive at,

$$e^{-i\varphi} \underline{\boldsymbol{O}}_{2\mathcal{N}}^{1/2} \underline{\boldsymbol{K}}^a \underline{\boldsymbol{Q}}_{2\mathcal{N}+1} \boldsymbol{B}_{EB}(\varphi) = \underline{\bar{\boldsymbol{I}}}_{2\mathcal{N}} \underline{\boldsymbol{O}}_{2\mathcal{N}}^{1/2} \underline{\boldsymbol{K}}^a \underline{\bar{\boldsymbol{I}}}_{2\mathcal{N}+1} \underline{\boldsymbol{Q}}_{2\mathcal{N}+1} \boldsymbol{B}_{EB}(\varphi). \tag{5.39}$$

Pre-multiplication of Eq. (5.39) with $\underline{\boldsymbol{Q}}_{2\mathcal{N}}^H$ and utilizing the fact that,

$$\underline{\boldsymbol{Q}}_{2\mathcal{N}}^H \underline{\bar{\boldsymbol{I}}}_{2\mathcal{N}} = \underline{\boldsymbol{Q}}_{2\mathcal{N}}^T \tag{5.40}$$

$$\underline{\bar{\boldsymbol{I}}}_{2\mathcal{N}+1} \underline{\boldsymbol{Q}}_{2\mathcal{N}+1} = \underline{\boldsymbol{Q}}_{2\mathcal{N}+1}^*, \tag{5.41}$$

yields,

$$\begin{aligned} e^{-i\varphi} \underline{\boldsymbol{Q}}_{2\mathcal{N}}^H \underline{\boldsymbol{O}}_{2\mathcal{N}}^{1/2} \underline{\boldsymbol{K}}^a \underline{\boldsymbol{Q}}_{2\mathcal{N}+1} \boldsymbol{B}_{EB}(\varphi) &= \underline{\boldsymbol{Q}}_{2\mathcal{N}}^T \underline{\boldsymbol{O}}_{2\mathcal{N}}^{1/2} \underline{\boldsymbol{K}}^a \underline{\boldsymbol{Q}}_{2\mathcal{N}+1}^* \boldsymbol{B}_{EB}(\varphi) \\ &= (\underline{\boldsymbol{Q}}_{2\mathcal{N}}^H \underline{\boldsymbol{O}}_{2\mathcal{N}}^{1/2} \underline{\boldsymbol{K}}^a \underline{\boldsymbol{Q}}_{2\mathcal{N}+1})^* \boldsymbol{B}_{EB}(\varphi). \end{aligned} \tag{5.42}$$

Therefore, it follows in analogy to Eq. (4.150) that,

$$\tan\left(\frac{\varphi}{2}\right) \underline{\tilde{\boldsymbol{G}}}_1 \boldsymbol{B}_{EB}(\varphi) = \underline{\tilde{\boldsymbol{G}}}_2 \boldsymbol{B}_{EB}(\varphi), \tag{5.43}$$

where,

$$\underline{\tilde{\boldsymbol{G}}}_1 \triangleq \mathrm{Re}\{\underline{\boldsymbol{Q}}_{2\mathcal{N}}^H \underline{\boldsymbol{O}}_{2\mathcal{N}}^{1/2} \underline{\boldsymbol{K}}^a \underline{\boldsymbol{Q}}_{2\mathcal{N}+1}\}, \tag{5.44}$$

$$\underline{\tilde{\boldsymbol{G}}}_2 \triangleq \mathrm{Im}\{\underline{\boldsymbol{Q}}_{2\mathcal{N}}^H \underline{\boldsymbol{O}}_{2\mathcal{N}}^{1/2} \underline{\boldsymbol{K}}^a \boldsymbol{Q}_{2\mathcal{N}+1}\}. \tag{5.45}$$

In order to obtain an estimate for the plane-wave's DOA, φ, the same mathematical operations that have been performed following Eq. (4.150) can be applied. Out of these operations, one very important intermediate step is reproduced here. As shown in Section 4.4.3, the estimation of the DOA using unitary ESPRIT involves an estimate of the eigenvectors of,

$$\begin{aligned}\underline{\tilde{\boldsymbol{U}}}_{EB} &\triangleq \begin{bmatrix}(\underline{\tilde{\boldsymbol{G}}}_1\hat{\boldsymbol{U}}_{S,EB})^T\\(\underline{\tilde{\boldsymbol{G}}}_2\hat{\boldsymbol{U}}_{S,EB})^T\end{bmatrix}\left[(\underline{\tilde{\boldsymbol{G}}}_1\hat{\boldsymbol{U}}_{S,EB})\;(\underline{\tilde{\boldsymbol{G}}}_2\hat{\boldsymbol{U}}_{S,EB})\right]\\ &= \begin{bmatrix}\hat{\boldsymbol{U}}_{S,EB}^T\underline{\tilde{\boldsymbol{G}}}_1^T\underline{\tilde{\boldsymbol{G}}}_1\hat{\boldsymbol{U}}_{S,EB} & \hat{\boldsymbol{U}}_{S,EB}^T\underline{\tilde{\boldsymbol{G}}}_1^T\underline{\tilde{\boldsymbol{G}}}_2\hat{\boldsymbol{U}}_{S,EB}\\ \hat{\boldsymbol{U}}_{S,EB}^T\underline{\tilde{\boldsymbol{G}}}_2^T\underline{\tilde{\boldsymbol{G}}}_1\hat{\boldsymbol{U}}_{S,EB} & \hat{\boldsymbol{U}}_{S,EB}^T\underline{\tilde{\boldsymbol{G}}}_2^T\underline{\tilde{\boldsymbol{G}}}_2\hat{\boldsymbol{U}}_{S,EB}\end{bmatrix},\end{aligned} \tag{5.46}$$

where $\hat{\boldsymbol{U}}_{S,EB}$ denotes the estimated signal subspace which is, for a single impinging plane-wave, the eigenvector corresponding to the largest eigenvalue of the modal correlation matrix, see Eq. (5.24). From Eq. (5.44), Eq. (5.45) and Eq. (5.31), it appears as if the estimates were frequency-dependent. However, as proven in Appendix E, the eigenvectors of Eq. (5.46) do not change if the frequency-dependent matrix $\underline{\boldsymbol{O}}_{2\mathcal{N}}^{1/2}$ is exchanged by an identity matrix, $\underline{\boldsymbol{I}}_{2\mathcal{N}}$. This fact, that can be attributed to the special structure of $\underline{\boldsymbol{O}}_{2\mathcal{N}}^{1/2}$, results in a frequency-independent DOA estimate, which is in contrast to the classical unitary ESPRIT algorithm.

The following list summarizes the steps required for estimating the DOAs of a single plane-wave using TLS EB-ESPRIT applied to a circular aperture.

1. Estimate the modal-spectral covariance matrix, $\hat{\underline{\boldsymbol{S}}}_{\boldsymbol{XX}}^{\mathrm{EB}}$, cf. Eq. (5.24), in analogy to Eq. (4.11).
2. Transform the complex modal-spectral covariance matrix into a real-valued one by, $\hat{\underline{\boldsymbol{S}}}_{\boldsymbol{XX},\mathrm{Re}}^{\mathrm{EB}} \triangleq \underline{\boldsymbol{Q}}_M^H\hat{\underline{\boldsymbol{S}}}_{\boldsymbol{XX}}^{\mathrm{EB}}\underline{\boldsymbol{Q}}_M$, cf. Eq. (4.112).
3. Perform a real-valued eigen-decomposition of $\hat{\underline{\boldsymbol{S}}}_{\boldsymbol{XX},\mathrm{Re}}^{\mathrm{EB}}$ and estimate the resulting modal signal subspace, $\hat{\boldsymbol{U}}_{S,\mathrm{EB}}$, cf. Eq. (5.27), which is a vector when considering a single impinging plane-wave.
4. Compute a real-valued eigen-decomposition of the matrix given in Eq. (5.46) to obtain a TLS solution of Eq. (4.137).
5. Denoting the eigenvalue of Eq. (4.137) corresponding to the impinging plane-wave by $\hat{\zeta}$ the DOA is given by, $\hat{\varphi} = 2\tan^{-1}\{\hat{\zeta}\}$.

Figure 5.5 – spanning two pages – shows the performance of the single source localization algorithm, i.e. EB-ESPRIT. The desired source is bandpass-filtered white noise where $f = 300\ldots3000$ Hz. The array comprises ten sensors sampling a continuous circular aperture of radius $R = 0.04$ m. As a result, $kR = 0.22\ldots2.22$. Both baffled and unbaffled apertures are considered. The wavefield is decomposed into seven circular harmonics, i.e up to and including order $\mathcal{N} = 3$. For comparison, not only spatially uncorrelated

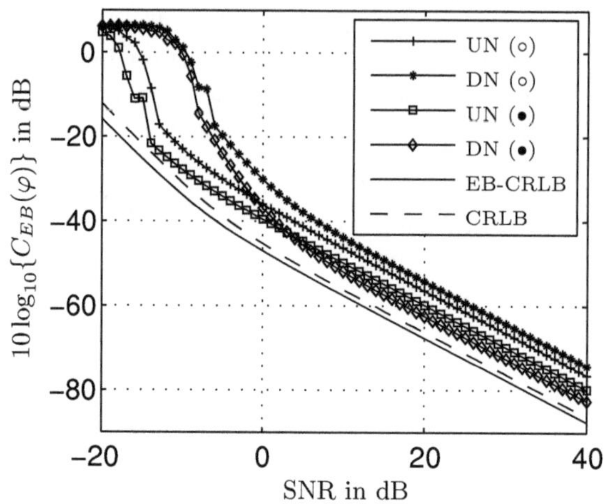

(a) Estimator variance in spatially uncorrelated noise (UN) and diffuse noise (DN)

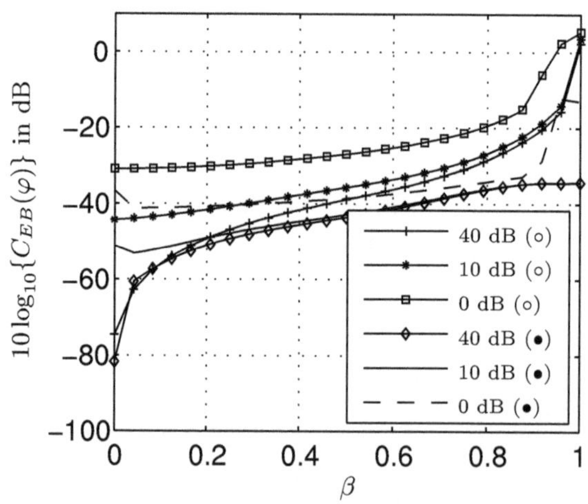

(b) Estimator variance in diffuse noise field

Fig. 5.5. Performance of circular EB-ESPRIT w.r.t. the EB-CRLB for a single plane-wave ($\varphi = 2\pi/9$), where $f = 300 \ldots 3000$ Hz, $R = 0.04$, $M = 10$, $\mathcal{N} = 3$, $K = 1024$, 300 independent trial runs ['○': unbaffled aperture '•': baffled aperture]

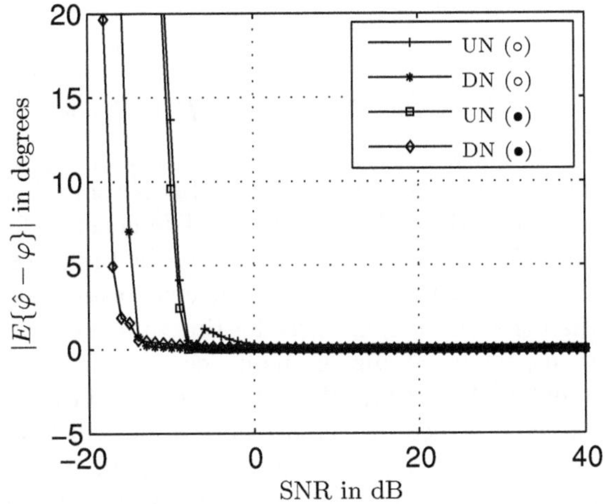

(c) Absolute value of mean of estimation error in spatially uncorrelated noise (UN) and diffuse noise (DN)

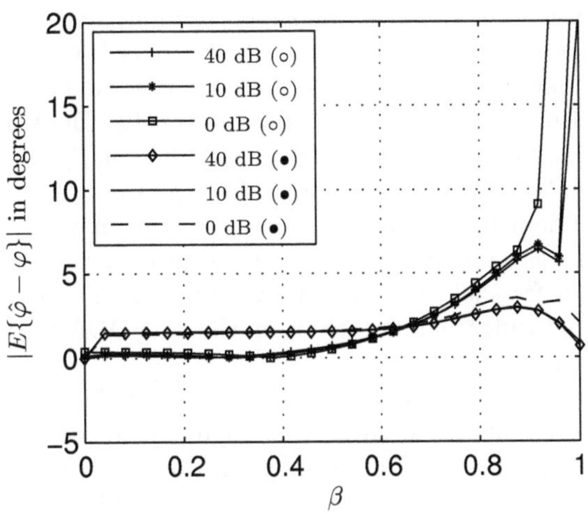

(d) Absolute value of mean of estimation error in diffuse noise field

Fig. 5.5. (cont'd) Performance of circular EB-ESPRIT w.r.t. the EB-CRLB for a single plane-wave ($\varphi = 2\pi/9$), where $f = 300 \ldots 3000$ Hz, $R = 0.04$, $M = 10$, $\mathcal{N} = 3$, $K = 1024$, 300 independent trial runs ['∘': unbaffled aperture '•': baffled aperture]

white noise but also cylindrically isotropic (diffuse) noise fields at the sensors are considered.

Since,

$$\sum_{n=-\mathcal{N}}^{\mathcal{N}} i^n B_n(kR)e^{-in\varphi} = \sum_{n=0}^{\mathcal{N}} \epsilon_n i^n B_n(kR)\cos n\varphi, \tag{5.47}$$

where $\epsilon_0 = 1$ and $\epsilon_n = 2$ for $n > 0$, the EB-CRLB is compared to the CRLB of a standard ULA comprising $\mathcal{N}+1$ sensors. The general characteristics of the curves in Fig. 5.5(a) parallel the ones shown in Fig. 4.18. A noticeable difference, however, is that the variance of the eigenbeam-based estimator is slightly higher than the sensor-space-based estimators. This can be explained by the fact that, especially at higher frequencies, modal aliasing, that was shown to be present in Fig. 3.11, affects the estimate in a negative way. A higher number of sensors used for the decomposition will lower the effect of modal aliasing in this frequency range and, as a result, the variance of the estimate. Furthermore, by increasing the maximum decomposition order $\mathcal{N}$, the effects due to series truncation will be reduced. An additional difference between EB-ESPRIT with circular arrays and ESPRIT using standard ULAs stems from the fact that the EB-ESPRIT algorithm allows for a full 2π field-of-view, i.e. EB-ESPRIT does not suffer from the ambiguity problem present in all algorithms employing linear sensor arrays.

As can be deduced from Fig. 5.5(a), a baffled circular sensor array exhibits lower estimation variance than its unbaffled counterpart. For this particular DOA of the wideband plane-wave, $\varphi = 2\pi/9$, the EB-CRLB is lower than the CRLB calculated for the same DOA and a standard ULA comprising $\mathcal{N}+1$ sensors. Recall that the CRLB depends on the DOA, exhibiting best performance for a plane-wave impinging perpendicularly w.r.t. the array's axis, while the EB-CRLB is independent of the DOA.

Figure 5.5(c) depicts the absolute value of the mean estimation error simulated for both spatially uncorrelated and diffuse noise, corresponding to Fig. 5.5(a), i.e. a free-field scenario. It can be seen that the mean estimation error does not differ significantly from the true DOA for an SNR>-5 dB in all cases considered.

Figure 5.5(b) exemplifies the performance of a sampled circular aperture in a simulated reverberant environment, making use of the image source method [AB79] with a single varying wall, ceiling, and floor reflection coefficient, β. Note that a circular array has no control over a wavefield that does not impinge from the horizontal plane. The influence of ceiling and floor reflections are therefore expected to be more dramatic than reflections off the walls. Although not easily quantifiable, variations in ceiling and floor reflection characteristics are not omitted here as they represent real acoustic environments. The room dimensions were chosen arbitrarily as $(W, D, H) = (6.5, 6.5, 3)$ m. Only the estimator variance of EB-ESPRIT, $C_{EB}(\varphi)$ cf. Eq. (4.64), evaluated for a cylindrically diffuse noise field where SNR=0,10,40 dB are shown since the performance for spatially uncorrelated noise fields are almost identical. As can be seen, especially for low SNR, the baffled aperture provides superior performance with respect to the unbaffled aperture. Figure 5.5(d) shows the absolute value of the mean estimation error corresponding to Fig. 5.5(b). Here, for small values of the reflection coefficient, the baffled circular array exhibits a slight offset with respect to the DOA estimate compared to the unbaffled circular array. It is believed that this offset is due to the existence of surface waves due to the cylindrical baffle. However, for large values of the reflection coefficient, the performance of the baffled circular array surpasses the one exhibited by the unbaffled circular array.

Note that the special structure of $\underline{\boldsymbol{O}}_{2\mathcal{N}}^{1/2}$ in Eq. (5.30), that leads to an inherently frequency-invariant algorithm for source localization, is only relevant to single-source scenarios. For multiple impinging sources, the algorithm needs to be modified slightly, as shown in the following.

ESPRIT-Based Algorithm for Multiple-Source Localization Tasks

As opposed to the procedure shown above, the DOA estimate can also be 'forced' to be frequency-independent by considering the modified entries of the eigenbeam array manifold vector as given by Eq. (5.21). In practice, this result can be obtained by applying an appropriate filter to each decomposed harmonic. Then, the eigenbeam invariance relationship becomes, cf. Eq. (4.150),

$$\tan\left(\frac{\varphi}{2}\right)\breve{\underline{\boldsymbol{G}}}_1\breve{\boldsymbol{V}}_{EB}(\varphi) = \breve{\underline{\boldsymbol{G}}}_2\breve{\boldsymbol{V}}_{EB}(\varphi), \tag{5.48}$$

where,

$$\breve{\underline{\boldsymbol{G}}}_1 \triangleq \mathrm{Re}\{\underline{\boldsymbol{Q}}_{2\mathcal{N}}^H \underline{\boldsymbol{K}}^a \underline{\boldsymbol{Q}}_{2\mathcal{N}+1}\}, \tag{5.49}$$

$$\breve{\underline{\boldsymbol{G}}}_2 \triangleq \mathrm{Im}\{\underline{\boldsymbol{Q}}_{2\mathcal{N}}^H \underline{\boldsymbol{K}}^a \boldsymbol{Q}_{2\mathcal{N}+1}\}. \tag{5.50}$$

As can be seen, Eq. (5.48) is now frequency-invariant, since the eigenbeam manifold vector $\breve{\boldsymbol{V}}_{EB}(\varphi)$ as defined in Eq. (5.20) does not depend on frequency. By defining an eigenbeam array manifold matrix for I impinging plane-waves as,

$$\breve{\underline{\boldsymbol{V}}}_{EB} \triangleq [\breve{\boldsymbol{V}}_{EB}(\varphi_1) \mid \breve{\boldsymbol{V}}_{EB}(\varphi_2) \mid \ldots \mid \breve{\boldsymbol{V}}_{EB}(\varphi_I)], \tag{5.51}$$

all further steps for obtaining estimates of the source DOAs are identical to the ones derived for the classical unitary-ESPRIT algorithm in Section 4.4.3, and are, therefore, not reproduced here.

A different approach for localizing multiple plane-wave signals using eigenbeams extracted by unbaffled circular sensor arrays is given in [MZ94], where a property of the recurrence relationship of Bessel function is utilized, cf. Eq. (B.14). However, the resulting algorithm is applicable to narrowband signals only and can therefore, in general, not be applied to acoustic signal processing tasks.

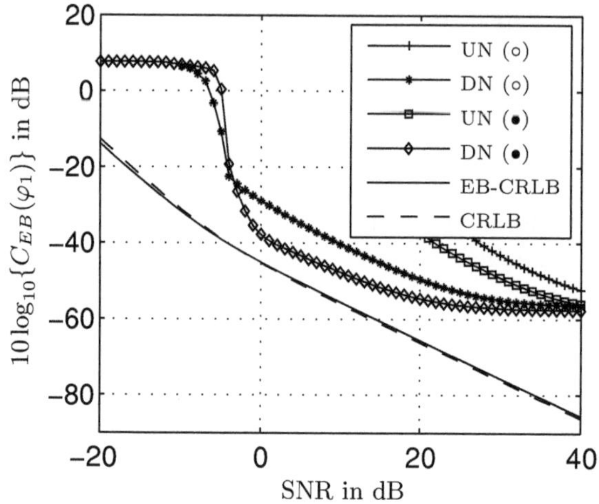

(a) Estimator variance in spatially uncorrelated noise (UN) and diffuse noise (DN)

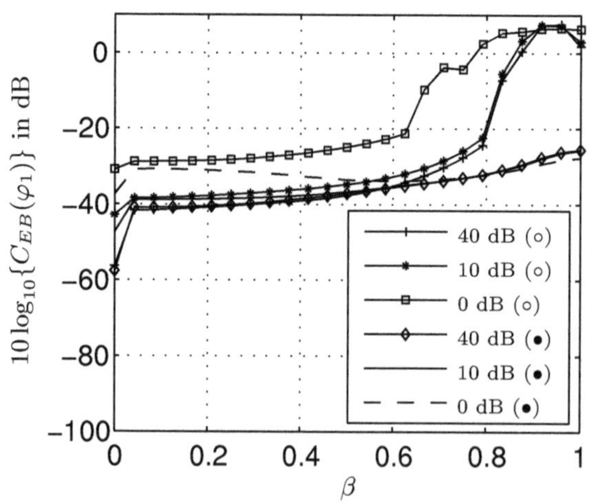

(b) Estimator variance in diffuse noise field

Fig. 5.6. Performance of circular EB-ESPRIT w.r.t. the EB-CRLB for two incident plane-waves ($\varphi = [2\pi/9, \pi/2]^T$, source one shown), where $f = 1000 \ldots 3000$ Hz, $R = 0.04$, $M = 10$, $\mathcal{N} = 3$, $K = 1024$, 300 independent trial runs ['○': unbaffled aperture '●': baffled aperture]

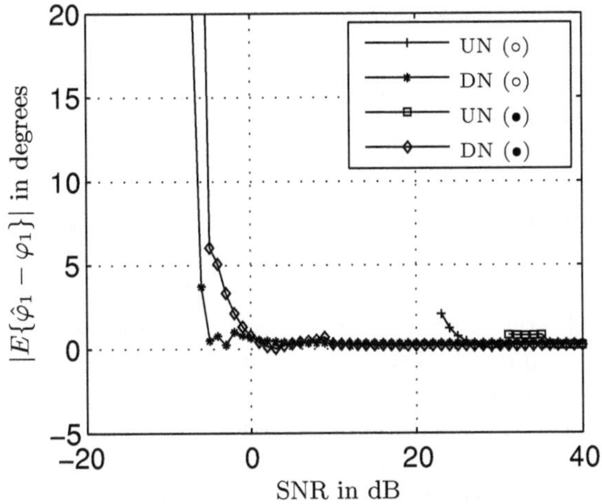

(c) Absolute value of mean of estimation error in spatially uncorrelated noise (UN) and diffuse noise (DN)

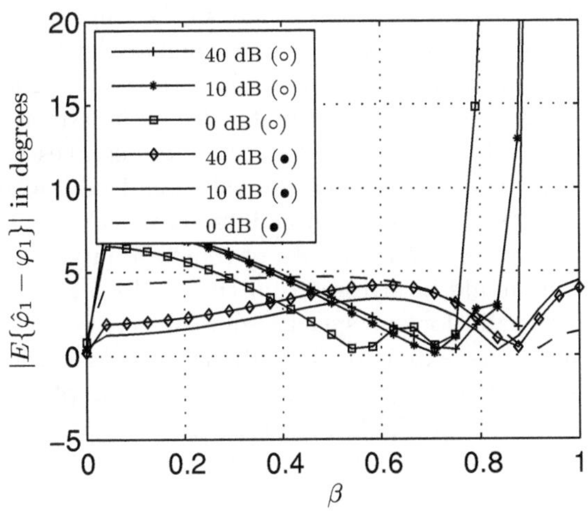

(d) Absolute value of mean of estimation error in diffuse noise field

Fig. 5.6. (cont'd) Performance of circular EB-ESPRIT w.r.t. the EB-CRLB for two incident plane-waves ($\varphi = [2\pi/9, \pi/2]^T$, source one shown), where $f = 1000 \ldots 3000$ Hz, $R = 0.04$, $M = 10$, $\mathcal{N} = 3$, $K = 1024$, 300 independent trial runs ['○': unbaffled aperture '●': baffled aperture]

Note that the algorithm for multiple plane-wave sources also applies to single plane-wave incidence. However, using the algorithm presented above, derived exclusively for single source scenarios, has advantages as will become clear from the following example.

Figure 5.6 – again spanning two pages – parallels the performance evaluation of EB-ESPRIT for a single impinging plane-wave as shown in Fig. 5.5. Here, however, two plane-wave signals, $\boldsymbol{\varphi} = [2\pi/9, \pi/2]^T$, impinge on the same sampled circular aperture as utilized before. Only the results for the first source, $\varphi_1 = 2\pi/9$, are shown. The results for the second source, $\varphi_2 = \pi/2$, are almost identical and are therefore, omitted. Performance degradation compared to the single-source scenario is obvious. This is in part due to the equalization procedure implied by Eq. (5.21). In particular, the presence of spatially uncorrelated white noise imposes the requirement of carefully calibrated sensors. The additional amplification of spatially uncorrelated white noise is the reason why the algorithm does not return any estimate in Fig. 5.6(a),(c) for SNR<20 dB. The absence of an estimate is due to an implementation detail realizing a reliability test described in [HN95]. This test verifies whether the eigenvalues of the matrix in Eq. (5.46) are all real, as anticipated due to purely real-valued operations, and does not return an estimate if this test fails.

Note that the eigenbeams and differential/superdirective sensor arrays share similar properties, especially their sensitivity to spatially uncorrelated noise which limits their usability at lower frequencies. Therefore, the lower cut-off frequency is chosen to be higher than in the single-source case.

Note that the degradation in performance compared with the single-source scenario is also due to the fact that, in general, the more signals are present, the more information needs to be extracted from the estimated correlation matrix. One can therefore conclude that, the more sources are present in a wavefield, the longer the observation interval should be selected for reliable estimates. Here, however, the observation interval was chosen to be as long as in the single-source scenario for the sake of comparability.

At this point, it should be stressed that the results for varying wall reflection coefficients are valid for one specific source-receiver-room scenario only. Due to space limitations, extensive simulations covering a wide range of different scenarios will not be presented. However, the example chosen here does reflect the general trend that baffled circular arrays yield better results than their unbaffled counterparts. More results using a system operating in a real acoustic environment are reproduced in Chapter 6.

5.2.5 Eigenbeam Processing Using Spherical Apertures

Spherical apertures and sensor arrays employing eigenbeams for the localization of multiple wideband sources need special considerations since the decomposed eigenbeams, cf. Section 3.2 and Section 5.1.2, depend on both the order n and the degree m. Hence, an invariance relationship as exploited by ESPRIT and EB-ESPRIT is not available. Therefore, a straightforward extension of the algorithm derived for circular sensor arrays is bound to fail. However, one can derive an algorithm that is, conceptually, similar to EB-ESPRIT, by employing a well-known recurrence relation for associated Legendre functions [Arf85], i.e.,

$$2m \cot\vartheta P_n^m(\cos\vartheta) = (m-n-1)(n+m)P_n^{m-1}(\cos\vartheta) - P_n^{m+1}(\cos\vartheta). \quad (5.52)$$

Note that this particular recurrence relation is based on a fixed order n of the associated Legendre functions. The idea of using recurrence relations for modal processing applied to circular apertures has been introduced in [MZ94]. The following discussion follows the path shown in [MZ94] and adapts it for its application to spherical apertures.

Now, by applying the basic idea of ESPRIT to form subarrays, three eigenbeam subarray vectors of length $2\mathcal{N}-1$ are extracted from Eq. (5.22) as,

$$\check{\boldsymbol{V}}_{EB}^{(l)}(\vartheta,\varphi) \triangleq \underline{\boldsymbol{\Delta}}^{(l)}\underline{\boldsymbol{D}}_0\check{\boldsymbol{V}}_{EB}(\vartheta,\varphi), \quad l=-1(1)1, \quad (5.53)$$

where $\underline{\boldsymbol{\Delta}}^{(-1)}$, $\underline{\boldsymbol{\Delta}}^{(0)}$, and $\underline{\boldsymbol{\Delta}}^{(1)}$ extract the first, middle, and last $2\mathcal{N}-1$ elements from $\underline{\boldsymbol{D}}_0\check{\boldsymbol{V}}_{EB}(\vartheta,\varphi)$, and where,

$$\underline{\boldsymbol{D}}_0 = \operatorname{diag}\{(-1)^{\mathcal{N}},\ldots,(-1)^0,1,\ldots,1^{\mathcal{N}}\}. \quad (5.54)$$

By considering I plane-waves impinging on the spherical aperture, the recurrence relation, Eq. (5.52) using Eq. (5.23), can be expressed as,

$$\underline{\boldsymbol{D}}_1\check{\underline{\boldsymbol{V}}}_{EB}^{(0)} = \underline{\boldsymbol{D}}_2\check{\underline{\boldsymbol{V}}}_{EB}^{(-1)}\boldsymbol{\Phi} + \underline{\boldsymbol{D}}_3\check{\underline{\boldsymbol{V}}}_{EB}^{(1)}\boldsymbol{\Phi}^*, \quad (5.55)$$

where,

$$\check{\underline{\boldsymbol{V}}}_{EB}^{(l)} = [\check{\boldsymbol{V}}_{EB}^{(l)}(\vartheta_1,\varphi_1) \mid \check{\boldsymbol{V}}_{EB}^{(l)}(\vartheta_2,\varphi_2) \mid \ldots \mid \check{\boldsymbol{V}}_{EB}^{(l)}(\vartheta_I,\varphi_I)], \quad l=-1(1)1, \quad (5.56)$$

and,

$$\boldsymbol{\Phi} = \operatorname{diag}\{\mu_1,\ldots,\mu_I\}, \quad (5.57)$$

where,

$$\mu_\iota = \tan\vartheta_\iota \cdot e^{-i\varphi_\iota}, \ \iota = 1(1)I. \quad (5.58)$$

Also,

$$\begin{aligned}
\underline{\boldsymbol{D}}_1 &= 2\operatorname{diag}\{(\mathcal{N}-1)/a_{\mathcal{N}}^{-(\mathcal{N}-1)},\ldots,1/a_{\mathcal{N}}^{-1},0,1/a_{\mathcal{N}}^{1},\ldots,(\mathcal{N}-1)/a_{\mathcal{N}}^{\mathcal{N}-1}\},\\
\underline{\boldsymbol{D}}_2 &= \operatorname{diag}\{(\nu-\mathcal{N}-1)\cdot(\mathcal{N}+\nu)/a_{\mathcal{N}}^{\nu-1}\}, \quad \nu=-(\mathcal{N}-1)(1)(\mathcal{N}-1),\\
\underline{\boldsymbol{D}}_3 &= \operatorname{diag}\{1/a_{\mathcal{N}}^{-(\mathcal{N}-2)},\ldots,1/a_{\mathcal{N}}^{0},-1/a_{\mathcal{N}}^{1},1/a_{\mathcal{N}}^{2},\ldots,1/a_{\mathcal{N}}^{\mathcal{N}}\},
\end{aligned}$$

where $a_{\mathcal{N}}^m$ is defined by Eq. (3.62) as the square-root term in Eq. (5.23) that depends on n and m.

Since the eigenbeam array manifold matrix,

$$\check{\underline{V}}_{EB} \triangleq [\check{\boldsymbol{V}}_{EB}(\vartheta_1, \varphi_1) \mid \check{\boldsymbol{V}}_{EB}(\vartheta_2, \varphi_2) \mid \ldots \mid \check{\boldsymbol{V}}_{EB}(\vartheta_I, \varphi_I)], \tag{5.59}$$

is related to the signal subspace matrix $\underline{\boldsymbol{U}}_S$ by a non-singular matrix $\underline{\boldsymbol{T}}$, cf. Eq. (4.124), it follows that $\underline{\boldsymbol{U}}_S^{(l)} = \underline{\boldsymbol{\Delta}}^{(l)} \underline{\boldsymbol{U}}_S, l = -1(1)1$.

As discussed earlier, the signal subspace matrix can be estimated by extracting the I principal eigenvalues from the modal correlation matrix, Eq. (5.24). This correlation matrix is estimated based on the decomposed output (spherical eigenbeams) of the spherical aperture. Equation (5.55) can then be expressed as,

$$\underline{\boldsymbol{D}}_1 \underline{\boldsymbol{U}}_S^{(0)} = \underline{\boldsymbol{E}} \begin{bmatrix} \underline{\boldsymbol{\Psi}}^T \\ \underline{\boldsymbol{\Psi}}^H \end{bmatrix}, \tag{5.60}$$

where,

$$\underline{\boldsymbol{E}} = [\underline{\boldsymbol{D}}_2 \underline{\boldsymbol{U}}_S^{(-1)} \mid \underline{\boldsymbol{D}}_3 \underline{\boldsymbol{U}}_S^{(1)}], \tag{5.61}$$

$$\underline{\boldsymbol{\Psi}} = \underline{\boldsymbol{T}}^{-1} \underline{\boldsymbol{\Phi}} \underline{\boldsymbol{T}}. \tag{5.62}$$

By solving Eq. (5.60) in a LS or TLS sense, an estimate for $\underline{\boldsymbol{\Psi}}$ can be obtained. Finally, by realizing that the complex eigenvalues of $\underline{\boldsymbol{\Psi}}$ are the entries of $\underline{\boldsymbol{\Phi}}$, the azimuth of the impinging plane-waves, $\varphi_\iota, \iota = 1(1)I$, can be readily identified by the phase value of these eigenvalues. Similarly, the direction-of-arrival in elevation, $\vartheta_\iota, \iota = 1(1)I$, simply correspond to the inverse tangent of the magnitude of the eigenvalues, see Eq. (5.58). Several observations are of interest. Firstly, as for EB-ESPRIT utilizing circular harmonics, the number of spherical harmonics to be extracted from the wavefield must satisfy the relation $\mathcal{N} \geq I + 1$. Secondly, the algorithm derived above can be identified as ESPRIT-like. Thirdly, source localization using eigenbeams obtained by wavefield decomposition along the surface of a spherical aperture is *inherently frequency-independent.* This is a direct consequence of the fact that the frequency-dependent terms in $\check{\underline{V}}$, i.e. $b_{\mathcal{N}}(kR)$ – cf. Eq. (5.23) –, cancel in Eq. (5.55).

Figure 5.7 and Fig. 5.8 – spanning multiple pages, respectively – exemplify the performance for one specific sources-sensors-room scenario of the DOA estimation algorithm using spherical arrays regarding azimuth and elevation, respectively. The sampling scheme used to produce these results was the t-design method, see Section 3.2.3, where $M = 32$. It can be seen that the variance of the estimator for spatially uncorrelated noise fields is significantly larger than the ideal estimator, i.e. the EB-CRLB for spherical apertures. This can be explained by the fact that the process of sampling spherical apertures introduces significant modal aliasing to the recorded spherical harmonics, cf. e.g. Fig. 3.22, which is not covered by the ideal model. This rather large amount of modal aliasing, at least compared to circular arrays, results in a biased estimate, especially for wall reflection coefficients, β, other than zero. Note that the EB-CRLB is really only meaningful for unbiased estimators in spatially uncorrelated white noise. Therefore, it can be deduced that while the estimation algorithm does work in principle, algorithms that incorporate

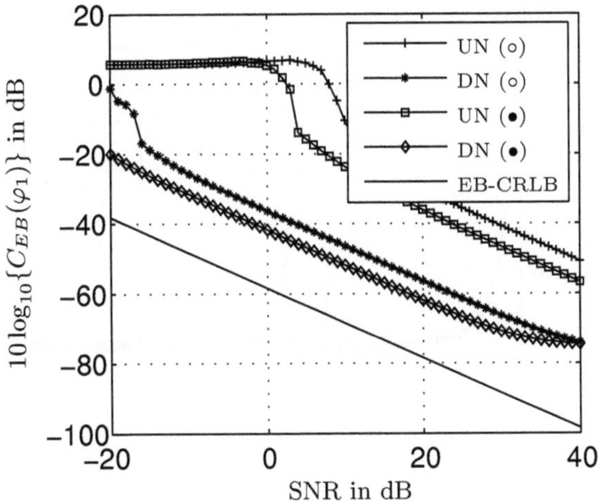

(a) Estimator variance in spatially uncorrelated noise (UN) and diffuse noise (DN)

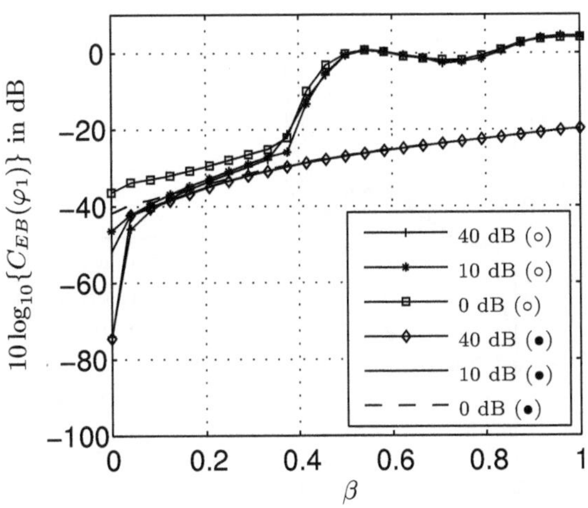

(b) Estimator variance in diffuse noise field

Fig. 5.7. Performance of spherical EB-ESPRIT w.r.t. the EB-CRLB in *azimuth* for two incident plane-waves ($\varphi = [\pi/6, \pi/2]^T$, source one shown), where $f = 1000 \ldots 3000$ Hz, $M = 32$, $\mathcal{N} = 3$, $K = 1024$, 300 independent trial runs ['∘': unbaffled spherical array '•': baffled spherical array]

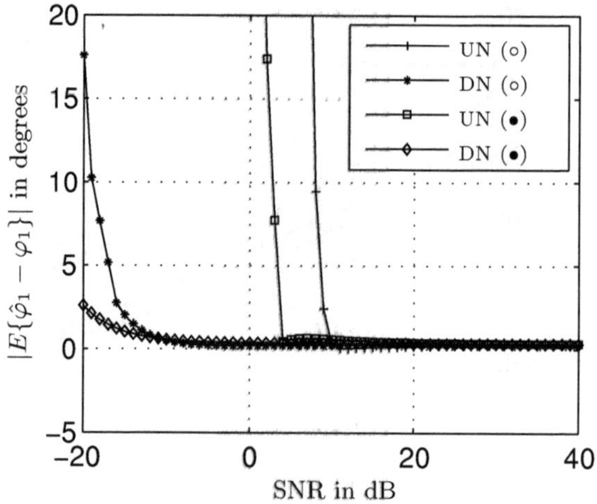

(c) Absolute value of mean of estimation error in spatially uncorrelated noise (UN) and diffuse noise (DN)

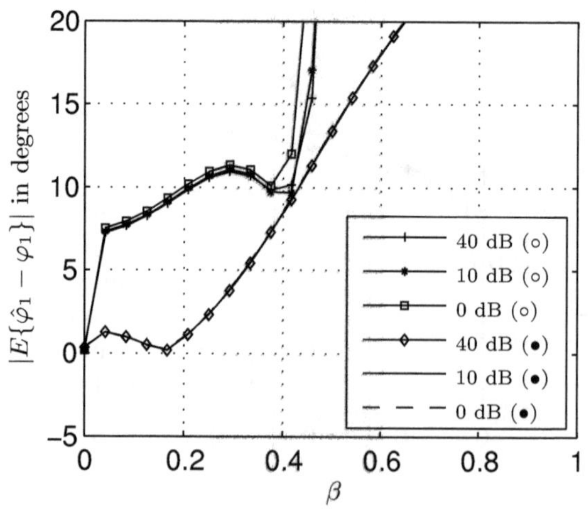

(d) Absolute value of mean of estimation error in diffuse noise field

Fig. 5.7. (cont'd) Performance of spherical EB-ESPRIT w.r.t. the EB-CRLB in *azimuth* for two incident plane-waves ($\boldsymbol{\varphi} = [\pi/6, \pi/2]^T$, source one shown), where $f = 1000 \ldots 3000$ Hz, $M = 32$, $\mathcal{N} = 3$, $K = 1024$, 300 independent trial runs ['∘': unbaffled spherical array '•': baffled spherical array]

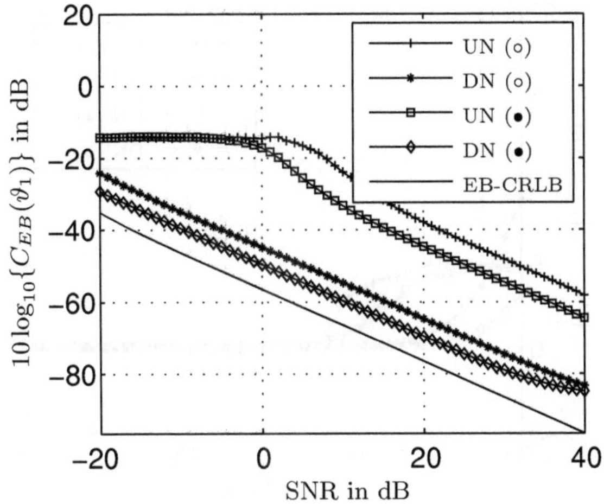

(a) Estimator variance in spatially uncorrelated noise (UN) and diffuse noise (DN)

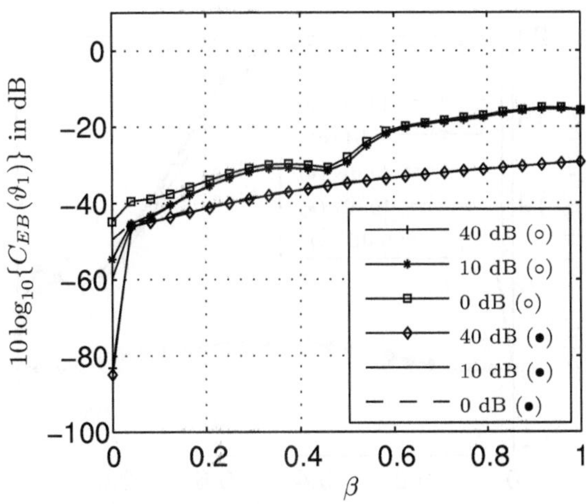

(b) Estimator variance in diffuse noise field

Fig. 5.8. Performance of spherical EB-ESPRIT w.r.t. the EB-CRLB in *elevation* for two incident plane-waves ($\vartheta = [\pi/9, 5\pi/18]^T$, source one shown), where $f = 1000 \ldots 3000$ Hz, $M = 32$, $\mathcal{N} = 3$, $K = 1024$, 300 independent trial runs ['○': unbaffled spherical array '●': baffled spherical array]

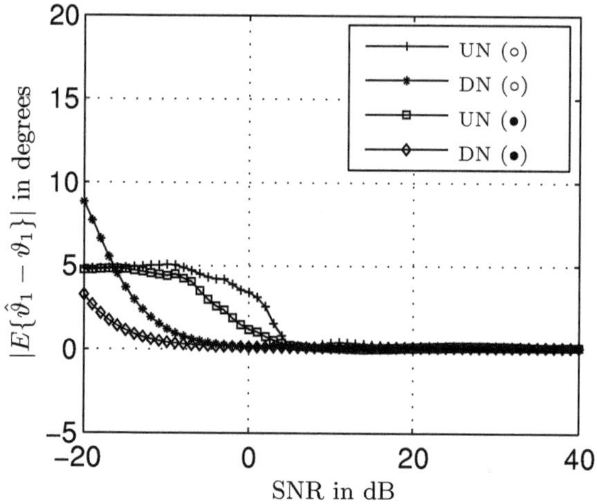

(c) Absolute value of mean of estimation error in spatially uncorrelated noise (UN) and diffuse noise (DN)

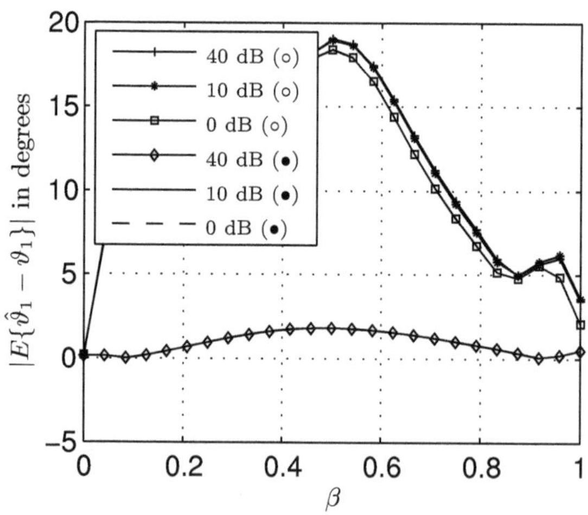

(d) Absolute value of mean of estimation error in diffuse noise field

Fig. 5.8. (cont'd) Performance of spherical EB-ESPRIT w.r.t. the EB-CRLB in *elevation* for two incident plane-waves ($\boldsymbol{\vartheta} = [\pi/9, 5\pi/18]^T$, source one shown), where $f = 1000 \ldots 3000$ Hz, $M = 32$, $\mathcal{N} = 3$, $K = 1024$, 300 independent trial runs ['○': unbaffled spherical array '•': baffled spherical array] a reverberated signal model should be considered for future developments. However, it is quite interesting that the estimation performance of the DOA in elevation using baffled spherical arrays is satisfactory, even for wall reflection coefficients considerably exceeding zero, cf. Figs. 5.8(b),(d). A satisfactory explanation for this phenomenon remains to be found.

At a first glance, the run of the curves in Fig. 5.8(d) appears to be erroneous. However, one has to keep in mind that for increasing boundary reflection coefficients, the variance increases considerably, cf. Fig. 5.8(b). This basically means that the estimates start to oscillate wildly around a value which, for this particular setup, happen to approach the true DOA for large values of β.

Note that, by using the algorithm described in this section, a signal impinging from $\vartheta = \pi/2$ cannot be localized because of the $\tan(\vartheta)$ term in Eq. (5.58). This problem can be alleviated by rotating the individual spherical harmonics in θ as described in Section 5.1.1.

5.2.6 Resolution Capacity and DOA Estimation of More than Two Wideband Sources

The resolution capacity of EB-ESPRIT, i.e. the minimum spatial distance between two sources that can be resolved is, in principle, zero. This means that, assuming a decomposition of the wavefield into an infinite number of harmonics, assuming absence of modal aliasing and infinite SNR, two sources can be spaced arbitrarily close together. However, as soon as the effects of wavefield truncation, modal aliasing due to the finite number of sensors, finite SNR, and finite observation intervals are present, the minimum spatial distance between two sources becomes non-zero. Exact predictions on how close two acoustic wideband sources can be positioned with respect to each other depend on too many parameters to be reproduced here and have to be examined individually for each scenario considered.

In this chapter, so far, the proof-of-concept concerning eigenbeam-based PE methods has been given for a maximum of two wideband sources only. It is well known that subspace-based PE methods require at least one more sensor than sources in order to form signal and noise subspaces from the estimated correlation matrices, see e.g. [MZ94] and cf. Eq. (4.12). Since the entire concept of eigenbeam-based PE methods stems from the fact that individual sensors are substituted by individual eigenbeams, it is clear that the requirement here is that a maximum of $2\mathcal{N}$ sources may be present if the wavefield is decomposed up to order $\mathcal{N}$, thereby yielding a total number of $2\mathcal{N} + 1$ eigenbeams, cf. Fig. 5.4. Qualitative examples for the performance of the EB-ESPRIT algorithm with respect to more than two sources are presented in Appendix F.

5.2.7 Detection of the Number of Active Sources

In this section, a detection algorithm based on eigenbeam processing for the number of active sources (NOS) present in a wavefield for both circular as well as for spherical apertures, is discussed.

Detection Algorithm for Circular Apertures

The following discussion details a detection algorithm for circular apertures which is based on the wavefield decomposition process outlined in Section 3.1. Note that the decomposition of a wavefield into circular harmonics is performed with respect to the aperture's perimeter. It can therefore be observed that after this transformation one obtains a single virtual receiver at the aperture's perimeter exhibiting multiple spatial selectivity characteristics, i.e. the harmonics of order n, simultaneously. As outlined in Section 3.1, $n = 0$ corresponds to an omnidirectional spatial characteristic, $n = 1$ corresponds to a dipole spatial characteristic, and so on.

First, the unbaffled circular aperture is considered. Then, some light is shed on how to modify the algorithm when baffled circular apertures are employed.

When I plane-waves with unity amplitude impinge on the circular aperture, Eq. (3.5) for $\vartheta = \pi/2$ becomes,

$$\begin{aligned} P(kR, \phi, \boldsymbol{\varphi}) &= \sum_{\iota=1}^{I} e^{ikR\cos(\phi-\varphi_\iota)} \\ &= \sum_{\iota=1}^{I} \sum_{n=-\infty}^{\infty} i^n J_n(kR) e^{in(\phi-\varphi_\iota)}, \end{aligned} \tag{5.63}$$

by virtue of the superposition principle. The symbol $\boldsymbol{\varphi}$ in Eq. (5.63) means that the angular dependency is due to the DOAs of all I impinging plane-waves, $\boldsymbol{\varphi} = [\varphi_1, \varphi_2, \ldots, \varphi_I]^T$. As a result, it follows for the circular harmonics corresponding to multiple plane-waves that, cf. Eq. (3.15),

$$\mathring{F}_n(kR, \boldsymbol{\varphi}) = \sum_{\iota=1}^{I} i^n J_n(kR) e^{-in\varphi_\iota}. \tag{5.64}$$

Up to this point, only continuous circular apertures have been considered. For actual implementations, however, the aperture has to be sampled at discrete sensor positions. In order to perform the sampling operation, Eq. (5.64) needs to be discretized to yield, for multiple plane-waves, the sampled circular harmonics, $\mathring{F}_n^s(kR, \boldsymbol{\varphi})$. Building upon the discussion presented in Section 3.1.3, it follows that,

$$\mathring{F}_n^s(kR, \boldsymbol{\varphi}) = \mathring{F}_n(kR, \boldsymbol{\varphi}) + \mathcal{E}_{\text{samp}}(n_a, kR, \boldsymbol{\varphi}), \tag{5.65}$$

where $\mathcal{E}_{\text{samp}}(n_a, kR, \varphi)$ is an additional term due to modal aliasing which, in essence, results in several modes of order $n_a > n$ leaking into mode n. This error can be controlled, although not eliminated, by appropriately choosing the number of microphones, M, the radius of the aperture, R, and the frequency range of interest, $kR = 2\pi f R/c$, see Section 3.1.3. For simplicity, it is assumed that $\mathcal{E}_{\text{samp}}(n_a, kR, \varphi)$ is sufficiently small in the following.

It is shown in Section 3.1.3 that by considering a real system using M microphones, a wavefield can be decomposed into a limited number of harmonics only, i.e.,

$$\mathcal{N} = \begin{cases} M/2, & M \text{ even} \\ (M-1)/2, & M \text{ odd} \end{cases}. \tag{5.66}$$

Note that *critical sampling*, cf. Section 3.1.3, is assumed here.

As a first step towards an algorithm that estimates the number of acoustic wideband sources, a wavefield synthesis operation that superimposes these limited number of circular harmonics is considered. This synthesis operation can be seen as an order-limited inverse spatial Fourier series expansion with respect to $\phi = \phi_0$, see Eq. (A.5), which gives with Eq. (5.64),

$$P(kR, \phi_0, \varphi) = \sum_{n=-\mathcal{N}}^{\mathcal{N}} \mathring{F}_n(kR, \varphi) e^{in\phi_0} + \mathcal{E}_{\text{trunc}}(kR, \phi_0, \varphi), \tag{5.67}$$

where,

$$\mathcal{E}_{\text{trunc}}(kR, \phi_0, \varphi) = \sum_{|n| > \mathcal{N}} \mathring{F}_n(kR, \varphi) e^{in\phi_0}, \tag{5.68}$$

is the error term due to the truncation of the infinite number of circular harmonics.

For the sake of simplicity, the truncation error, $\mathcal{E}_{\text{trunc}}(kR, \phi_0, \varphi)$, is assumed to be sufficiently small in the frequency range of interest in the following. See Section 3.1.4 for more details on the wavefield truncation error. Then, Eq. (5.67) with Eq. (5.63) and $\phi_0 = 0$ can be expressed as,

$$P(kR, 0, \varphi) = \sum_{\iota=1}^{I} e^{ikR\cos\varphi_\iota} \approx \sum_{n=-\mathcal{N}}^{\mathcal{N}} \mathring{F}_n(kR, \varphi). \tag{5.69}$$

Transformation of Eq. (5.69) into the time-domain gives the *modal impulse response*,

$$p(t, \boldsymbol{\tau}) = \sum_{\iota=1}^{I} \delta(t - \tau_\iota), \tag{5.70}$$

where the symbol $\boldsymbol{\tau}$ combines the delays of all I impinging plane-waves, $\boldsymbol{\tau} = [\tau_1, \tau_2, \ldots, \tau_I]^T$, and where,

$$\tau_\iota = \frac{R}{c} \cos \varphi_\iota, \qquad \iota = 1(1)L, \tag{5.71}$$

is a delay caused by the ι-th plane-wave.

A rule for the system detection problem for wave-fronts propagating in free-field can now be derived by observing that the number of Dirac impulses in Eq. (5.70) correspond to the number of sources in the impinging wavefield.

The next step involves an adaptive identification of the modal impulse response, $p(t, \boldsymbol{\tau})$, from a block of observed data. In order to transform this task into a classical system identification problem [Hay02], a reference signal is required. This signal should be independent of the direction of the incoming plane-waves as well as of the number of sources. A suitable choice is to utilize the zeroth-order circular harmonic for this purpose since it exhibits an omnidirectional spatial characteristic. This signal can be found in Eq. (5.64) for $n = 0$. For the purpose of system identification, the frequency-dependence of the zeroth-order harmonic needs to be equalized. The modified reference signal is therefore,

$$\mathring{\mathcal{F}}_0(kR) \triangleq \frac{\mathring{F}_0(kR)}{J_0(kR)} = I, \tag{5.72}$$

which in the time domain equals,

$$f_0(t) = I \cdot \delta(t). \tag{5.73}$$

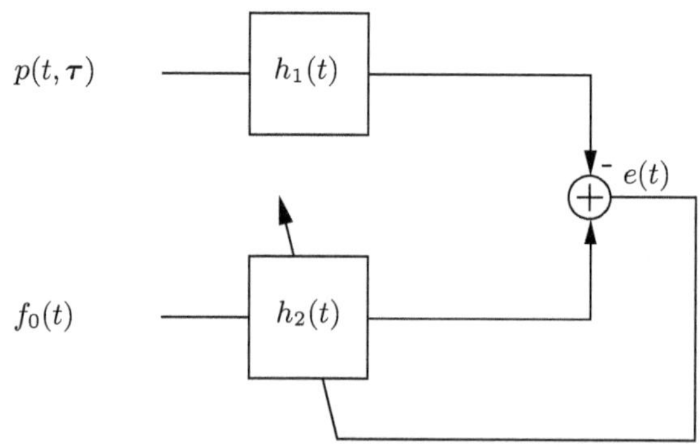

Fig. 5.9. Equivalent system identification problem for white noise input signals

A schematic of the resulting system identification problem is shown in Fig. 5.9, where the filter $h_1(t)$ is simply a delay that compensates for the delay introduced by the adaptive filter $h_2(t)$, which should be identical to a delayed version of $p(t, \boldsymbol{\tau})$ after convergence.

In practice, the algorithm can be implemented by using any adaptive filtering technique such as LMS-based or RLS-based algorithms [Hay02].

Figure 5.10 depicts the discretized modal impulse responses $p(t = \nu T, \boldsymbol{\tau})$, where ν is an integer and T is the sampling interval, and $f_0(t = \nu T)$ for a single band-limited plane-wave incident on a continuous aperture in free-field

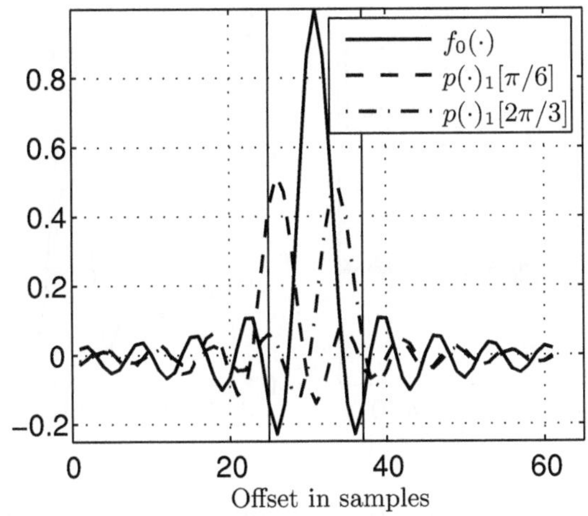

Fig. 5.10. Normalized modal impulse responses corresponding to a wavefield containing a single plane-wave $\varphi_1 = \pi/6$ and $\varphi_1 = 2\pi/3$

for two cases, $\varphi = \varphi_1 = \pi/6$ and $\varphi = \varphi_1 = 2\pi/3$, respectively. The parameters are chosen as $R = 0.04$ m, $\mathcal{N} = 5$, $f_s = 1/T = 48$ kHz, and $c = 340$ m/s, where f_s denotes the sampling frequency. The sensor signals were bandlimited to $f_{\max} = 7$ kHz, which corresponds to $k_{\max}R \approx 5.2$.

Also shown are two vertical lines that represent the maximum delays $|\tau_{\max}| = R/c$, cf. Eq. (5.71), that need to be considered in this detection scheme. This observation is of some importance since it is expected that moderately reverberant environments only have a limited negative impact on the estimation of the modal impulse response in this very narrow section of the impulse response. That means that even though the signal model is based on a free-field model, moderate reverberation should not seriously affect the performance of the algorithm.

This claim is now illustrated by an example. The parameters, especially the DOA, are the same as the ones that were used to produce Fig. 5.10 for $\varphi_1 = \pi/6$.

Here, however, the actual modal impulse response to be identified, $h_2(\cdot)$, is estimated by a frequency-domain algorithm controlling an 512-tap adaptive FIR filter. The result is plotted in Fig. 5.11. Both an anechoic as well as a reverberated environment, simulated by applying the image method [AB79], where the reflection coefficients are chosen to be $\beta = 0.8$, are considered. The effect of reverberation is twofold. First, additional peaks, indicated by arrows in Fig. 5.11(a), can be observed to the far left and to the far right with respect to the area of interest, denoted by the two vertical lines. These peaks correspond to the individual reflections that appear in the identified modal impulse response. Second, the peaks in the area of interest are now

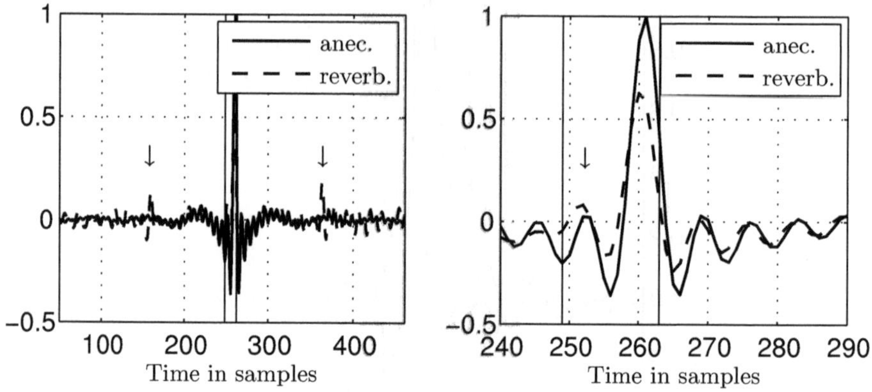

(a) Estimated modal impulse responses

(b) Excerpt detailing the area of interest

Fig. 5.11. Estimation of $h_2(\cdot)$ in anechoic (anec.) and reverberated (reverb.) environments

somewhat less prevalent, i.e. the power of the highest peak relative to the second highest peak – which does not correspond to an impinging plane-wave, cf. arrow in Fig. 5.11(b) – is reduced. This effect is depicted in Fig. 5.12, where the relative power of first to second peak with increasing reverberation, i.e. increasing wall reflection coefficient, is shown. As can be seen, for moderate reflection coefficients, the power ratio is quite high which, therefore, enables reliable source/no-source decision.

Next, two simultaneously impinging plane-waves are considered. The parameters chosen were the same ones that were used to produce Fig. 5.13, the difference being that $\boldsymbol{\varphi} = [\pi/6, 2\pi/3]^T$, i.e. $\varphi_1 = \pi/6$ and $\varphi_2 = 2\pi/3$, indicating simultaneously arriving wave-fronts. Two distinct peaks in the modal impulse response appear which correspond to the two sources. All results obtained for a single impinging plane-wave can also be extended to the two-source case. However, as can be deduced from the cosine term in Eq. (5.71), the two sources cannot be located arbitrarily along the circle. For instance, a plane-wave impinging from $\varphi_1 = \pi/6$ cannot be distinguished from a second plane-wave impinging from $\varphi_2 = 5\pi/6$ since they cause an identical delay in Eq. (5.71). This ambiguity problem can be alleviated by also performing the wavefield synthesis operation in Eq. (5.67) with respect to $\phi_0 \neq 0$, e.g. $\phi_0 = \pi/2$, and use both detection results for an overall estimate. A value of $\phi_0 \neq 0$ basically is equivalent to rotating the individual harmonics by ϕ_0, thereby resolving the ambiguity problem.

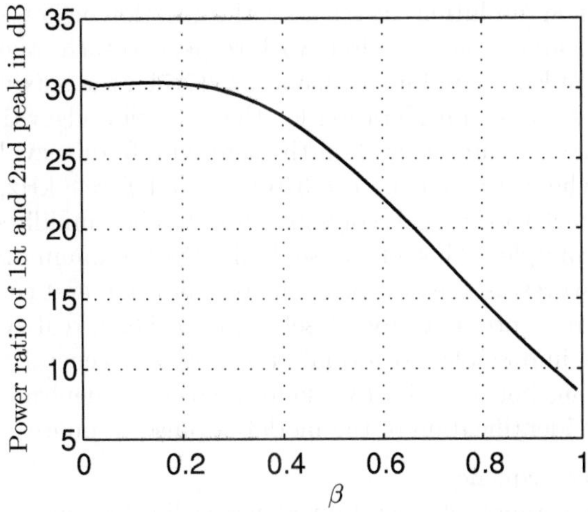

Fig. 5.12. Relative power of first to second peak with increasing reverberation

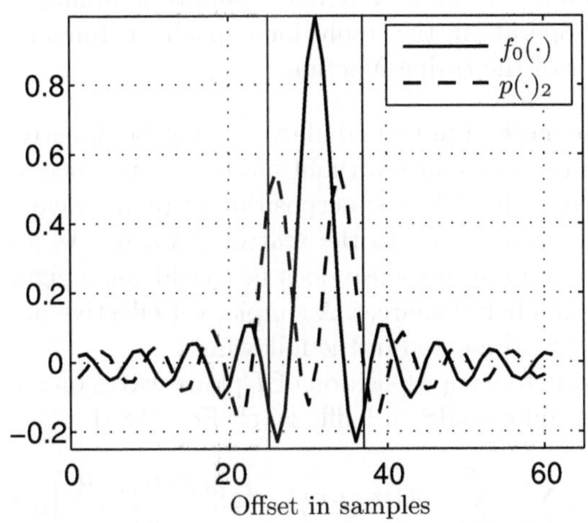

Fig. 5.13. Normalized modal impulse responses corresponding to a wavefield containing two plane-waves $\varphi = [\pi/3, 2\pi/3]^T$

A few words on resolution capacity, i.e. the question on how close the plane-waves can be located along the circle with respect to each other, are in order. As can be seen in Fig. 5.10, Fig. 5.13, and Eq. (5.71), the maximum delay that needs to be taken into consideration for the detection algorithm depends on the radius of the aperture as well on the sampling frequency. In the example shown above, where $R = 0.04$ m, $c = 340$ m/s, and f_s=48 kHz, the maximum delay with respect to the peak corresponding to the omnidirectional term is only about ± 6 samples. This, of course, limits the minimum distance between the two plane-waves as the two peaks merge into one and can therefore not be resolved if the sources are too closely spaced. For a real system it is very difficult, if not impossible, to formulate criteria of general validity on the minimum spacing between the two sources since its determination, i.e. the accuracy of the identification of the modal impulse responses, depend on

- the sampling frequency,
- the frequency range of operation which is limited due to the effects of wavefield truncation and aperture sampling,
- the position of the two sources due to the properties of the cosine function.

Note that even for an idealized system, i.e. infinite sampling frequency and frequency range of operation, the resolution capacity is fundamentally bounded by the properties of the cosine function.

Note that the method presented above cannot be directly applied to apertures mounted into rigid baffles without modifications since re-synthesis of the wavefield does not only yield a superposition of plane-waves, but also results in components corresponding to the scattered waves. As a consequence, its estimated modal impulse response does not yield unambiguous peaks corresponding to the number of sources. A simple, yet effective method that solves this problem will be presented in the following.

A wavefield due to a superposition of I plane-waves on a circular aperture mounted into a rigid cylindrical baffle is, cf. Eq. (2.83),

$$P(kR, \phi, \varphi) = \sum_{\iota=1}^{I} \sum_{n=-\infty}^{\infty} i^n \left[J_n(kR) - \frac{J_n'(kR) H_n(kR)}{H_n'(kR)} \right] e^{in(\phi - \varphi_\iota)}. \tag{5.74}$$

Applying the decomposition into circular harmonics, as described in Section 3.1.1, yields,

$$\overset{\bullet}{F}_n (kR, \varphi) = \sum_{\iota=1}^{I} i^n \left[J_n(kR) - \frac{J_n'(kR) H_n(kR)}{H_n'(kR)} \right] e^{-in\varphi_\iota}. \tag{5.75}$$

A finite sum of these circular harmonics, does not yield an estimate for the superposition of plane-waves at the center of the aperture. In contrast to unbaffled circular apertures, the circular harmonics derived from baffled apertures not only describe a wavefield containing plane-waves, but also an additional wavefield corresponding to the scattered wave on the surface of the rigid cylinder. Therefore, the superimposed circular harmonics do not correspond to individual peaks in the time-domain. Rather, additional peaks occur due to the scattered wavefield. This fact turns the detection task into a very difficult problem. One possibility to overcome this problem is to 'undo' the wave scattering effect by a filtering operation. Note that the scattered wavefield only appears in the frequency-dependent component of the circular harmonics, whereas the angular-dependent component remains unaffected. By defining the modified circular harmonics of a baffled circular aperture as,

$$\overset{\bullet}{\mathcal{F}}_n(kR,\varphi) = \frac{\overset{\bullet}{F}_n\,(kR,\varphi)}{\left[J_n(kR) - \dfrac{J_n^{'}(kR)H_n(kR)}{H_n^{'}(kR)}\right]}\,J_n(kR), \tag{5.76}$$

and choosing,

$$\overset{\bullet}{\mathcal{F}}_0(kR,\varphi) = \frac{\overset{\bullet}{F}_0\,(kR,\varphi)}{\left[J_0(kR) - \dfrac{J_0^{'}(kR)H_0(kR)}{H_0^{'}(kR)}\right]} = I, \tag{5.77}$$

the same algorithm as presented above for unbaffled circular apertures can then be directly transferred to the baffled circular aperture scenario. All results obtained for the unbaffled circular aperture, in essence, remain valid and are therefore not reproduced here.

Note that the simulations presented in this section can be easily repeated for more than two impinging sources. However, the radius of the circular aperture as well as the sampling frequency and the bandwidth of operation have to be chosen such that the desired number of peaks can be detected in the significant part of the estimated modal impulse response $h_2(\cdot)$, cf. Fig. 5.9.

More examples obtained with a baffled circular microphone array fed by data recorded in real acoustic environments are presented in Chapter 6.

An estimate of the number of sources can in fact be obtained by simply considering two microphones along a circular aperture after applying the AED algorithm, see [Ben00] and Section 4.4.2, for an estimate of the impulse response "between" the two microphones. However, the main advantage of the method described in this section is the fact that baffled apertures can be rather easily included in the framework. In the sensor-space, the effect of scattering depends on the direction-of-arrival of the impinging plane-waves, cf. Fig. 2.9. Therefore, it is impossible to compensate for this effect by a digital filtering operation. On the other hand, by applying the notion of eigenbeams, the effect of scattering is independent of the plane-waves' direction-of-arrival and the frequency-dependence of each individual component can be rather easily be compensated for, as shown above. In summary, only by using an eigenbeam-based method baffled apertures can be employed for the task of source detection.

Further advantages include a higher SNR with respect to uncorrelated white noise. It is well-known that an increase of the number of microphones by a factor of two results in an SNR gain of 3 dB. Essentially, the omnidirectional eigenbeam, $n = 0$, is obtained by summing the contribution of *all* individual microphones. For example, by utilizing ten microphones for the wavefield decomposition process, a gain of about 10 dB can be achieved compared to a single omnidirectional microphone. Therefore, the more microphones are used to sample the continuous circular aperture, the higher the resulting SNR of the decomposed wavefield.

A third advantage, which is mainly of academic interest, is that the angle by which the eigenbeams can be rotated (ϕ_0) is continuous, while in the sensor-space approach only discrete microphone pairs can be selected in order to solve the ambiguity problem.

Detection Algorithm for Spherical Apertures

The following discussion transfers the system detection algorithm for circular apertures to spherical apertures. In fact, the algorithm can be applied almost directly, provided several details are taken into consideration.

First, of course, the algorithm is now based on the wavefield decomposition process outlined in Section 3.2. Therefore, it follows for the superposition of plane-waves on an unbaffled spherical aperture with Eq. (3.59), Eq. (3.60), and Eq. (3.66) that,

$$\begin{aligned} P(kR,\theta,\phi,\boldsymbol{\vartheta},\boldsymbol{\varphi}) &= \sum_{\iota=1}^{I} e^{ikR[\sin\theta\sin\vartheta_\iota\cos(\phi-\varphi_\iota)+\cos\theta\cos\vartheta_\iota]} \\ &= 4\pi\sum_{\iota=1}^{I}\sum_{n=0}^{\infty} i^n j_n(kR)\sum_{m=-n}^{n} Y_n^m(\theta,\phi)Y_n^m(\vartheta_\iota,\varphi_\iota)^*, \end{aligned} \tag{5.78}$$

where the symbol $\boldsymbol{\vartheta}$ in Eq. (5.78) means that the angular dependency in elevation is due to the DOAs of I impinging plane-waves, $\boldsymbol{\vartheta} = [\vartheta_1, \vartheta_2, \ldots, \vartheta_I]^T$.

After performing the decomposition step, essentially a spherical harmonics transform of Eq. (5.78), see Section 3.2.1, one obtains for the wavefield due to I incoming plane-waves decomposed by an unbaffled spherical aperture,

$$\mathring{G}_n^m(kR,\boldsymbol{\vartheta},\boldsymbol{\varphi}) = \sqrt{4\pi}\sum_{\iota=1}^{I} i^n j_n(kR) Y_n^m(\vartheta_\iota,\varphi_\iota)^*. \tag{5.79}$$

Now, applying the same ideas as presented for circular apertures, a re-synthesis operation with respect to a reference point on the aperture, (θ_0, ϕ_0), is applied. This corresponds to a spherical harmonics transform with respect to this reference point. The result, after the sampling and wavefield truncation operations is,

$$P(kR,\theta_0,\phi_0,\boldsymbol{\vartheta},\boldsymbol{\varphi}) = 4\pi \sum_{\iota=1}^{I}\sum_{n=0}^{\mathcal{N}} i^n j_n(kR) \sum_{m=-n}^{n} Y_n^m(\theta_0,\phi_0) Y_n^m(\vartheta_\iota,\varphi_\iota)^* + \mathcal{E}_{\text{samp}}(n_a,m_a,kR,\theta_0,\phi_0,\boldsymbol{\vartheta},\boldsymbol{\varphi}) + \mathcal{E}_{\text{trunc}}(kR,\theta_0,\phi_0,\boldsymbol{\vartheta},\boldsymbol{\varphi}). \tag{5.80}$$

Again, for the sake of clarity, the error due to the truncation, $\mathcal{E}_{\text{trunc}}(\cdot)$, and the error due to sampling, $\mathcal{E}_{\text{samp}}(\cdot)$, are assumed to be sufficiently small in the following. More details on these errors can be found in Section 3.2.4 and Section 3.2.3, respectively. Then, Eq. (5.80) can be written as,

$$\begin{aligned} P(kR,\theta_0,\phi_0,\boldsymbol{\vartheta},\boldsymbol{\varphi}) &= \sum_{\iota=1}^{I} e^{ikR[\sin\theta_0 \sin\vartheta_\iota \cos(\phi_0-\varphi_\iota)+\cos\theta_0\cos\vartheta_\iota]} \\ &\approx 4\pi \sum_{\iota=1}^{I}\sum_{n=0}^{\mathcal{N}} i^n j_n(kR) \sum_{m=-n}^{n} Y_n^m(\theta_0,\phi_0) Y_n^m(\vartheta_\iota,\varphi_\iota)^*. \end{aligned} \tag{5.81}$$

In the time-domain, Eq. (5.81) can be written as,

$$p(t,\boldsymbol{\tau}) = \sum_{\iota=1}^{I} \delta(t-\tau_\iota), \tag{5.82}$$

where,

$$\tau_\iota = \frac{R}{c}\Big[\sin\theta_0 \sin\vartheta_\iota \cos(\phi_0-\varphi_\iota) + \cos\theta_0\cos\vartheta_\iota\Big], \qquad \iota = 1(1)I. \tag{5.83}$$

As before, in order to formulate an adaptive algorithm, a reference signal needs to be identified. This is, again, the omnidirectional component, which is,

$$\mathring{\mathcal{G}}\,{}_0^0(kR) = \frac{\mathring{G}\,{}_0^0(kR)}{j_0(kR)} = I, \tag{5.84}$$

where $\mathring{G}\,{}_0^0(kR)$ is given by Eq. (5.79) with $n = m = 0$. In the time-domain Eq. (5.84) reads,

$$g_0^0(t) = I \cdot \delta(t). \tag{5.85}$$

As an example, Fig. 5.14 shows the modal impulse responses corresponding to a wave field containing two plane-waves for $\boldsymbol{\vartheta} = [\pi/3, 4\pi/9]^T$ and $\boldsymbol{\varphi} = [\pi/6, 2\pi/3]^T$. The reference point on the aperture was chosen to be $(\theta_0,\phi_0) = (\pi/2, 0)$. Two distinct peaks corresponding to the two sources are clearly visible, which facilitate the detection algorithm for spherical apertures.

It is not difficult to verify that expressions involving baffled spherical apertures can be derived by appropriately modifying Eqs. (5.74)–(5.77). For the sake of brevity, explicit mathematical expressions are omitted here.

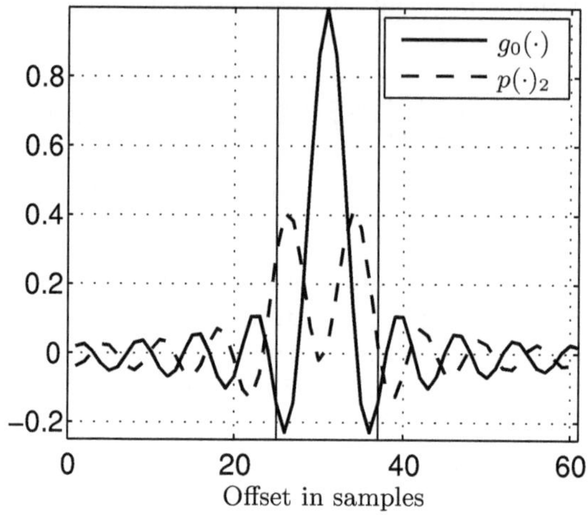

Fig. 5.14. Normalized modal impulse responses corresponding to a wavefield containing two plane-waves where $\boldsymbol{\vartheta} = [\pi/3, 4\pi/9]^T$ and $\boldsymbol{\varphi} = [\pi/6, 2\pi/3]^T$

6

A Practical Acoustic Scene Analysis System

This chapter describes a practical real-time capable system designed to verify the claim developed throughout the previous chapters that acoustic scene analysis tasks can be tackled by wavefield decomposition methods using microphone arrays. In this chapter and in Appendix F, it will be shown that a circular microphone array mounted into a rigid cylindrical baffle is capable of performing waveform estimation and parameter estimation in the sense of Chapter 5 in real acoustic environments. The remainder of this chapter is organized as follows. Section 6.1 discusses hardware, software, and algorithmic details utilized throughout the evaluation process. The measurements to be presented in Section 6.2 were performed in a real room allowing for varying degrees of reverberation.

6.1 System Details

This section presents a few details regarding the hardware and software used during the evaluation process. Also, some details on the implementation of the algorithms presented in Chapter 5 are given.

6.1.1 Hardware

As shown in Chapter 3, the two array design parameters, the radius of the circle, R, and the number of microphones, M, are intimately related to each other and to the desired frequency range of operation, $kR = 2\pi f R/c$. The lower the lowest desired frequency is chosen, the larger the radius of the circle has to be selected to be able to extract higher-order components of the wavefield with significant strength, see e.g. Fig. 3.5. On the other hand, a large radius requires a large number of microphones in order to decompose the wavefield into components yielding as little modal aliasing as possible at the highest desired frequency, see Section 3.1. However, a large number of microphones may be prohibitive for applications requiring real-time operation. Note that for many applications, a small array diameter may be desirable and may constitute a major advantage over traditional microphone array geometries, most notably linear arrays.

The main focus of attention that motivated the work presented in this book is scene analysis of an acoustic environment containing speech signals. Traditionally, the frequency bandwidth of operation is selected to equal the bandwidth that can be found in standard telephony, i.e. $f = 300 \ldots 3000$ Hz.

In order to adhere to these requirements and to the desire to keep the array as compact as possible, ten microphones were chosen to sample a continuous aperture of $R = 0.04$ m. In fact, this particular setup has already been used in previous chapters for the simulations.

Figure 6.1 shows the ten-sensor circular microphone array mounted into a finite-length rigid cylindrical baffle. The baffle can be removed for the purpose of comparing the results of a baffled to an unbaffled aperture. The omnidirec-

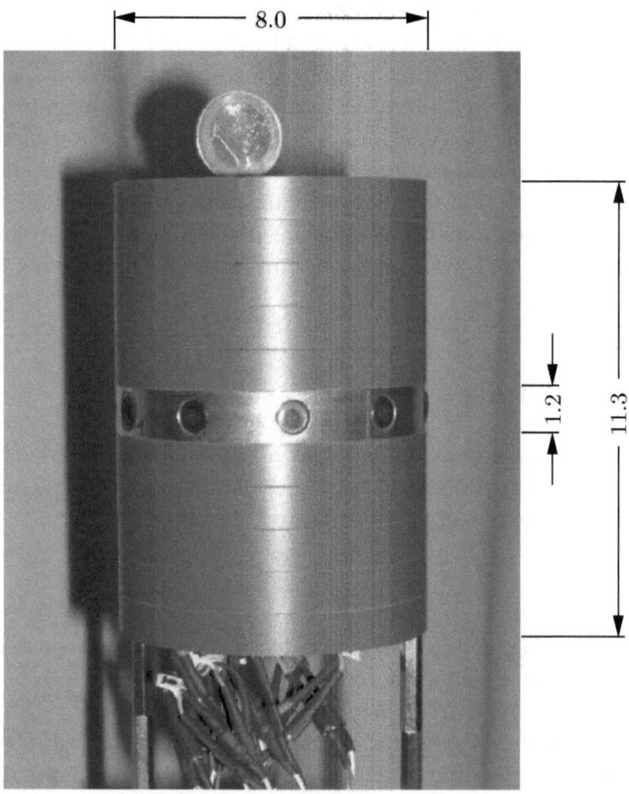

Fig. 6.1. Ten-sensor circular microphone array mounted into a rigid cylindrical baffle (dimensioning in centimeters)

tional quarter-inch electret microphone capsules are of type AKG 2112Z0017, which can be found in several commercially available AKG microphones [AKG]. The microphones used exhibit omnidirectional spatial characteristics since it is well known that omnidirectional microphones can be calibrated more effectively than directional microphones, such as dipoles and cardioids.

The multi-channel microphone preamplifier and AD-converter used is a non-commercial custom-made solution that also supplies the phantom power required by the electret capsules in a single unit.

The digitized microphone signals are presented to an RME HDSP9652 Hammerfall multi-channel sound-card system [RME] built into a modern PC.

The sampling rate used throughout the evaluation procedure was fixed to 48 kHz.

6.1.2 Algorithm Implementation

For the most part, the algorithms presented in Chapter 5 can be implemented directly in MATLAB [MAT] or C/C++ without any modifications. However, there are a few subtleties that need to be considered for best performance.

First, as opposed to a continuous-time formulation of the algorithms presented above, any digital computer implementation requires a discrete-time representation of the signals. In contrast to spatial sampling along curved lines, sampling of continuous time-domain signals is well understood and the effects of aliasing can be avoided almost completely by careful anti-aliasing filter design, e.g. [Lam79]. In the following, therefore, it is assumed that the discrete-time signals represent the continuous-time signals with sufficient accuracy. The transition from continuous-time signals and systems to their respective discrete-time counterparts is standard and can be found in any textbook on digital signal processing, e.g. [OS75]. Further details are, therefore, omitted here.

For an efficient implementation of EB-ESPRIT as described in Section 5.2.4 using circular microphone arrays, and in order to be able to estimate the modal correlation matrix given by Eq. (5.24), in a statistically significant way, the discrete-time signals are evaluated in a block-based fashion. In addition, the modal array correlation matrix is estimated recursively for being able to cope with non-stationary signals, such as speech.

The basis of all algorithms is the decomposition of a wavefield containing plane-waves into a set of eigen-solutions corresponding to the acoustic wave equation in circular or spherical coordinates. For circular apertures, the procedure that yields this result is given by Eq. (3.6), which corresponds to a spatial Fourier series expansion along the perimeter of the circular aperture. This spatial Fourier transform can be implemented efficiently by an FFT, applied to each block of signals along the microphones.

As insinuated in previous chapters, a prerequisite for good performance are microphones that are matched in magnitude and phase. FIR filters of order

128, serving as calibration filters, were obtained by measuring the impulse-responses of all ten microphones in a low-reverberation chamber and calculating filters that match each microphone to a reference sensor in magnitude and phase using a least-squares optimization procedure. A more efficient method of microphone calibration that does not require quasi-anechoic environments is described in [EKM03]. It has been found during the course of this investigation, however, that unless the microphones are operated in an adverse environment, i.e. where significant temperature/humidity changes must be expected, it is possible to keep the calibration filters fixed for at least a year without significant degradation in performance.

Finally, several algorithms presented in Chapter 5 require filters that compensate for the frequency-dependence of the eigenbeams, see e.g. Eq. (5.21). For the purpose of system validation, this non-trivial filter design problem was attacked by applying a standard least-squares optimization criterion which optimizes filters corresponding to the inverse frequency-dependence of the eigenbeams.

6.1.3 Software

The offline evaluation software is implemented in MATLAB while the real-time code is implemented in C/C++ under Linux. Since the algorithms make extensive use of matrix algebra and fast Fourier transforms, a large part of the algorithmic implementation utilizes the Intel Integrated Performance Primitives library [IPP] of version 4.0 for fast code execution. [PTVF02] provides a library offering routines for eigen-decomposition of matrices that has also been used in the real-time implementation of the algorithms.

The sound-card is accessed via the high-level JACK library [JAC] which itself is based on the low-level ALSA sound architecture [ALS].

A custom-designed MATLAB-based impulse-response measurement system was also used extensively during the evaluation process.

6.2 Evaluations

This section describes the measurements performed to validate the applicability of eigenbeam processing to acoustic scene analysis in terms of beamforming and source localization of possibly multiple acoustic sources.

6.2.1 Evaluation Setup

Figure 6.2 shows the setup utilized for evaluating the parameter estimation algorithms developed in Section 5.2. The microphone array, as shown in Fig. 6.1, was placed in a room with variable acoustics by means of retractable curtains, see Fig. 6.2. In order to keep the number of possible acoustic scenarios

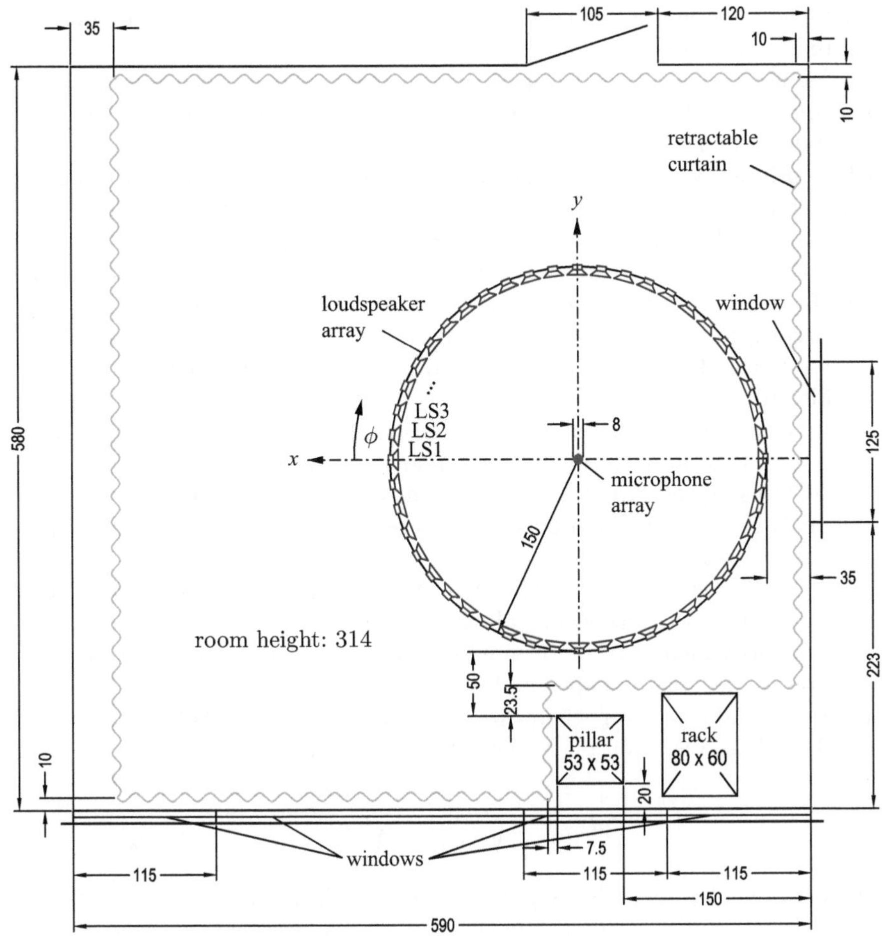

Fig. 6.2. Setup of the evaluation system in a real acoustic environment (dimensioning in centimeters)

tractable, only two extreme cases were considered, i.e. all curtains closed on the one hand and all curtains open on the other hand. In the following, the former scenario will be denoted as ACC (all curtains closed) and the latter one as ACO (all curtains open). By keeping all curtains closed, an approximate reverberation time was determined to be $T_{60,\mathrm{ACC}} \approx 200$ ms [1]. By opening all curtains, thereby exposing concrete walls and windows leading to hard reflec-

[1] The reverberation time T_{60} is defined as the time required for the spatial average of the sound energy to drop by 60 dB after the sound source has been switched off [Pie89].

tions, the reverberation time increases to $T_{60,ACO} \approx 400$ ms. The floor was covered by a thin carpet.

The microphone array was positioned in the center of a loudspeaker array of radius 1.5 m. This loudspeaker array, comprising 48 dual-diaphragm ELAC 301 loudspeakers [ELA], is part of a laboratory setup comprising a wave-field synthesis system (WFS) under investigation at the Chair of Multimedia Communications and Signal Processing, University of Erlangen-Nuremberg, Germany [SBR04, LMS]. Here, however, the loudspeakers were not jointly operated in WFS-mode but were driven individually to simulate a large number of possible source positions covering the full 2π field-of-view offered by the circular microphone array. Note that, for practical reasons, the coordinate system indicated in Fig. 6.2 is a mirror image of the coordinate system considered in the previous chapters.

For evaluation purposes, the impulse-responses of each individual loudspeaker with respect to all ten microphones were measured for both ACC and ACO using the MATLAB-based impulse-response measurement tool described above. Unreverberated signals were then convolved with the respective impulse-responses and superimposed with cylindrically isotropic noise of varying SNR, which can be generated by using the impulse-responses of all 48 loudspeakers with respect to all ten microphones.

A quantity that may be of interest when considering the effect of reverberation is the *critical distance*, $d_{c,T_{60}}$. It is defined as the distance between the source and a receiver where the direct sound level equals the reverberant sound level, and is given by [You59, Pie89, Bal91],

$$d_{c,T_{60}} = \sqrt{\frac{3 \ln 10}{2\pi c}} \cdot \sqrt{\frac{V_r Q_d}{T_{60}}}, \tag{6.1}$$

where V_r denotes the volume of the room and Q_d is the source's directional factor. It is assumed that the loudspeakers can be modeled as omnidirectionally radiating sources. This assumption is reasonable for dynamic loudspeakers mounted into closed cabinets when considering low to moderately high frequencies, as is done here. Therefore, $Q_d = 1$ [Pie89], and it follows for this particular setup with $c = 340$ m/s,

$$d_{c,T_{60}} \approx \begin{cases} 1.32\ m, \text{ for } T_{60,\mathrm{ACC}} \approx 200 \text{ ms} \\ 0.93\ m, \text{ for } T_{60,\mathrm{ACO}} \approx 400 \text{ ms} \end{cases}, \tag{6.2}$$

which is in both cases smaller than the distance between the microphones and the loudspeakers. In other words, the source is located far enough with respect to the sensors such that the reverberation takes full effect.

The performance regarding waveform estimation techniques, see Section 5.1, was evaluated using a slightly different setup than the one depicted in Fig. 6.2. In order to exclude the effects of multipath propagation, the array was placed into a low-reverberation chamber of size 2.5 m×2.7 m, yielding a reverberation time of about $T_{60} \approx 50$ ms. The microphone array was attached to a stepper motor that was controlled via MATLAB by the impulse-response measurement software described above. With the help of the stepper motor, impulse responses at a spatial resolution of 1° were recorded using a loudspeaker emitting maximum-length sequences (MLS). These impulse-responses were used to evaluate modal magnitude and spatial, i.e. multipole, responses. They were also used for calibrating the individual microphones.

6.2.2 Waveform Estimation

Figure 6.3(b) shows the measured *free-field* modal magnitude response of the baffled circular microphone array as depicted in Fig. 6.1. For reference, the *simulated* modal magnitude response for this array configuration is depicted in Fig 6.3(a). As can be seen, the measurements are in good agreement with the simulations, especially for moderately large to large values of kR. Modal aliasing can be identified in the simulations and in the measurements, for instance, as the region where the modal magnitude response of the fifth-order harmonic exceeds the lower-order modes for $kR > 3.5$. The difference between the simulations and the measurements with respect to small values of kR can be attributed to the fact that the microphones are not calibrated perfectly and that, starting from about -60 dB, the system's self-noise dominates over the decomposed eigenbeams for $n > 2$ at low frequencies. The slight deviations obvious in the zero-order mode throughout the frequency band are a result from the presence of boundary conditions at the endcaps of the finite-length cylinder that were not taken into account in Fig. 6.3(a). Note that Fig. 6.3(a) shows simulations of a circular microphone array mounted into an *infinite-length* cylindrical baffle. Since contributions from a farfield source impinging from $\vartheta = 0$, i.e. along the baffle's axis, only show up in the omnidirectional mode, components resulting from endcap boundary conditions only couple into the zero-order mode with significant strength. Note that this phenomenon may change in nature as the microphones are positioned closer to the endcaps of the cylindrical baffle.

Figure 6.4(b) shows the *measured* modal magnitude response of an *unbaffled* circular microphone array. The array, referred to as unbaffled here, is the one depicted in Fig. 6.1 where the finite cylinders had been removed prior to the measurements, leaving only the aluminum ring holding the microphones. However, this setup cannot be truly called 'unbaffled' since the aluminum ring has a width of 1.2 cm and this thin ring itself is an object subject to scattering phenomena. It has been shown in Section 3.1.1 that the effects of scattering are present even when the ratio of the length of the scattering object with respect to radius of the circular aperture is small. This effect is therefore also represented in Fig. 6.4(b). Note that Fig. 6.4(a) depicts the modal magnitude response of a truly unbaffled circular aperture, obtained via simulations.

In the following, only the baffled circular microphone array is considered for all further evaluations.

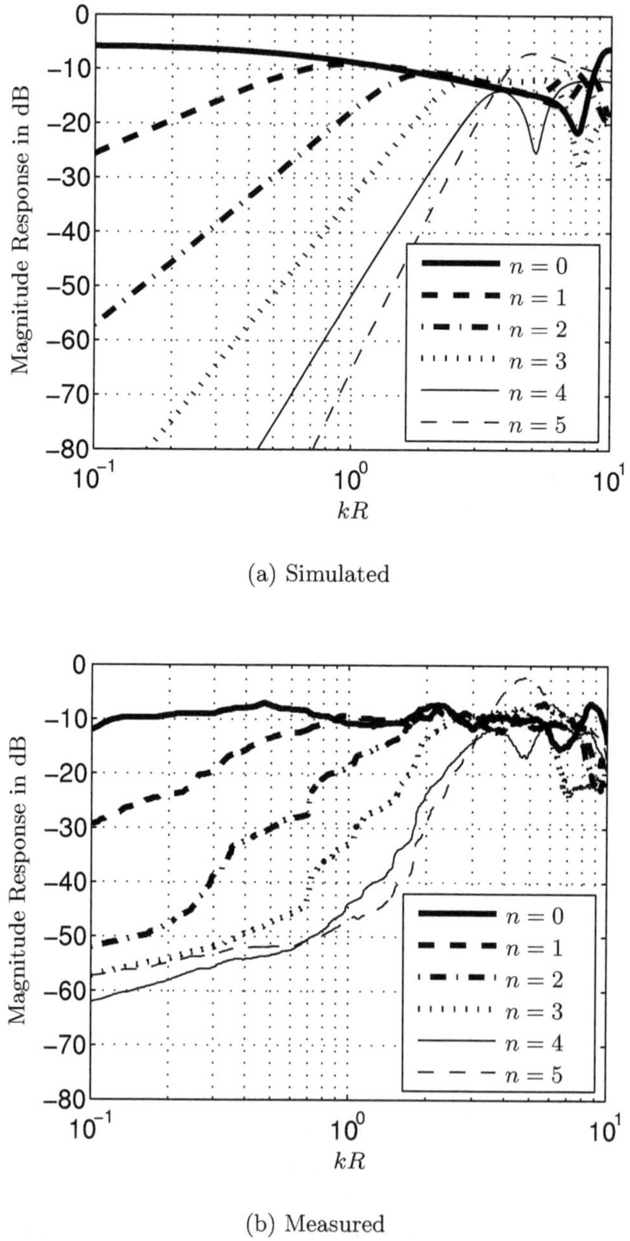

(a) Simulated

(b) Measured

Fig. 6.3. Modal magnitude response of a ten-sensor baffled circular microphone array

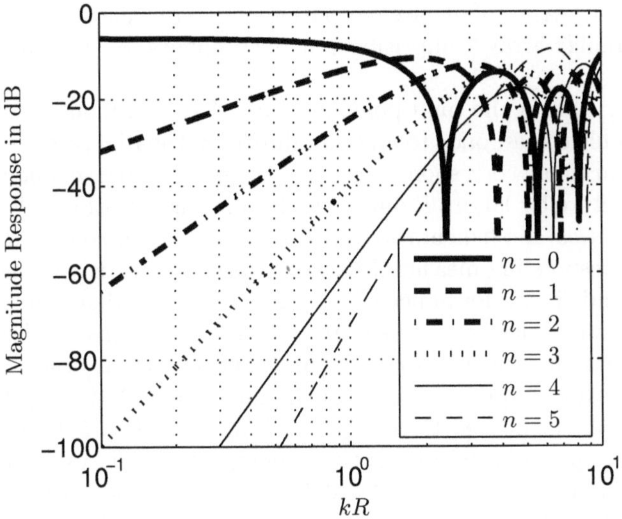

(a) Simulated

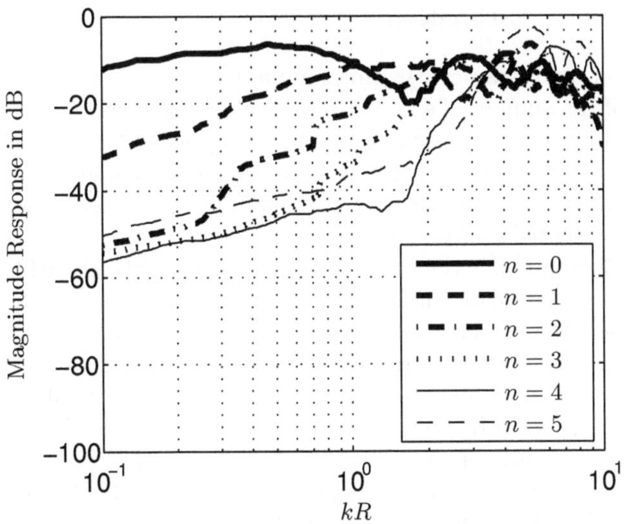

(b) Measured

Fig. 6.4. Modal magnitude response of a ten-sensor unbaffled circular microphone array

The simulated/measured spatial response with respect to a plane-wave with varying DOA, φ, of all circular harmonics – or eigenbeams – corresponding to a circular array comprising ten microphones are shown in Fig. 6.5.

Clearly visible is the multipole structure of the harmonics, i.e. the zero-order harmonic corresponds to an omni-directional spatial response, while the first-order harmonic corresponds to a dipole spatial response, and so on. Even the fine structure for large kR due to modal aliasing can still be demonstrated with the practical implementation. In summary, comparison of the simulated with the corresponding measured harmonics suggests satisfying agreement of the theory and the performance obtained by the realized system.

As an example for optimum beamforming in the sense of Section 5.1.2, Fig. 6.6 shows the unconstrained optimum beampattern design using measured eigenbeams of order up to and including $\mathcal{N} = 3$. Note that Fig. 6.6 can be, in principle, compared to the lower left figure in Fig. 5.2 for $0 \leq \varphi \leq \pi$. In order to keep the WNG higher than about -20 dB, the values for kR were chosen not to be smaller than $kR \approx 0.74$ which corresponds to a frequency of about $f = 1000$ Hz. The usable upper frequency bound is limited by modal aliasing and is chosen to be $kR \approx 2.2$ which corresponds to a frequency of $f = 3000$ Hz.

6.2.3 Parameter Estimation

In this section, the parameter estimation methods, i.e. source localization and estimation of the number of sources in a wavefield, as presented in Section 5.2, are validated by real data obtained by the setup shown in Fig. 6.2. Here, for consistency with Section 5.2 the excitation signal was chosen to be bandpass-filtered white noise. Examples involving more realistic excitation signals, i.e. speech, are presented in Appendix F.

Figure 6.7 exemplifies the performance of circular EB-ESPRIT for a single plane-wave in two acoustic environments, i.e. all curtains open (ACO) and all curtains closed (ACC), cf. Fig. 6.2. The source signal was emitted by loudspeaker LS 7 which was determined to correspond to $\varphi \approx 2\pi/9$ here. The frequency range was chosen to be $f = 300 \ldots 3000$ Hz. Applying the arguments brought forward in Section 3.1.5, i.e. considering the distance from the loudspeaker to the microphones and the frequency range, this source can be considered planar at the microphone positions. The wavefield was decomposed up to third order. Additionally, cylindrically diffuse noise, generated by all 48 loudspeakers, was added to each microphone covering an SNR range from -20 dB to 40 dB.

Figure 6.7(a) shows the estimator variance while Figure 6.7(b) shows the absolute value of the mean estimation error in both acoustic environments. It can be concluded that reliable estimates in both environments can be obtained for an SNR of down to about 3 dB. For even lower values of the SNR, the

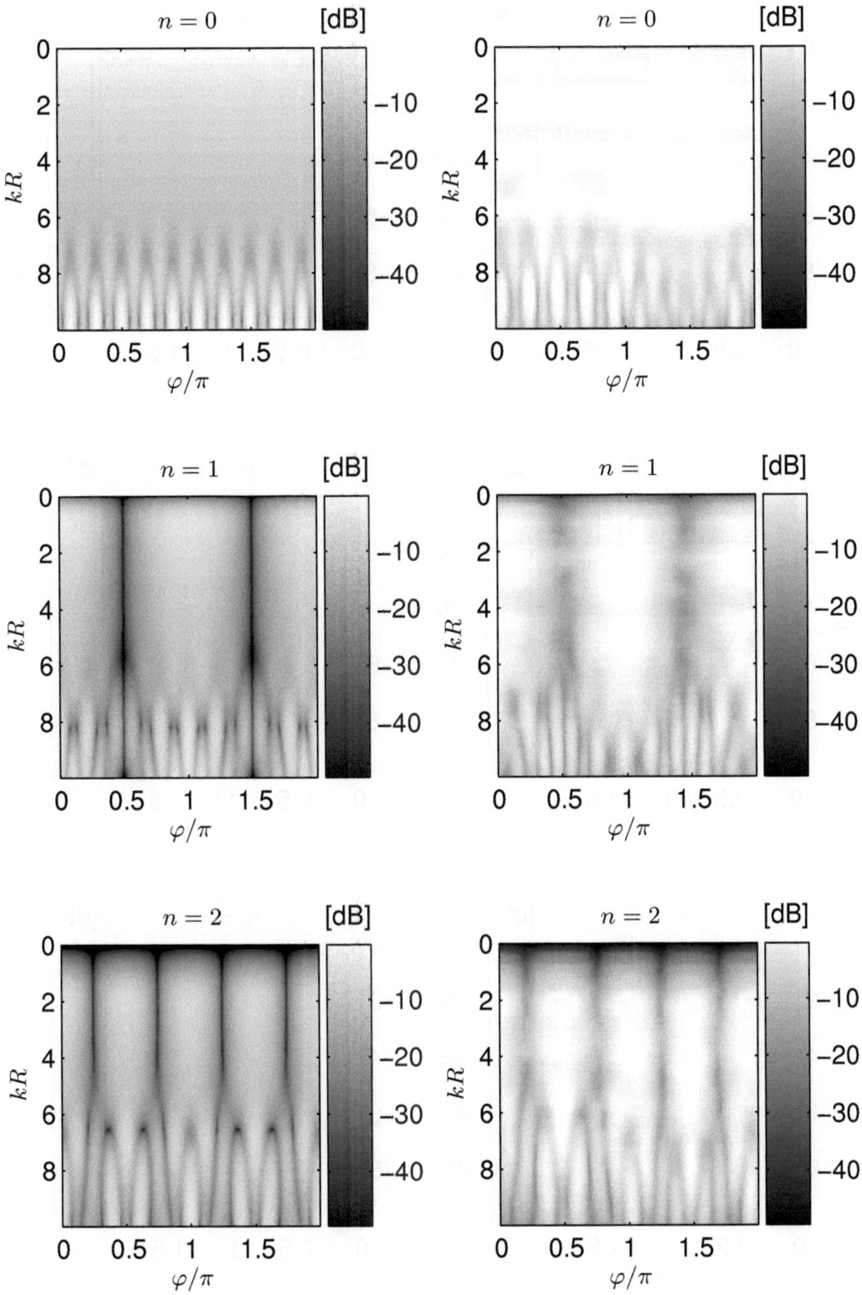

Fig. 6.5. Simulated (left column) and measured (right column) circular harmonics – or eigenbeams – using a baffled circular microphone array

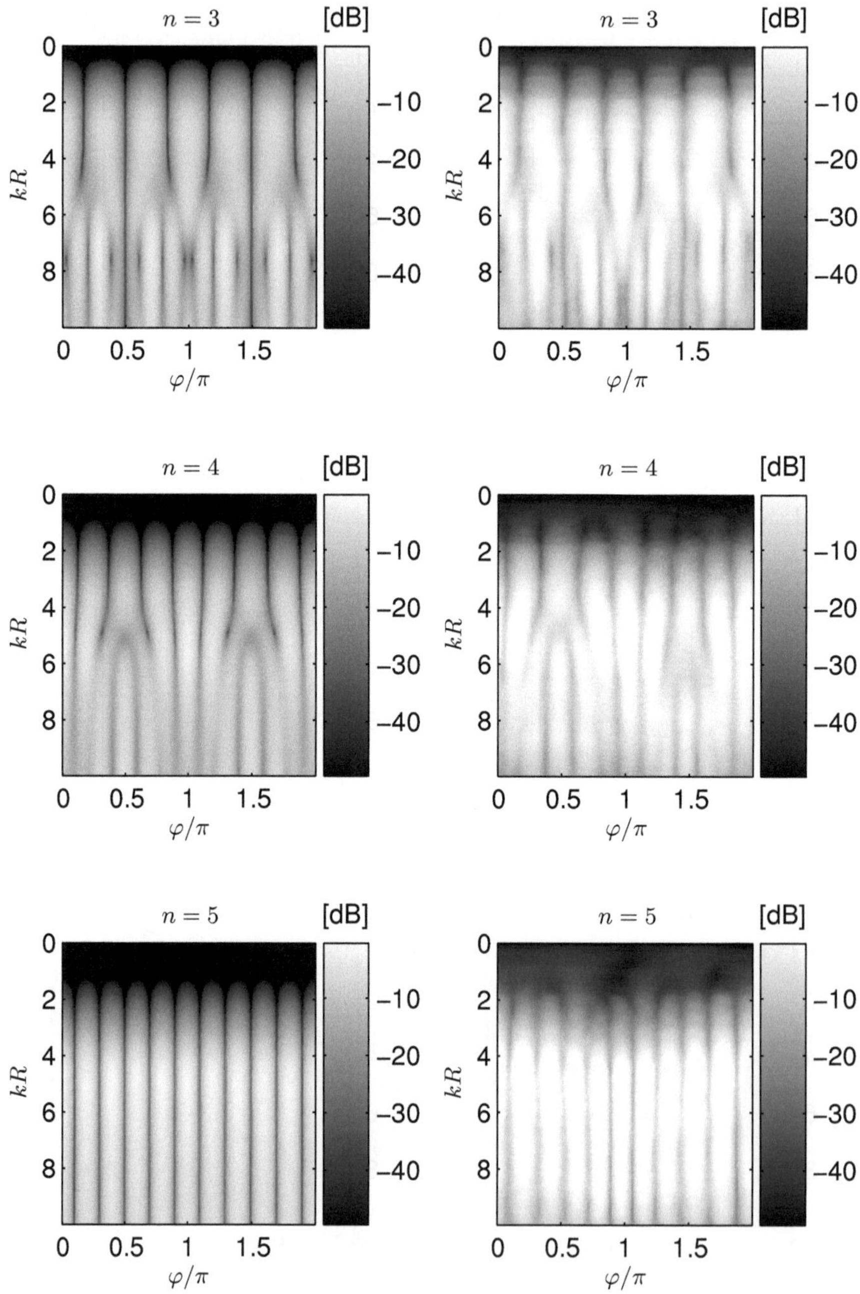

Fig. 6.5. (cont'd) Simulated (left column) and measured (right column) circular harmonics – or eigenbeams – using a baffled circular microphone array

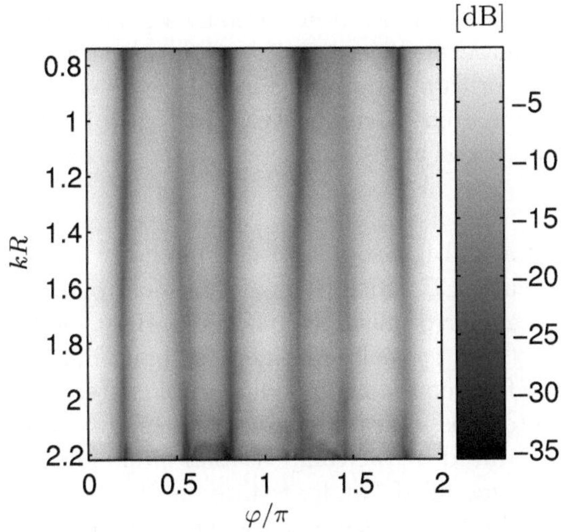

Fig. 6.6. Unconstrained optimum beampattern design using measured eigenbeams of order up to $\mathcal{N} = 3$

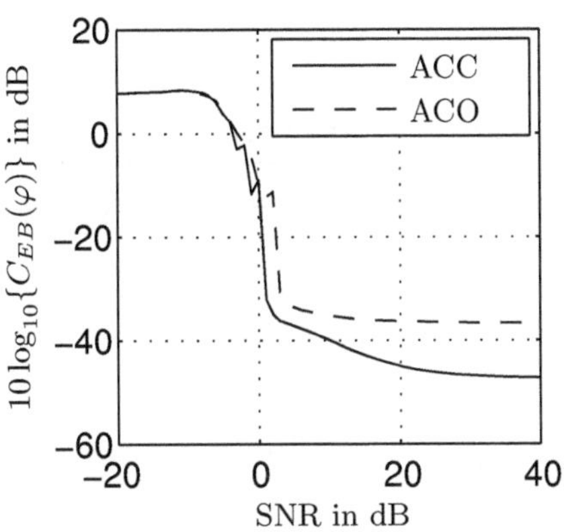

(a) Estimator variance in diffuse noise field

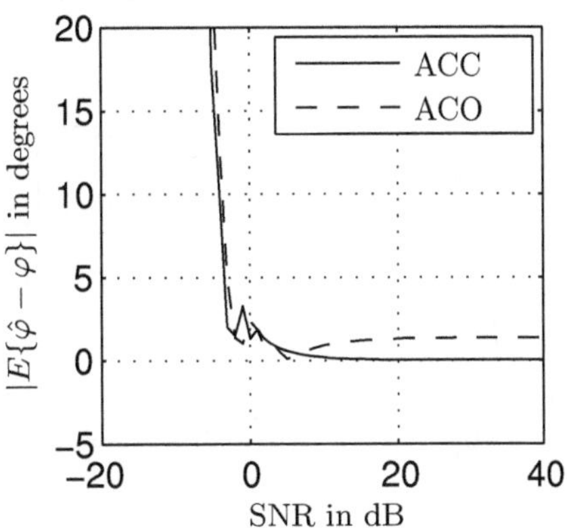

(b) Absolute value of mean of estimation error in diffuse noise field

Fig. 6.7. Performance of circular EB-ESPRIT for a single plane-wave ($\varphi \approx 2\pi/9$) in two acoustic environments, where $f = 300 \dots 3000$ Hz, $M = 10$, $\mathcal{N} = 3$, $K = 1024$, 200 trial runs threshold effect that was discussed in Section 4.4.3 makes the estimate unreliable. As expected, however, the estimation performance does degrade to some extend when the algorithm is evaluated in a more reverberated environment, cf. curve 'ACO' compared to curve 'ACC' in Fig. 6.7. More results involving other source positions and scenarios using speech signals can be found in Appendix F.

An example of a multiple-source scenario is given in Fig. 6.8, where the performance of circular EB-ESPRIT for two incident plane-waves, in terms of estimator variance and error mean, is considered. Here, loudspeakers LS 7 and LS 14 were used to emit spatially uncorrelated band-limited white noise signals. Note that the frequency corresponding to the lower edge of the passband of the bandpass filter was increased to keep the amplification of uncorrelated white noise as introduced by the compensation filters low, cf. Eq. (5.21). The introduction of the modified bandpass filter is the reason why the performance of EB-ESPRIT for two sources appears to be better than the performance of EB-ESPRIT for a single source, see Fig. 6.7. The bandpass filter 'removes' the reverberation adversely affecting the algorithm for $f < 1000$ Hz in the single-source case, where no compensation filters are required, cf. Section 5.2.4. Again, cylindrically isotropic noise was added to each microphone. In Fig. 6.8, the result corresponding to source 1, $\varphi_1 = 2\pi/9$, is shown. The

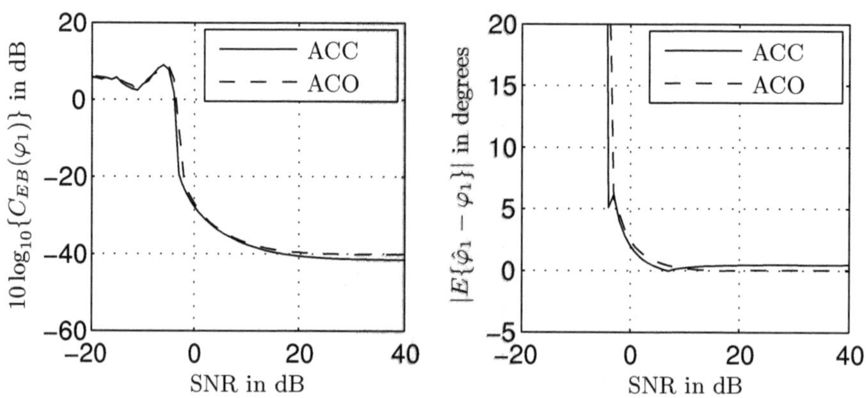

(a) Estimator variance in diffuse noise field

(b) Absolute value of mean of estimation error in diffuse noise field

Fig. 6.8. Performance of circular EB-ESPRIT for two incident plane-waves ($\boldsymbol{\varphi} \approx [2\pi/9, \pi/2]^T$, source one shown) in two acoustic environments, where $f = 1000\ldots3000$ Hz, $M = 10$, $\mathcal{N} = 3$, $K = 1024$, 200 trial runs respective results for source 2, $\varphi_2 = \pi/2$, are very similar and are, hence, not reproduced here. As in the single-source scenario, it follows that reliable estimates in both environments can be obtained for an SNR of down to about 3 dB. More results involving a wide variety of relative source positions are given in Appendix F.

Finally, some results on the estimation of the number of active sources in a real acoustic wavefield are given in Fig. 6.9. The number of sources was de-

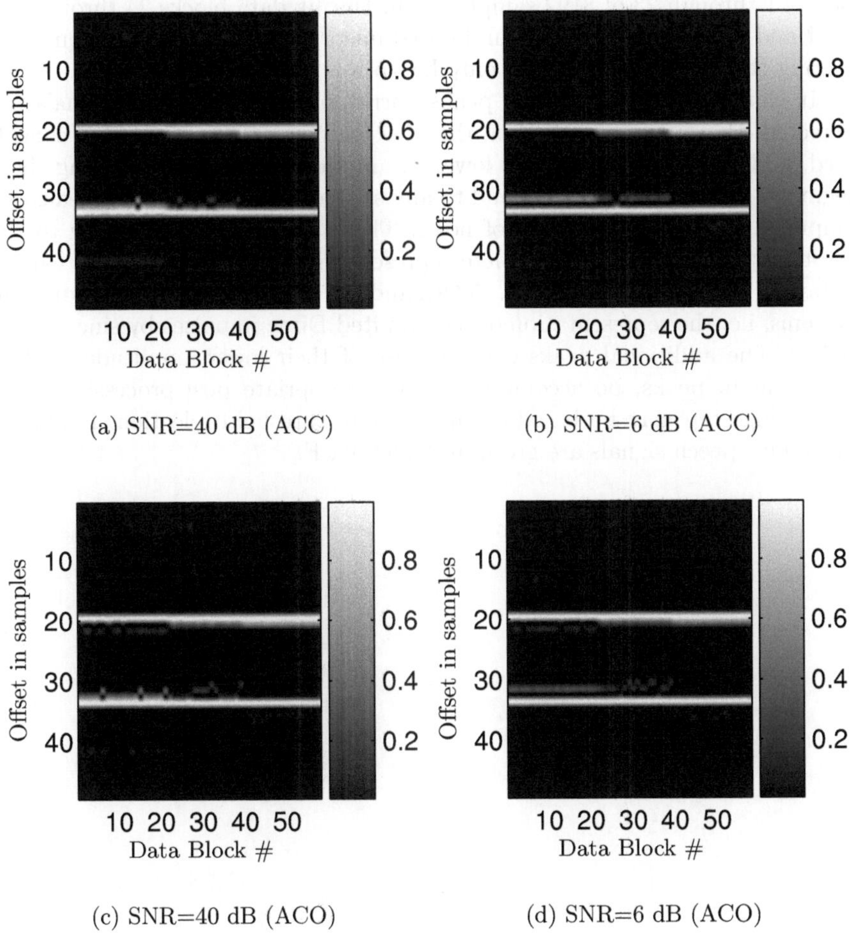

Fig. 6.9. Number of active noise signals in real acoustic environments using loudspeakers LS 1 and LS 25, where $f_{\max} = 7$ kHz, $M = 10$, $\mathcal{N} = 5$, data blocklength is 8192 samples

termined using the algorithm derived in Section 5.2.7 as follows. The impulse-response $h_2(t)$ in Fig. 5.9 was estimated by applying a frequency-domain LMS algorithm [Hay02]. The resulting impulse-response, represented by an 256-tap adaptive filter, was then examined for peaks in the range defined by the radius of the circular aperture. In Fig. 6.9, the significant range is indicated by two white horizontal lines, spanning 12 samples. All positions in the impulse-response that correspond to a local maximum were marked and weighted by the actual value of the peak. The image-plots represent the results for both acoustic environments as well as for high and moderately low SNR scenarios. Loudspeaker LS 1 was emitting a bandlimited white noise signal during data blocks 1 through 23 of 8192 samples each. During data blocks 24 through 40, LS 1 and LS 25 were playing bandlimited uncorrelated white noise sequences. During the last 20 data blocks, only LS 25 was active.

It can be seen that strong peaks corresponding to the two signals are detected. However, depending on the SNR and acoustic environment considered, additional spurious peaks lower in amplitude also appear during these segments which do not correspond to any source. In addition, these peaks get amplified due to the presence of noise. These peaks are present due to the error introduced by the incomplete representation of a wavefield that can be obtained by real systems, cf. Eq. (5.69), and the inherent properties of discrete systems, i.e. the representation of bandlimited Dirac impulses by sinc-shaped pulses. The additional peaks can, because of their lower amplitude relative to the main peaks, be accounted for by appropriate post-processing. This processing step may be based on appropriate threshold selection. Examples involving speech signals are given in Appendix F.

7

Summary and Conclusions

Modal array signal processing, as treated in this book, comprises waveform estimation tasks, i.e. beamforming, as well as parameter estimation tasks, i.e. source localization and the estimation of the number of active sources. Modal array signal processing has important applications in the areas of teleconferencing, surveillance, and seamless acoustic human-machine interfaces.

Waveform estimation is typically performed by applying the principles of DSB or FSB. Thereby, the microphone signals are aligned, i.e. steered, toward the desired acoustic source. Then, depending on the optimization procedure, scalars or FIR filters are applied independently to each microphone signal, see Section 4.3.

Acoustic source localization is often based on the evaluation of the TDOA between microphone pairs. A fundamental disadvantage of TDOA-based techniques, such as (weighted) GCC and AED, is the fact that the underlying signal model only considers a single impinging wave-front, see Section 4.4.2. A source localization technique that, in contrast to TDOA-based methods, is able to handle multiple simultaneously active sources are the subspace-based methods introduced in Section 4.4.3. However, a fundamental limitation of the subspace-based techniques, such as MUSIC and ESPRIT, is that they are formulated only for narrowband signals [Sch81, RK89].

All of the above mentioned techniques have in common that they do not take advantage of the underlying physics of wave propagation and scattering in two- or three-dimensional space.

The main motivation that ignited this work was the desire to establish links between classical acoustics and classical array signal processing to arrive at novel solutions for tasks associated with acoustic scene analysis, in particular for dealing with multiple wideband acoustic sources.

Classical acoustics, as considered in this book, is the science of acoustic wave propagation and wave scattering in two and three spatial dimensions, see Chapter 2. Wavefields propagating in two spatial dimensions can be used as reasonable models for propagating acoustic wavefields in closed rooms where ceiling and floor reflections are sufficiently attenuated. It was shown that a natural way of analyzing a two-dimensional wavefield is to decompose it into an orthogonal set of eigenfunctions of the acoustic wave equation in cylindrical coordinates, i.e. the circular harmonics or circular eigenbeams, utilizing circular apertures and circular microphone arrays. Similarly, a three-dimensional wavefield can be analyzed by its decomposition into an orthogonal set of eigenfunctions of the acoustic wave equation in spherical coordinates, i.e. the spherical harmonics or spherical eigenbeams, utilizing spherical apertures and spherical microphone arrays, see Chapter 3.

One of the key observations and results of this work, which builds the bridge between wavefield decomposition on the one hand and array signal processing on the other hand, is the fact that a single microphone of an array in the classical array signal processing sense, can be associated with a single eigenbeam resulting from the decomposition process. Note that *all* microphones comprising the circular or spherical array are used in the decomposition process to obtain the eigenbeams. Having established this link, parameter estimation algorithms known from classical array signal processing, i.e. MUSIC and ESPRIT, have been formulated by using the notion of eigenbeams, resulting in EB-MUSIC and EB-ESPRIT, see Chapter 5. A significant advantage of this new paradigm is the fact that eigenbeam-based parameter estimation methods are inherently frequency-independent. This previously undiscovered observation allows the application of parameter estimation methods, which are able to deal with multiple simultaneously active sources, for scenarios requiring wideband acoustic signal processing, such as scenarios involving speech signals.

It has been shown, see Chapter 6 and Appendix F, that the eigenbeam-based algorithms derived in Chapter 5 can be applied to wavefields that can be found in real-world acoustic environments. A real-time capable system comprising a ten-element circular microphone array mounted into a finite-length rigid cylindrical baffle has been constructed for the purpose of extensive performance evaluations. In order to demonstrate the frequency-independence of the parameter estimation algorithms, both bandpass-filtered white noise signals as well as speech signals were used as excitation signals. Robust parameter estimation results have been obtained for low- to moderately reverberated environments, with reverberation times up to $T_{60} = 400$ ms. This reverberation time can be expected in a typical office environment. The concept of eigenbeams applied to compact (baffled) circular sensor arrays may therefore be advantageously applied to applications aiming at teleconferencing or acoustic surveillance scenarios.

In summary, the main contributions of this work can be stated as follows.

- A rigorous investigation of *wavefield decomposition using circular and spherical apertures/arrays* has been conducted which finally led to the notion of *eigenbeams*.

- The novel concept of applying eigenbeams to the task of *localizing multiple wideband acoustic sources* has been described in some depth. As a result of the investigations, an inherent wideband formulation of classical subspace-based source localization techniques, most notably ESPRIT, has been found.
- The applicability of eigenbeams to the task of *estimating the number of active sources* in a wavefield has been shown.
- Performance evaluations of a *compact-size* and *real-time* capable array signal processing system based on the paradigm of eigenbeams have been presented that prove its applicability to scenarios which can be found in real acoustic environments.

In the presented initial system implementation, however, the performance must be expected to deteriorate in highly reverberated environments since the wavefield decomposition process, as detailed in Chapter 3, is based on the free-field propagation model introduced in Chapter 2. Future work on eigenbeam-based signal processing should address this problem by trying to find a framework that also explicitly models reverberation.

An interesting extension to the wavefield decomposition process using baffled circular apertures may be to stack individual circular arrays of several

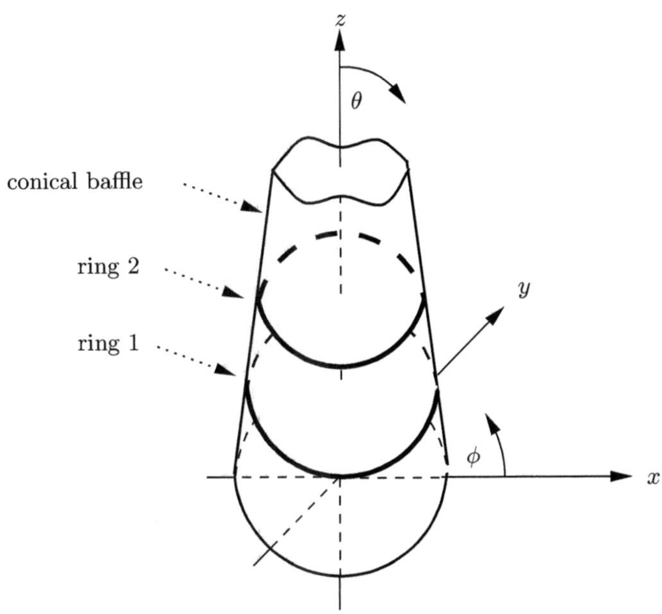

Fig. 7.1. Geometric model of two circular apertures mounted into a rigid cone

radii and mount them into a *rigid cone* [Kel05]. A possible geometry for two circular apertures mounted into a rigid cone is sketched in Fig. 7.1. A potential advantage of this arrangement is an extension of the usable bandwidth of the array, which is especially interesting at low frequencies.

Further work may also include,

- a performance evaluation of waveform estimation as well as parameter estimation methods utilizing baffled *spherical* microphone arrays in real acoustic environments,
- studies on whether other signal processing tasks often performed by multiple microphone configurations, such as (multichannel) acoustic echo cancellation [SK92, BBK03], (blind) source separation [BAK04, BAK05a], and noise reduction methods [EM84, Doc03] are applicable to the eigenbeam framework,
- examination of the applicability of eigenbeam-based methods to the analysis of room acoustics, for example by identifying the first relevant reflections in a room.
- examination of the applicability of eigenbeam-based methods to the analysis of the radiation behavior of vibrating bodies, for instance by considering a technique called *nearfield acoustic holography* (NAH) [Wil99].

A

Signal Transforms

A.1 One- and Multi-dimensional Fourier Transforms

In this book, the temporal Fourier transform pair is defined as [JD93],

$$G(\omega) = \int_{-\infty}^{\infty} g(t)\, e^{i\omega t}\, dt, \tag{A.1}$$

$$g(t) = \frac{1}{2\pi} \int_{-\infty}^{\infty} G(\omega)\, e^{-i\omega t}\, d\omega. \tag{A.2}$$

The spatial Fourier transform pair in d spatial coordinates is defined as,

$$G(\boldsymbol{k}) = \int_{\mathbb{R}^d} g(\boldsymbol{r}) e^{-i\boldsymbol{k}^T\boldsymbol{r}}\, d\boldsymbol{r}, \tag{A.3}$$

$$g(\boldsymbol{r}) = \frac{1}{(2\pi)^d} \int_{\mathbb{R}^d} G(\boldsymbol{k}) e^{i\boldsymbol{k}^T\boldsymbol{r}}\, d\boldsymbol{k}, \tag{A.4}$$

Note the different sign in the temporal and spatial Fourier transforms which is due to the notion of traveling waves [Wil99]. This convention is reasonably standard in many textbooks dealing with acoustics and apertures [MF53, JD93, Wil99].

A.2 Fourier Series

Assuming circular symmetry, the equivalent to the one-dimensional spatial Fourier transform is the Fourier series. A function on the unit circle in polar coordinates, $g(\phi)$, can be represented by a Fourier series as [Wil99],

$$g(\phi) = \sum_{n=-\infty}^{\infty} g_n \, e^{in\phi}, \tag{A.5}$$

where, due to the completeness relation of the exponential function, the Fourier coefficients, g_n, are given by,

$$g_n = \frac{1}{2\pi} \int_0^{2\pi} g(\phi) \, e^{-in\phi} \, d\phi. \tag{A.6}$$

A.3 Spherical Harmonics Transform

Any function $f(\theta, \phi)$ that is square-integrable over the unit sphere, i.e. $0 \leq \theta \leq \pi$ and $0 \leq \phi \leq 2\pi$, can be expanded into a series of spherical harmonics as [DJ94],

$$f(\theta, \phi) = \sqrt{4\pi} \sum_{n=0}^{\infty} \sum_{m=-n}^{n} f_{mn} Y_n^m(\theta, \phi), \tag{A.7}$$

where, due to the completeness relation of the spherical harmonics, the coefficients f_{mn} are given by,

$$f_{mn} = \frac{1}{\sqrt{4\pi}} \int_0^{2\pi} \int_0^{\pi} f(\theta, \phi) Y_n^m(\theta, \phi)^* \sin\theta d\theta d\phi. \tag{A.8}$$

A spherical harmonics transform on the sphere is the equivalent of the Fourier series of periodic functions.

B

Special Functions

This section defines and shows properties of special functions frequently used in this book. All results, unless noted otherwise, are adopted from [AS72] and [Wil99].

B.1 Bessel Functions

Bessel functions of the first kind, $J_n(z)$, Bessel functions of the second kind, $Y_n(z)$, and Bessel functions of the third kind, $H_n^{(1)}(z), H_n^{(2)}(z)$ are all solutions of Bessel's differential equation,

$$z^2 \frac{\partial^2 w}{\partial z^2} + z \frac{\partial w}{\partial z} + (z^2 - n^2) w = 0. \tag{B.1}$$

$Y_n(z)$ is called a Neumann function, and $H_n^{(1)}(z), H_n^{(2)}(z)$ are denoted as Hankel functions of the first and second kind, respectively. All functions are of order $n \in \mathbb{Z}$. The Bessel function (of the first kind) is implicitly defined as,

$$J_n(z) = \frac{1}{2\pi i^n} \int_0^{2\pi} e^{i(z \cos \phi + n\phi)} \, d\phi. \tag{B.2}$$

The Neumann function is defined as,

$$Y_n(z) = \frac{J_n(z) \cos(n\pi) - J_{-n}(z)}{\sin(n\pi)}. \tag{B.3}$$

Figure B.1 shows the Bessel functions of the first and second kind for $n = 0, 1, 2, 3$. The Hankel functions of the first and second kind are defined as,

$$H_n^{(1)}(z) = J_n(z) + iY_n(z), \tag{B.4}$$

and,

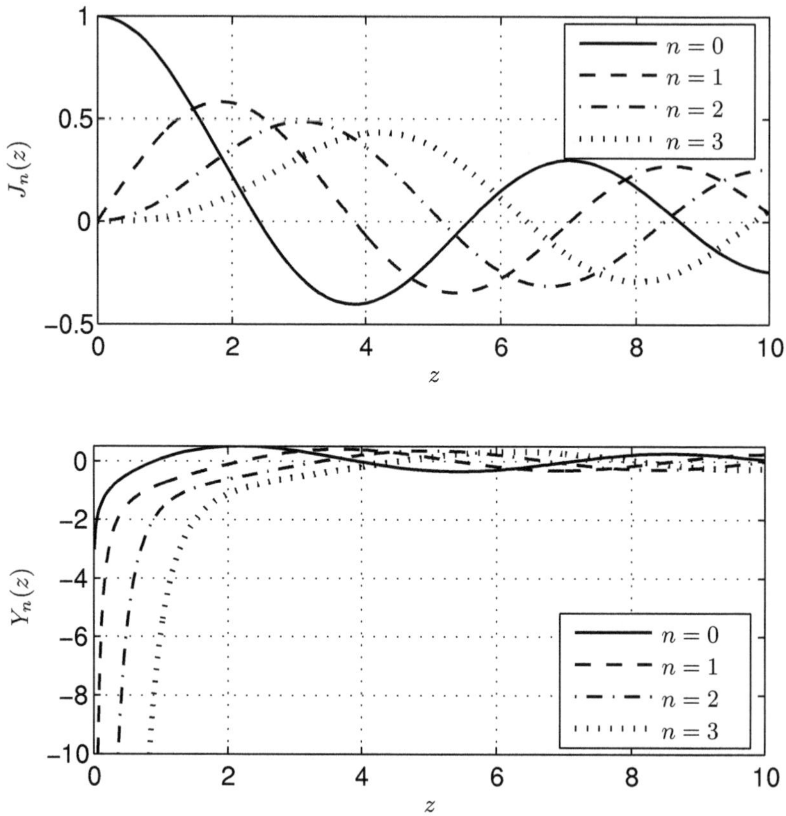

Fig. B.1. Bessel functions of the first and second kind for $n = 0, 1, 2, 3$

$$H_n^{(2)}(z) = J_n(z) - iY_n(z), \tag{B.5}$$

respectively. Further important relationships are,

$$J_{-n}(z) = (-1)^n J_n(z), \tag{B.6}$$
$$Y_{-n}(z) = (-1)^n Y_n(z), \tag{B.7}$$

and,

$$H_{-n}^{(1)}(z) = e^{in\pi} H_n^{(1)}(z), \tag{B.8}$$
$$H_{-n}^{(2)}(z) = e^{-in\pi} H_n^{(2)}(z). \tag{B.9}$$

Also,

$$J_n(-z) = (-1)^n J_n(z), \tag{B.10}$$
$$Y_n(-z) = (-1)^n Y_n(z), \tag{B.11}$$

and,

$$H_n^{(1)}(-z) = e^{in\pi} H_n^{(1)}(z), \tag{B.12}$$
$$H_n^{(2)}(-z) = e^{-in\pi} H_n^{(2)}(z). \tag{B.13}$$

The recurrence relations are,

$$K_{n-1}(z) + K_{n+1}(z) = \frac{2n}{z} K_n(z), \tag{B.14}$$
$$K_{n-1}(z) - K_{n+1}(z) = 2K_n'(z), \tag{B.15}$$
$$K_{n-1}(z) - \frac{n}{z} K_n(z) = K_n'(z), \tag{B.16}$$
$$-K_{n+1}(z) + \frac{n}{z} K_n(z) = K_n'(z), \tag{B.17}$$

where $K_\nu(z)$ denotes any of the above introduced Bessel functions or linear combinations thereof. Small argument approximations, $z \ll n$, for the Bessel and Neumann functions are, respectively,

$$J_n(z) \approx \frac{(z)^n}{2^n n!}, \tag{B.18}$$

$$Y_n(z) \approx -\frac{(n-1)!}{\pi} \left(\frac{2}{z}\right)^n, \tag{B.19}$$

A small argument approximation for the derivative of the Hankel function is,

$$H_n^{'(1,2)}(z) \approx \pm \frac{in!}{\pi \epsilon_n} \left(\frac{2}{z}\right)^{n+1}, \tag{B.20}$$

where,

$$\epsilon_n = \begin{cases} 1, \ n = 0 \\ 2, \ n > 0 \end{cases}. \tag{B.21}$$

The asymptotic behavior of the Bessel functions for large arguments are,

$$J_n(z) \approx \sqrt{\frac{2}{\pi z}} \cos(z - n\pi/2 - \pi/4), \tag{B.22}$$

$$Y_n(z) \approx \sqrt{\frac{2}{\pi z}} \sin(z - n\pi/2 - \pi/4), \tag{B.23}$$

$$H_n^{(1)}(z) \approx \sqrt{\frac{2}{\pi z}} e^{i(z - n\pi/2 - \pi/4)}, \tag{B.24}$$

$$H_n^{(2)}(z) \approx \sqrt{\frac{2}{\pi z}} e^{-i(z - n\pi/2 - \pi/4)}. \tag{B.25}$$

B.2 Spherical Bessel Functions

Spherical Bessel functions of the first kind, $j_n(z)$, spherical Bessel functions of the second kind, $y_n(z)$, and spherical Bessel functions of the third kind, i.e. spherical Hankel functions, $h_n^{(1)}(z), h_n^{(2)}(z)$ are all solutions of the differential equation,

$$z^2 \frac{\partial^2 w}{\partial z^2} + 2z \frac{\partial w}{\partial z} + [z^2 - n(n+1)]w = 0. \tag{B.26}$$

$y_n(z)$ is denoted as a spherical Neumann function, and $h_n^{(1)}(z), h_n^{(2)}(z)$ are spherical Hankel functions of the first and second kind, respectively. All functions are of integer order, n.

The spherical Bessel functions are related to the Bessel functions by the following equivalents,

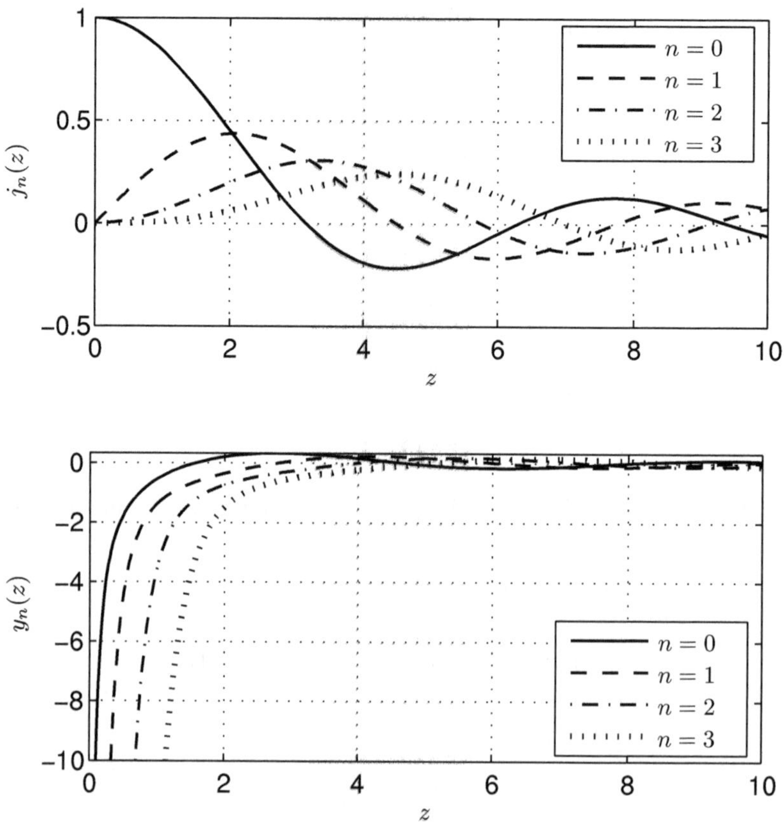

Fig. B.2. Spherical Bessel functions of the first and second kind for $n = 0, 1, 2, 3$

$$j_n(z) = \sqrt{\frac{\pi}{2z}} J_{n+1/2}(z), \tag{B.27}$$

$$y_n(z) = \sqrt{\frac{\pi}{2z}} Y_{n+1/2}(z), \tag{B.28}$$

$$h_n^{(1)}(z) = j_n(z) + iy_n(z) = \sqrt{\frac{\pi}{2z}}[J_{n+1/2}(z) + iY_{n+1/2}(z)], \tag{B.29}$$

$$h_n^{(1)}(z) = j_n(z) - iy_n(z) = \sqrt{\frac{\pi}{2z}}[J_{n+1/2}(z) - iY_{n+1/2}(z)]. \tag{B.30}$$

In contrast to Bessel functions, spherical Bessel functions can be written in terms of trigonometric functions as,

$$j_n(z) = (-z)^n \left(\frac{1}{z}\frac{d}{dz}\right)^n \left(\frac{\sin z}{z}\right), \tag{B.31}$$

$$y_n(z) = -(-z)^n \left(\frac{1}{z}\frac{d}{dz}\right)^n \left(\frac{\cos z}{z}\right), \tag{B.32}$$

$$h_n^{(1,2)}(z) = (-z)^n \left(\frac{1}{z}\frac{d}{dz}\right)^n \left(\frac{e^{\pm iz}}{\pm iz}\right). \tag{B.33}$$

Specifically for $n = 0$ on obtains,

$$j_0(z) = \frac{\sin z}{z} = \operatorname{sinc}(z), \tag{B.34}$$

$$h_0^{(1,2)}(z) = \frac{e^{\pm iz}}{\pm iz}. \tag{B.35}$$

Some recurrence relations are,

$$\frac{2n+1}{z} k_n(z) = k_{n-1}(z) + k_{n+1}(z), \tag{B.36}$$

$$k_n'(z) = k_{n-1}(z) - \frac{n+1}{z} k_n(z), \tag{B.37}$$

where $k_n(z)$ denotes any of the above introduced spherical Bessel functions or linear combinations thereof. Small argument approximations, $z \ll n$, for the spherical Bessel, Neumann, and Hankel functions are, respectively,

$$j_n(z) \approx \frac{(z)^n}{(2n+1)!!}, \tag{B.38}$$

$$y_n(z) \approx -\frac{(2n-1)!!}{(z)^{n+1}}, \tag{B.39}$$

and,

$$h_n^{(1,2)}(z) \approx \mp i \frac{(2n-1)!!}{z^{n+1}}. \tag{B.40}$$

A small argument approximation for the derivative of the spherical Hankel function is,

$$h_n^{\prime(1,2)}(z) \approx \pm i \frac{(n+1)(2n-1)!!}{z^{n+2}}, \tag{B.41}$$

where the double factorial is defined as,

$$n!! = \begin{cases} n \cdot (n-2) \cdot \cdots \cdot 3 \cdot 1, & n > 0 \text{ odd} \\ n \cdot (n-1) \cdot \cdots \cdot 4 \cdot 2, & n > 0 \text{ even} \\ 1, & n = -1, 0 \end{cases}. \tag{B.42}$$

For large arguments an asymptotic expression for the spherical Hankel function of the first kind can be formulated as,

$$h_n^{(1,2)}(z) \approx (\mp i)^{n+1} \frac{e^{\pm iz}}{z}. \tag{B.43}$$

B.3 Legendre Polynomials, Associated Legendre Functions, and Spherical Harmonics

Legendre polynomials are defined as,

$$P_n(z) = \frac{1}{2^n n!} \frac{d^n}{dz^n} (z^2 - 1)^n, \tag{B.44}$$

where $n \in \mathbb{Z}$. An important property of Legendre polynomials is that they are mutually orthogonal, i.e. they satisfy,

$$\int_{-1}^{1} P_n(z) P_m(z) \, dz = \frac{2}{2n+1} \delta_{mn}, \tag{B.45}$$

where $m \in \mathbb{Z}$ and δ_{mn} denotes the Kronecker symbol which is unity for $m = n$ and zero otherwise. The first few Legendre polynomials are,

$$P_0(z) = 1, \tag{B.46}$$

$$P_1(z) = z, \tag{B.47}$$

$$P_2(z) = (3z^2 - 1)/2, \tag{B.48}$$

$$P_3(z) = (5z^3 - 3z)/2, \tag{B.49}$$

which are often plotted as a function of θ where $z = \cos\theta$.

Associated Legendre functions are defined as,

$$P_n^m(z) = (-1)^m (1 - z^2)^{m/2} \frac{d^m}{dz^m} P_n(z). \tag{B.50}$$

For each m, the associated Legendre functions form a complete set of orthogonal functions as,

$$\int_{-1}^{1} P_{n'}^{m}(z) P_{n}^{m}(z)\, dz = \frac{2}{2n+1} \frac{(n+m)!}{(n-m)!} \delta_{n'n}. \tag{B.51}$$

The associated Legendre functions are one of the angle functions as part of the solution of the acoustic wave equation in spherical coordinates with $z = \cos\theta$. Associated Legendre functions of negative degree are defined as,

$$P_{n}^{-m}(z) = (-1)^{m} \frac{(n-m)!}{(n+m)!} P_{n}^{m}(z). \tag{B.52}$$

Table B.1. Associated Legendre functions for $n = 0(1)2$ and $m = -2(1)2$ where $z = \cos\theta$

$P_n^m(\cos\theta)$	$n=0$	$n=1$	$n=2$
m=-2	0	0	$(\sin^2\theta)/8$
m=-1	0	$(\sin\theta)/2$	$(\cos\theta\sin\theta)/2$
m=0	1	$\cos\theta$	$(1+3\cos 2\theta)/4$
m=1	0	$-\sin\theta$	$-3\cos\theta\,\sin\theta$
m=2	0	0	$3\sin^2\theta$

Solutions of the acoustic wave equation in spherical coordinates involve separate solutions for the angles θ and ϕ. These solutions can be conveniently combined by using the notation of spherical harmonics as,

$$Y_{n}^{m}(\theta,\phi) \triangleq \sqrt{\frac{(2n+1)}{4\pi} \frac{(n-m)!}{(n+m)!}}\, P_{n}^{m}(\cos\theta)\, e^{im\phi}. \tag{B.53}$$

They satisfy,

$$Y_{n}^{-m}(\theta,\phi) = (-1)^{m} Y_{n}^{m}(\theta,\phi)^{*}, \tag{B.54}$$

and,

$$Y_{n}^{0}(\theta,\phi) = \sqrt{\frac{2n+1}{4\pi}}\, P_{n}(\cos\theta). \tag{B.55}$$

The spherical harmonics are mutually orthonormal, i.e.,

$$\int_0^{2\pi}\int_0^{\pi} Y_n^m(\theta,\phi)Y_{n'}^{m'}(\theta,\phi)^* \sin\theta\, d\theta d\phi = \delta_{nn'}\delta_{mm'}. \tag{B.56}$$

Furthermore, they satisfy the following completeness relation,

$$\sum_{n=0}^{\infty}\sum_{m=-n}^{n} Y_n^m(\theta,\phi)Y_n^m(\theta',\phi')^* = \delta(\phi-\phi')\delta(\cos\theta-\cos\theta'). \tag{B.57}$$

The first few spherical harmonics are given in Tab. B.2.

Table B.2. Spherical harmonics for $n = 0(1)2$ and $m = -2(1)2$

$Y_n^m(\theta,\phi)$	$n=0$	$n=1$	$n=2$
m=-2	0	0	$3e^{-2i\phi}\sqrt{\frac{5}{96\pi}}\sin^2\theta$
m=-1	0	$e^{-i\phi}\sqrt{\frac{3}{8\pi}}\sin\theta$	$\frac{3}{2}e^{-i\phi}\sqrt{\frac{5}{24\pi}}\sin 2\theta$
m=0	$\sqrt{\frac{1}{4\pi}}$	$\sqrt{\frac{3}{4\pi}}\cos\theta$	$\sqrt{\frac{5}{16\pi}}(3\cos^2\theta-1)$
m=1	0	$-e^{i\phi}\sqrt{\frac{3}{8\pi}}\sin\theta$	$-\frac{3}{2}e^{i\phi}\sqrt{\frac{5}{24\pi}}\sin 2\theta$
m=2	0	0	$3e^{2i\phi}\sqrt{\frac{5}{96\pi}}\sin^2\theta$

Let (θ_1,ϕ_1) and (θ_2,ϕ_2) define two points on a sphere in a spherical coordinate system, separated by an angle γ, where,

$$\cos\gamma = \cos\theta_1\cos\theta_2 + \sin\theta_1\sin\theta_2\cos(\phi_1-\phi_2). \tag{B.58}$$

Then, an addition theorem for spherical harmonics can be formulated as [Arf85],

$$P_n(\cos\gamma) = \frac{4\pi}{2n+1}\sum_{m=-n}^{n} Y_n^m(\theta_1,\phi_1)Y_n^m(\theta_2,\phi_2)^*, \tag{B.59}$$

In particular, for $\theta_1=\theta_2$ and $\phi_1=\phi_2$, it follows that,

$$\sum_{m=-n}^{n} |Y_n^m(\theta,\phi)|^2 = \frac{2n+1}{4\pi}. \tag{B.60}$$

Recurrence relations for the spherical harmonics can be derived from a plethora of recurrence relations for associated Legendre functions. A few important ones used in this book are [Arf85],

$$\cos\theta\, Y_n^m(\theta,\phi) = \sqrt{\frac{(n-m+1)(n+m+1)}{(2n+1)(2n+3)}}\, Y_{n+1}^m(\theta,\phi) + \sqrt{\frac{(n-m)(n+m)}{(2n-1)(2n+1)}}\, Y_{n-1}^m(\theta,\phi), \tag{B.61a}$$

$$e^{i\phi}\sin\theta\, Y_n^m(\theta,\phi) = -\sqrt{\frac{(n+m+1)(n+m+2)}{(2n+1)(2n+3)}}\, Y_{n+1}^{m+1}(\theta,\phi) + \sqrt{\frac{(n-m)(n-m-1)}{(2n-1)(2n+1)}}\, Y_{n-1}^{m+1}(\theta,\phi), \tag{B.61b}$$

$$e^{-i\phi}\sin\theta\, Y_n^m(\theta,\phi) = \sqrt{\frac{(n-m+1)(n-m+2)}{(2n+1)(2n+3)}}\, Y_{n+1}^{m-1}(\theta,\phi) - \sqrt{\frac{(n+m)(n+m-1)}{(2n-1)(2n+1)}}\, Y_{n-1}^{m-1}(\theta,\phi). \tag{B.61c}$$

C

Microphone Arrays and Nearfield/Farfield Sources

Figure C.1 adumbrates a microphone array comprising M sensors placed into a wavefield due to nearfield and farfield sources. The microphone positioned at $\boldsymbol{r}_{\mathcal{M}_1}$ is chosen to be located at the origin of an, in general, arbitrary orthogonal coordinate system. For simplicity of presentation, a two-dimensional Cartesian coordinate system is depicted in Fig. C.1. It is further assumed that the origin of the coordinate system is also the phase center of the array, i.e.,

$$\sum_{\iota=1}^{M} \boldsymbol{r}_{\mathcal{M}_\iota} = \boldsymbol{0}, \tag{C.1}$$

which can be considered the center of gravity of the array [JD93].

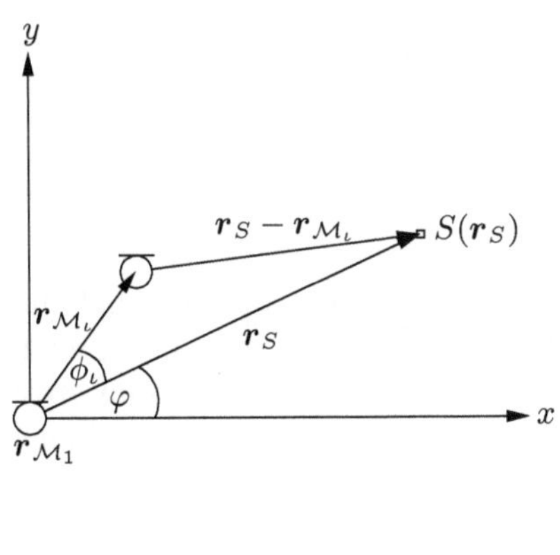

(a) nearfield

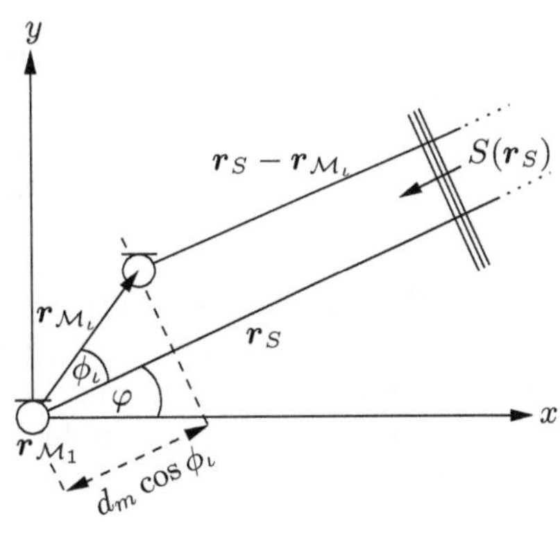

(b) farfield

Fig. C.1. Microphone array with respect to a nearfield/farfield source

The microphone positioned at $\boldsymbol{r}_{\mathcal{M}_\iota}$ is assumed to have the largest Euclidean distance from the origin of the coordinate system. Further symbols are defined as, $d_\iota = \|\boldsymbol{r}_{\mathcal{M}_\iota}\|, \tilde{d} = \|\boldsymbol{r}_S - \boldsymbol{r}_{\mathcal{M}_\iota}\|$, and $d = \|\boldsymbol{r}_S\|$.

In the farfield, cf. Fig. C.1(b), it follows that,

$$\tilde{d}_{\text{ff}} = d - d_\iota \cos\phi_\iota, \tag{C.2}$$

where the subscript 'ff' represents farfield conditions.

If the source is located in the nearfield, cf. Fig. C.1(a), it follows that,

$$\tilde{d}_{\text{nf}}^2 = d^2 + d_\iota^2 - 2dd_\iota \cos\phi_\iota, \tag{C.3}$$

where the subscript 'nf' represents nearfield conditions. Then,

$$\tilde{d}_{\text{nf}} = d\sqrt{1 - 2\frac{d_\iota}{d}\cos\phi_\iota + \left(\frac{d_\iota}{d}\right)^2}. \tag{C.4}$$

A Taylor series expansion of Eq. (C.4) yields,

$$\begin{aligned}\tilde{d}_{\text{nf}} = d\Bigg\{1 &+ \frac{1}{2}\left[2\frac{d_\iota}{d}\cos\phi_\iota - \left(\frac{d_\iota}{d}\right)^2\right] \\ &- \frac{1}{8}\left[2\frac{d_\iota}{d}\cos\phi_\iota - \left(\frac{d_\iota}{d}\right)^2\right]^2 + \mathcal{O}\left[\left(\frac{d_\iota}{d}\right)^3\right]\Bigg\}.\end{aligned} \tag{C.5}$$

Evaluating Eq. (C.5) gives, after neglecting all terms d_ι/d of power greater than 2,

$$\begin{aligned}\tilde{d}_{\text{nf}} &\approx d - d_\iota \cos\phi_\iota + \frac{d_\iota^2}{2d} - \frac{d_\iota^2}{2d}\cos^2\phi_\iota \\ &= \underbrace{d - d_\iota\cos\phi_\iota}_{\tilde{d}_{\text{ff}}} + \underbrace{\frac{d_\iota^2}{2d}\sin^2\phi_\iota}_{\epsilon_\iota}.\end{aligned} \tag{C.6}$$

Here, ϵ_ι denotes the approximate error made by assuming planar wavefield incidence while the impinging wavefield should be considered spherically spreading with respect to the microphone positioned at $\boldsymbol{r}_{\mathcal{M}_\iota}$. The maximum error occurs when $\phi_\iota = \pi/2$ and it follows for a general aperture with maximum linear dimension $D = 2d_\iota$ that,

$$\epsilon_{\iota_{max}} = \frac{D^2}{8d}. \tag{C.7}$$

Equation (C.7) can be used to finding a distance from the aperture where the maximum error is defined in terms of a phase error quantified as $1/\kappa$-th of a wavelength. Therefore, it follows from Eq. (C.7) that,

$$\hat{d} \triangleq d|_{\epsilon_{\iota max}=\lambda/\kappa} = \frac{D^2}{8\frac{\lambda}{\kappa}} = \frac{kD^2}{16\pi}\kappa, \qquad \kappa > 0. \tag{C.8}$$

In the antenna literature, κ is often chosen as 16 [Sko70], therefore allowing for a maximum phase error of 22.5°. For this special value, the region $d < \hat{d}$ is also called the Fresnel region, and the region $d > \hat{d}$ is called Fraunhofer region [Sko70].

By using Eq. (C.8), a transition from the nearfield to the farfield can thus be specified given a maximum allowable error. The distance from the aperture to the point where farfield conditions can be assumed with a given error can also be written as a ratio of the distance of the source to the most distant sensor of the aperture as,

$$\hat{d}_D \triangleq \frac{\hat{d}}{D} = \frac{\kappa}{16\pi} \cdot kD. \tag{C.9}$$

Figure C.2 shows $\hat{d}_D$ for the examples of $\kappa = 16$ and $\kappa = 36$, which correspond to a maximum phase error of 22.5° and 10°, respectively.

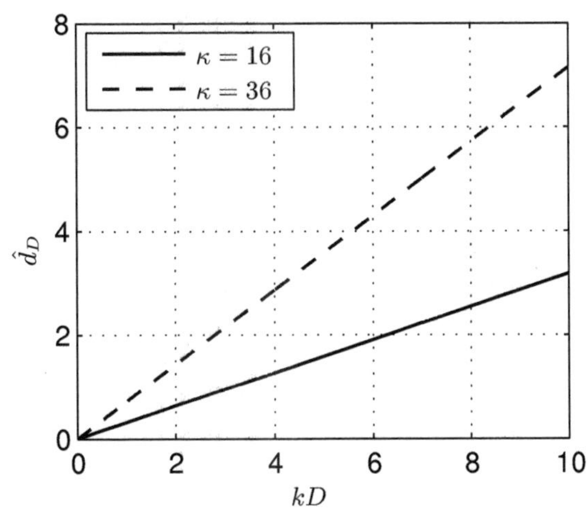

Fig. C.2. Relative distance from the source to an aperture with maximum linear dimension D

D

Eigenbeam-CRLB for a Single Source

In this section, the EB-CRLB for a single source, Eq. (5.29), is derived. First, it is noted that with $\underline{\boldsymbol{S}}_{\boldsymbol{SS}}(\omega) = \sigma_s^2(\omega)$, $\mathrm{SNR}(\omega) \triangleq \sigma_s^2(\omega)/\sigma_w^2$, the EB-CRLB follows from Eq. (4.66) as,

$$\begin{aligned}\underline{\boldsymbol{C}}_{EB-CRLB}(\varphi,\omega) = &\frac{1}{2K}\Bigg\{\mathrm{Re}\Bigg\{\mathrm{SNR}(\omega)\Bigg[\left(\underline{\boldsymbol{I}} + \underline{\check{\boldsymbol{V}}}_{\mathrm{EB}}^H(\varphi)\underline{\check{\boldsymbol{V}}}_{\mathrm{EB}}(\varphi)\mathrm{SNR}(\omega)\right)^{-1} \\ &\times \left(\underline{\check{\boldsymbol{V}}}_{\mathrm{EB}}^H(\varphi)\underline{\check{\boldsymbol{V}}}_{\mathrm{EB}}(\varphi)\mathrm{SNR}(\omega)\right)\Bigg] \odot \underline{\boldsymbol{H}}_{\mathrm{EB}}^T(\varphi)\Bigg\}\Bigg\}^{-1},\end{aligned} \tag{D.1}$$

where with Eqs. (4.67) and Eq. (5.28),

$$\underline{\boldsymbol{H}}_{\mathrm{EB}} = \underline{\boldsymbol{D}}_{\mathrm{EB}}^H \left[\underline{\boldsymbol{I}} - \underline{\check{\boldsymbol{V}}}_{\mathrm{EB}}(\underline{\check{\boldsymbol{V}}}_{\mathrm{EB}}^H\underline{\check{\boldsymbol{V}}}_{\mathrm{EB}})^{-1}\underline{\check{\boldsymbol{V}}}_{\mathrm{EB}}^H\right]\underline{\boldsymbol{D}}_{\mathrm{EB}}, \tag{D.2a}$$

$$\underline{\boldsymbol{D}}_{\mathrm{EB}} = \frac{\partial \underline{\check{\boldsymbol{V}}}_{\mathrm{EB}}}{\partial \boldsymbol{\varphi}} = \left[\frac{\partial \check{\boldsymbol{V}}_{\mathrm{EB}}(\varphi_1)}{\partial \varphi_1} \,\middle|\, \frac{\partial \check{\boldsymbol{V}}_{\mathrm{EB}}(\varphi_2)}{\partial \varphi_2} \,\middle|\, \cdots \,\middle|\, \frac{\partial \check{\boldsymbol{V}}_{\mathrm{EB}}(\varphi_I)}{\partial \varphi_I}\right] \tag{D.2b}$$

Note that, where possible, the dependencies on $\boldsymbol{\varphi}$ and ω have been omitted for better readability.

For a single source it follows with Eq. (5.21) that,

$$\underline{\check{\boldsymbol{V}}}_{\mathrm{EB}}(\boldsymbol{\varphi}) = \check{\boldsymbol{V}}_{\mathrm{EB}}(\varphi_1) \triangleq \check{\boldsymbol{V}}_{\mathrm{EB}}(\varphi) = [e^{i\mathcal{N}\varphi}, e^{i(\mathcal{N}-1)\varphi}, \ldots, 1, e^{-i\varphi}, \ldots, e^{-i\mathcal{N}\varphi}]^T, \tag{D.3}$$

and,

$$\underline{\boldsymbol{D}}_{\mathrm{EB}}(\boldsymbol{\varphi}) = \boldsymbol{D}_{\mathrm{EB}}(\varphi_1) \triangleq \boldsymbol{D}_{\mathrm{EB}}(\varphi) = i\underline{\boldsymbol{M}}\check{\boldsymbol{V}}_{\mathrm{EB}}(\varphi), \tag{D.4}$$

where,

$$\underline{\boldsymbol{M}} \triangleq \mathrm{diag}\{\mathcal{N}, (\mathcal{N}-1), \ldots, 0, -1, \ldots, -\mathcal{N}\}. \tag{D.5}$$

It can readily be verified that,

$$\boldsymbol{V}_{\mathrm{EB}}^{H}(\varphi)\boldsymbol{V}_{\mathrm{EB}}(\varphi) = \sum_{n=-\mathcal{N}}^{\mathcal{N}} n = 2\mathcal{N}+1, \tag{D.6}$$

$$\boldsymbol{D}_{\mathrm{EB}}^{H}(\varphi)\boldsymbol{D}_{\mathrm{EB}}(\varphi) = \sum_{n=-\mathcal{N}}^{\mathcal{N}} n^2 = \frac{1}{3}\mathcal{N}(\mathcal{N}+1)(2\mathcal{N}+1), \tag{D.7}$$

and,

$$\boldsymbol{V}_{\mathrm{EB}}^{H}(\varphi)\boldsymbol{D}_{\mathrm{EB}}(\varphi) = 0. \tag{D.8}$$

Then, by using these expressions and after omitting the angular dependency for notational convenience, Eq. (D.2a) becomes,

$$\begin{aligned}
\underline{\boldsymbol{H}}_{\mathrm{EB}} = \boldsymbol{H}_{\mathrm{EB}} &= -i\left(\underline{\boldsymbol{M}}\breve{\boldsymbol{V}}_{\mathrm{EB}}\right)^{H}\left(\boldsymbol{I} - \frac{\breve{\boldsymbol{V}}_{\mathrm{EB}}\breve{\boldsymbol{V}}_{\mathrm{EB}}^{H}}{2\mathcal{N}+1}\right) i\underline{\boldsymbol{M}}\breve{\boldsymbol{V}}_{\mathrm{EB}} \\
&= \breve{\boldsymbol{V}}_{\mathrm{EB}}^{H}\underline{\boldsymbol{M}\boldsymbol{M}}\breve{\boldsymbol{V}}_{\mathrm{EB}} - \frac{1}{2\mathcal{N}+1}\left(\breve{\boldsymbol{V}}_{\mathrm{EB}}^{H}\underline{\boldsymbol{M}}\breve{\boldsymbol{V}}_{\mathrm{EB}}\right)^{2} \\
&= \boldsymbol{D}_{\mathrm{EB}}^{H}\boldsymbol{D}_{\mathrm{EB}} + \frac{i}{2\mathcal{N}+1}\left(\breve{\boldsymbol{V}}_{\mathrm{EB}}^{H}\boldsymbol{D}_{\mathrm{EB}}\right)^{2} \\
&= \frac{1}{3}\mathcal{N}(\mathcal{N}+1)(2\mathcal{N}+1).
\end{aligned} \tag{D.9}$$

With this result, the EB-CRLB, Eq. (D.1), for a single source can finally be written as,

$$C_{\text{EB-CRLB}}(\omega) = 3\left[2K\mathcal{N}(\mathcal{N}+1)(2\mathcal{N}+1)^2\frac{\mathrm{SNR}^2(\omega)}{1+(2\mathcal{N}+1)\mathrm{SNR}(\omega)}\right]^{-1}. \tag{D.10}$$

E

Frequency-Independence of EB-ESPRIT for a Single Source

In this section, a rigorous proof of the claim stated in Section 5.2.4 – i.e. that the DOA estimation algorithm for a single impinging plane-wave is inherently frequency-independent – is given.

The proof starts with a Lemma, reviewing a known property of skew-symmetric matrices.

Lemma E.1 *Let* $\underline{\boldsymbol{A}} \in \mathbb{R}^{M\times M}$ *be a real-valued skew-symmetric matrix, satisfying* $\underline{\boldsymbol{A}}^T = -\underline{\boldsymbol{A}}$, *and* $\boldsymbol{B} \in \mathbb{R}^{M\times 1}$ *a real-valued vector, then the quadratic form* f *vanishes, i.e.* $f \triangleq \boldsymbol{B}^T\underline{\boldsymbol{A}}\boldsymbol{B} = 0 \in \mathbb{R}^{M\times 1}$.

Proof. $f = f^T = (\boldsymbol{B}^T\underline{\boldsymbol{A}}\boldsymbol{B})^T = \boldsymbol{B}^T\underline{\boldsymbol{A}}^T\boldsymbol{B} = -\boldsymbol{B}^T\underline{\boldsymbol{A}}\boldsymbol{B} = -f \;\leftrightarrow\; f = 0.$ □

In the following, the notation is taken directly from the text in Section 4.4.3 and Section 5.2.4. In particular, $\hat{\boldsymbol{U}}_S \in \mathbb{R}^{M\times 1}$ denotes the estimated signal subspace, i.e. the eigenvector corresponding to the largest eigenvalue of the modal correlation matrix. Here, subscripted square matrices containing the symbol 'M' define their dimension. $\underline{\boldsymbol{O}}_{M-1}^{1/2}$ is defined by Eq. (5.31), $\underline{\boldsymbol{Q}}$ is an unitary matrix defined by Eq. (4.114) of dimension $M-1$ or M. $\underline{\boldsymbol{K}}$ is a selection matrix defined by Eq. (4.117), and $\underline{\bar{\boldsymbol{I}}}$ denotes the exchange matrix given by Eq. (4.106). Furthermore,

$$\underline{\tilde{\boldsymbol{G}}}_1 \triangleq \mathrm{Re}\{\underline{\boldsymbol{Q}}_{M-1}^H\underline{\boldsymbol{O}}_{M-1}^{1/2}\underline{\boldsymbol{K}}\underline{\boldsymbol{Q}}_M\} \triangleq \mathrm{Re}\{\underline{\boldsymbol{Q}}_{M-1}^H\underline{\tilde{\boldsymbol{K}}}\underline{\boldsymbol{Q}}_M\} \triangleq \frac{1}{2}(\underline{\tilde{\boldsymbol{Z}}} + \underline{\tilde{\boldsymbol{Z}}}^*), \tag{E.1}$$

$$\underline{\tilde{\boldsymbol{G}}}_2 \triangleq \mathrm{Im}\{\underline{\boldsymbol{Q}}_{M-1}^H\underline{\boldsymbol{O}}_{M-1}^{1/2}\underline{\boldsymbol{K}}\underline{\boldsymbol{Q}}_M\} \triangleq \mathrm{Im}\{\underline{\boldsymbol{Q}}_{M-1}^H\underline{\tilde{\boldsymbol{K}}}\underline{\boldsymbol{Q}}_M\} \triangleq \frac{1}{2i}(\underline{\tilde{\boldsymbol{Z}}} - \underline{\tilde{\boldsymbol{Z}}}^*), \tag{E.2}$$

and,

$$\underline{\boldsymbol{G}}_1 \triangleq \mathrm{Re}\{\underline{\boldsymbol{Q}}_{M-1}^H\underline{\boldsymbol{K}}\underline{\boldsymbol{Q}}_M\} \triangleq \frac{1}{2}(\underline{\boldsymbol{Z}} + \underline{\boldsymbol{Z}}^*), \tag{E.3}$$

$$\underline{\boldsymbol{G}}_2 \triangleq \mathrm{Im}\{\underline{\boldsymbol{Q}}_{M-1}^H\underline{\boldsymbol{K}}\underline{\boldsymbol{Q}}_M\} \triangleq \frac{1}{2i}(\underline{\boldsymbol{Z}} - \underline{\boldsymbol{Z}}^*), \tag{E.4}$$

where,

$$\underline{\boldsymbol{Z}} \triangleq \underline{\boldsymbol{Q}}_{M-1}^{H} \underline{\boldsymbol{K}} \underline{\boldsymbol{Q}}_{M}, \tag{E.5}$$

$$\underline{\tilde{\boldsymbol{Z}}} \triangleq \underline{\boldsymbol{Q}}_{M-1}^{H} \underline{\tilde{\boldsymbol{K}}} \underline{\boldsymbol{Q}}_{M}, \tag{E.6}$$

and,

$$\underline{\tilde{\boldsymbol{K}}} \triangleq \underline{\boldsymbol{O}}_{M-1}^{1/2} \underline{\boldsymbol{K}}. \tag{E.7}$$

Theorem E.2 *The eigenvectors of a matrix defined by Eq. (5.46),*

$$\underline{\tilde{\boldsymbol{U}}} \triangleq \begin{bmatrix} \hat{\boldsymbol{U}}_S^T \underline{\tilde{\boldsymbol{G}}}_1^T \underline{\tilde{\boldsymbol{G}}}_1 \hat{\boldsymbol{U}}_S & \hat{\boldsymbol{U}}_S^T \underline{\tilde{\boldsymbol{G}}}_1^T \underline{\tilde{\boldsymbol{G}}}_2 \hat{\boldsymbol{U}}_S \\ \hat{\boldsymbol{U}}_S^T \underline{\tilde{\boldsymbol{G}}}_2^T \underline{\tilde{\boldsymbol{G}}}_1 \hat{\boldsymbol{U}}_S & \hat{\boldsymbol{U}}_S^T \underline{\tilde{\boldsymbol{G}}}_2^T \underline{\tilde{\boldsymbol{G}}}_2 \hat{\boldsymbol{U}}_S \end{bmatrix}, \tag{E.8}$$

are identical to the eigenvectors of the matrix,

$$\underline{\boldsymbol{U}} \triangleq \begin{bmatrix} \hat{\boldsymbol{U}}_S^T \underline{\boldsymbol{G}}_1^T \underline{\boldsymbol{G}}_1 \hat{\boldsymbol{U}}_S & \hat{\boldsymbol{U}}_S^T \underline{\boldsymbol{G}}_1^T \underline{\boldsymbol{G}}_2 \hat{\boldsymbol{U}}_S \\ \hat{\boldsymbol{U}}_S^T \underline{\boldsymbol{G}}_2^T \underline{\boldsymbol{G}}_1 \hat{\boldsymbol{U}}_S & \hat{\boldsymbol{U}}_S^T \underline{\boldsymbol{G}}_2^T \underline{\boldsymbol{G}}_2 \hat{\boldsymbol{U}}_S \end{bmatrix}, \tag{E.9}$$

Proof. A sufficient condition for this statement is that,

$$\underline{\boldsymbol{U}} - \underline{\tilde{\boldsymbol{U}}} = \begin{bmatrix} \hat{\boldsymbol{U}}_S^T \underline{\boldsymbol{A}}_1 \hat{\boldsymbol{U}}_S & \hat{\boldsymbol{U}}_S^T \underline{\boldsymbol{A}}_2 \hat{\boldsymbol{U}}_S \\ \hat{\boldsymbol{U}}_S^T \underline{\boldsymbol{A}}_3 \hat{\boldsymbol{U}}_S & \hat{\boldsymbol{U}}_S^T \underline{\boldsymbol{A}}_4 \hat{\boldsymbol{U}}_S \end{bmatrix} \stackrel{!}{=} \varkappa \underline{\boldsymbol{I}}, \quad \varkappa \in \mathbb{R} \tag{E.10}$$

where,

$$\underline{\boldsymbol{A}}_1 \triangleq \underline{\boldsymbol{G}}_1^T \underline{\boldsymbol{G}}_1 - \underline{\tilde{\boldsymbol{G}}}_1^T \underline{\tilde{\boldsymbol{G}}}_1 \qquad \underline{\boldsymbol{A}}_2 \triangleq \underline{\boldsymbol{G}}_1^T \underline{\boldsymbol{G}}_2 - \underline{\tilde{\boldsymbol{G}}}_1^T \underline{\tilde{\boldsymbol{G}}}_2$$

$$\underline{\boldsymbol{A}}_3 \triangleq \underline{\boldsymbol{G}}_2^T \underline{\boldsymbol{G}}_1 - \underline{\tilde{\boldsymbol{G}}}_2^T \underline{\tilde{\boldsymbol{G}}}_1 \qquad \underline{\boldsymbol{A}}_4 \triangleq \underline{\boldsymbol{G}}_2^T \underline{\boldsymbol{G}}_2 - \underline{\tilde{\boldsymbol{G}}}_2^T \underline{\tilde{\boldsymbol{G}}}_2.$$

Theorem E.3 *A sufficient condition for the main diagonal entries in Eq. (E.10) to be equal is* $\underline{\boldsymbol{A}}_1 = \underline{\boldsymbol{A}}_4$.

Proof.

$$\begin{aligned}
\underline{\boldsymbol{A}}_1 - \underline{\boldsymbol{A}}_4 &= \frac{1}{4}\left[(\underline{\boldsymbol{Z}}+\underline{\boldsymbol{Z}}^*)^T(\underline{\boldsymbol{Z}}+\underline{\boldsymbol{Z}}^*) - (\underline{\tilde{\boldsymbol{Z}}}+\underline{\tilde{\boldsymbol{Z}}}^*)^T(\underline{\tilde{\boldsymbol{Z}}}+\underline{\tilde{\boldsymbol{Z}}}^*)\right] \\
&\quad + \frac{1}{4}\left[(\underline{\boldsymbol{Z}}-\underline{\boldsymbol{Z}}^*)^T(\underline{\boldsymbol{Z}}-\underline{\boldsymbol{Z}}^*) - (\underline{\tilde{\boldsymbol{Z}}}-\underline{\tilde{\boldsymbol{Z}}}^*)^T(\underline{\tilde{\boldsymbol{Z}}}-\underline{\tilde{\boldsymbol{Z}}}^*)\right] \\
&= \frac{1}{2}\left[\underline{\boldsymbol{Z}}^T\underline{\boldsymbol{Z}} + \underline{\boldsymbol{Z}}^H\underline{\boldsymbol{Z}}^* - \underline{\tilde{\boldsymbol{Z}}}^T\underline{\tilde{\boldsymbol{Z}}} - \underline{\tilde{\boldsymbol{Z}}}^H\underline{\tilde{\boldsymbol{Z}}}^*\right] \\
&= \frac{1}{2}\Big[(\underline{\boldsymbol{Q}}_{M-1}^H\underline{\boldsymbol{K}}\underline{\boldsymbol{Q}}_M)^T(\underline{\boldsymbol{Q}}_{M-1}^H\underline{\boldsymbol{K}}\underline{\boldsymbol{Q}}_M) \\
&\quad + (\underline{\boldsymbol{Q}}_{M-1}^H\underline{\boldsymbol{K}}\underline{\boldsymbol{Q}}_M)^H(\underline{\boldsymbol{Q}}_{M-1}^H\underline{\boldsymbol{K}}\underline{\boldsymbol{Q}}_M)^* \\
&\quad - (\underline{\boldsymbol{Q}}_{M-1}^H\underline{\tilde{\boldsymbol{K}}}\underline{\boldsymbol{Q}}_M)^T(\underline{\boldsymbol{Q}}_{M-1}^H\underline{\tilde{\boldsymbol{K}}}\underline{\boldsymbol{Q}}_M) \\
&\quad - (\underline{\boldsymbol{Q}}_{M-1}^H\underline{\tilde{\boldsymbol{K}}}\underline{\boldsymbol{Q}}_M)^H(\underline{\boldsymbol{Q}}_{M-1}^H\underline{\tilde{\boldsymbol{K}}}\underline{\boldsymbol{Q}}_M)^*\Big] \\
&= \frac{1}{2}\Big[\underline{\boldsymbol{Q}}_M^T\underline{\boldsymbol{K}}^T\underbrace{\underline{\boldsymbol{Q}}_{M-1}^*\underline{\boldsymbol{Q}}_{M-1}^H}_{\underline{\bar{\boldsymbol{I}}}_{M-1}\triangleq\underline{\bar{\boldsymbol{I}}}}\underline{\boldsymbol{K}}\underline{\boldsymbol{Q}}_M + \underline{\boldsymbol{Q}}_M^H\underline{\boldsymbol{K}}^T\underbrace{\underline{\boldsymbol{Q}}_{M-1}\underline{\boldsymbol{Q}}_{M-1}^T}_{\underline{\bar{\boldsymbol{I}}}}\underline{\boldsymbol{K}}\underline{\boldsymbol{Q}}_M^* \\
&\quad - \underline{\boldsymbol{Q}}_M^T\underline{\tilde{\boldsymbol{K}}}^T\underline{\boldsymbol{Q}}_{M-1}^*\underline{\boldsymbol{Q}}_{M-1}^H\underline{\tilde{\boldsymbol{K}}}\underline{\boldsymbol{Q}}_M - \underline{\boldsymbol{Q}}_M^H\underline{\tilde{\boldsymbol{K}}}^T\underline{\boldsymbol{Q}}_{M-1}\underline{\boldsymbol{Q}}_{M-1}^T\underline{\tilde{\boldsymbol{K}}}\underline{\boldsymbol{Q}}_M^*\Big] \\
&= \frac{1}{2}\left[\underline{\boldsymbol{Q}}_M^T(\underline{\boldsymbol{K}}^T\underline{\bar{\boldsymbol{I}}}\underline{\boldsymbol{K}} - \underline{\tilde{\boldsymbol{K}}}^T\underline{\bar{\boldsymbol{I}}}\underline{\tilde{\boldsymbol{K}}})\underline{\boldsymbol{Q}}_M + \underline{\boldsymbol{Q}}_M^H(\underline{\boldsymbol{K}}^T\underline{\bar{\boldsymbol{I}}}\underline{\boldsymbol{K}} - \underline{\tilde{\boldsymbol{K}}}^T\underline{\bar{\boldsymbol{I}}}\underline{\tilde{\boldsymbol{K}}})\underline{\boldsymbol{Q}}_M^*\right] \\
&= 0,
\end{aligned}$$

since,

$$\begin{aligned}
\underline{\tilde{\boldsymbol{K}}}^T\underline{\bar{\boldsymbol{I}}}_{M-1}\underline{\tilde{\boldsymbol{K}}} &= (\underline{\boldsymbol{O}}_{M-1}^{1/2}\underline{\boldsymbol{K}})^T\underline{\bar{\boldsymbol{I}}}_{M-1}\underline{\boldsymbol{O}}_{M-1}^{1/2}\underline{\boldsymbol{K}} = \underline{\boldsymbol{K}}^T\underline{\boldsymbol{O}}_{M-1}^{1/2^T}\underline{\bar{\boldsymbol{I}}}_{M-1}\underline{\boldsymbol{O}}_{M-1}^{1/2}\underline{\boldsymbol{K}} \\
&= \overline{\underline{\boldsymbol{K}}^T\underline{\bar{\boldsymbol{I}}}_{M-1}\underline{\boldsymbol{K}}}',
\end{aligned} \tag{E.11}$$

where in the last step Eq. (5.32) has been used. □

Theorem E.4 *A sufficient condition that the anti-diagonal entries in the matrix defined by Eq. (E.10) are zero is that* $\underline{\boldsymbol{A}}_2$ *and* $\underline{\boldsymbol{A}}_3$ *are skew-symmetric.*

Proof.

$$\begin{aligned}
\underline{\boldsymbol{A}}_2 + \underline{\boldsymbol{A}}_2^T &= \frac{1}{4i}\left[(\underline{\boldsymbol{Z}}+\underline{\boldsymbol{Z}}^*)^T(\underline{\boldsymbol{Z}}-\underline{\boldsymbol{Z}}^*) - (\underline{\tilde{\boldsymbol{Z}}}+\underline{\tilde{\boldsymbol{Z}}}^*)^T(\underline{\tilde{\boldsymbol{Z}}}-\underline{\tilde{\boldsymbol{Z}}}^*)\right] \\
&\quad + \frac{1}{4i}\left[(\underline{\boldsymbol{Z}}+\underline{\boldsymbol{Z}}^*)^T(\underline{\boldsymbol{Z}}-\underline{\boldsymbol{Z}}^*) - (\underline{\tilde{\boldsymbol{Z}}}+\underline{\tilde{\boldsymbol{Z}}}^*)^T(\underline{\tilde{\boldsymbol{Z}}}-\underline{\tilde{\boldsymbol{Z}}}^*)\right]^T \\
&= \frac{1}{2i}\left[\underline{\boldsymbol{Z}}^T\underline{\boldsymbol{Z}} - \underline{\boldsymbol{Z}}^H\underline{\boldsymbol{Z}}^* - \underline{\tilde{\boldsymbol{Z}}}^T\underline{\tilde{\boldsymbol{Z}}} + \underline{\tilde{\boldsymbol{Z}}}^H\underline{\tilde{\boldsymbol{Z}}}^*\right] \\
&= \frac{1}{2}\left[\underline{\boldsymbol{Q}}_M^T(\underline{\boldsymbol{K}}^T\underline{\bar{\boldsymbol{I}}}\underline{\boldsymbol{K}} - \underline{\tilde{\boldsymbol{K}}}^T\underline{\bar{\boldsymbol{I}}}\underline{\tilde{\boldsymbol{K}}})\underline{\boldsymbol{Q}}_M - \underline{\boldsymbol{Q}}_M^H(\underline{\boldsymbol{K}}^T\underline{\bar{\boldsymbol{I}}}\underline{\boldsymbol{K}} - \underline{\tilde{\boldsymbol{K}}}^T\underline{\bar{\boldsymbol{I}}}\underline{\tilde{\boldsymbol{K}}})\underline{\boldsymbol{Q}}_M^*\right] \\
&= \underline{\boldsymbol{0}},
\end{aligned}$$

where intermediate steps were taken from the proof following Corollary E.3. Similarly,

$$\begin{aligned}\underline{\boldsymbol{A}}_3 + \underline{\boldsymbol{A}}_3^T &= \frac{1}{4i}\left[(\underline{\boldsymbol{Z}} - \underline{\boldsymbol{Z}}^*)^T(\underline{\boldsymbol{Z}} + \underline{\boldsymbol{Z}}^*) - (\tilde{\underline{\boldsymbol{Z}}} - \tilde{\underline{\boldsymbol{Z}}}^*)^T(\tilde{\underline{\boldsymbol{Z}}} + \tilde{\underline{\boldsymbol{Z}}}^*)\right] \\ &+ \frac{1}{4i}\left[(\underline{\boldsymbol{Z}} + \underline{\boldsymbol{Z}}^*)^T(\underline{\boldsymbol{Z}} - \underline{\boldsymbol{Z}}^*) - (\tilde{\underline{\boldsymbol{Z}}} + \tilde{\underline{\boldsymbol{Z}}}^*)^T(\tilde{\underline{\boldsymbol{Z}}} - \tilde{\underline{\boldsymbol{Z}}}^*)\right]^T \\ &= \frac{1}{2i}\left[\underline{\boldsymbol{Z}}^T\underline{\boldsymbol{Z}} - \underline{\boldsymbol{Z}}^H\underline{\boldsymbol{Z}}^* - \tilde{\underline{\boldsymbol{Z}}}^T\tilde{\underline{\boldsymbol{Z}}} + \tilde{\underline{\boldsymbol{Z}}}^H\tilde{\underline{\boldsymbol{Z}}}^*\right] \\ &= \underline{\boldsymbol{A}}_2 + \underline{\boldsymbol{A}}_2^T = \underline{\boldsymbol{0}}.\end{aligned} \tag{E.12}$$

□

This concludes the proof of Theorem E.2.

F

A Practical Acoustic Scene Analysis System – Further Results

In this appendix, further results on the evaluation of an acoustic scene analysis system based on wavefield decomposition methods using a real microphone array are presented. Several of the experiments involve speech as excitation signals. The speech signals used are plotted in Fig. F.1. On the left-hand side, a four-second segment of male speech, used for single-source scenarios, is depicted. The right-hand side of Fig. F.1 depicts two speech signals used for the evaluation of the NOS (number of sources) estimator.

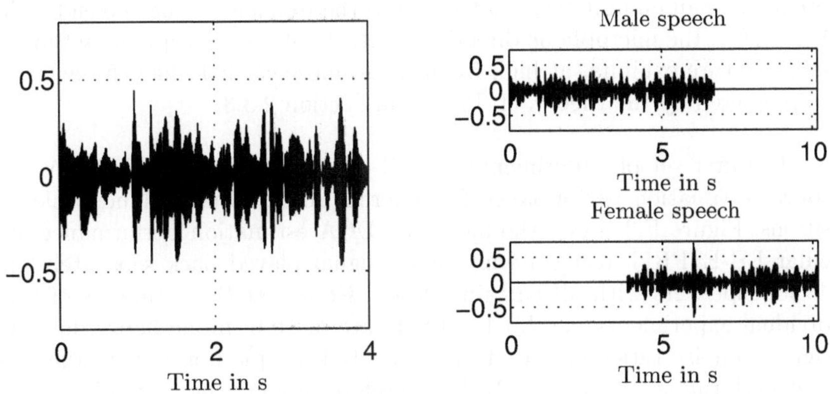

Fig. F.1. Speech signals used for several of the system evaluation procedures

A result that is very interesting from the standpoint of both WE and PE methods is the measured directivity pattern of an individual microphone set into either the baffled or the unbaffled aperture, which is shown in Fig. F.2. Note that the microphones used for the hardware setup are quarter-inch om-

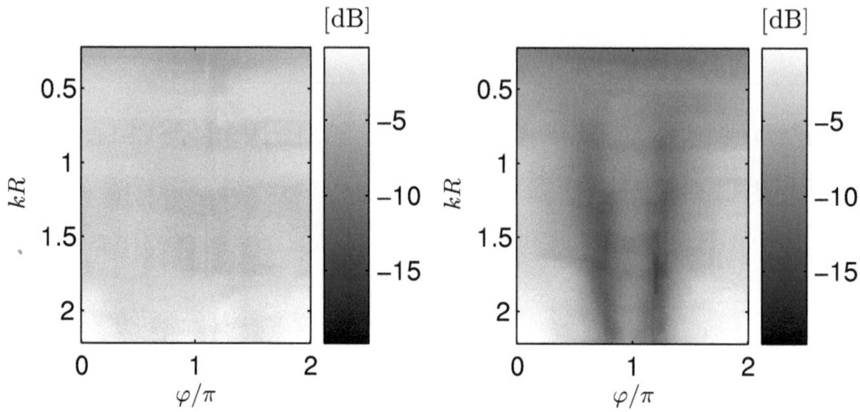

(a) Microphone set in unbaffled aperture

(b) Microphone set in baffled aperture

Fig. F.2. Directivity pattern of a single microphone for $kR = 0.22 \ldots 2.2$

nidirectional AKG 2112Z0017 electret capsules. While the microphone directivity pattern of the microphone set in an unbaffled aperture roughly approximates an omnidirectional pattern in the frequency range considered, see Fig F.2(a), the microphone directivity pattern of the microphone set in a baffled aperture exhibits, at high frequencies, more cardioid-like characteristics, as predicted by the theory, cf. Fig 2.9 in Section 2.3.3.

The first set of experiments for PE evaluates the system's DOA single-source estimation performance for different source, that is loudspeaker, positions. Figure F.3 shows the measured DOA estimation performance of circular EB-ESPRIT using a white noise signal played back separately on all 48 loudspeakers with all curtains closed. Note that the estimates of the individual experiments, i.e. LS 1... LS 48 are plotted in one figure, for several signal-to-noise ratios based on cylindrically isotropic noise. Note the consistent performance for all 48 DOAs, therefore proving the system's capability to cover the array's entire field-of-view, i.e. 2π.

Figure F.4 shows the same scenario in the ACO acoustic environment. Note that, putting the slight overall degradation due to stronger multipath aside, consistent estimation performance for all 48 DOAs can be observed in this environment, too.

For the evaluations represented by Fig. F.5 and Fig. F.6, the white noise excitation signal was replaced by the speech signal shown on the left-hand side in Fig. F.1. Figure F.5 depicts the ACC scenario while Fig. F.6 deals

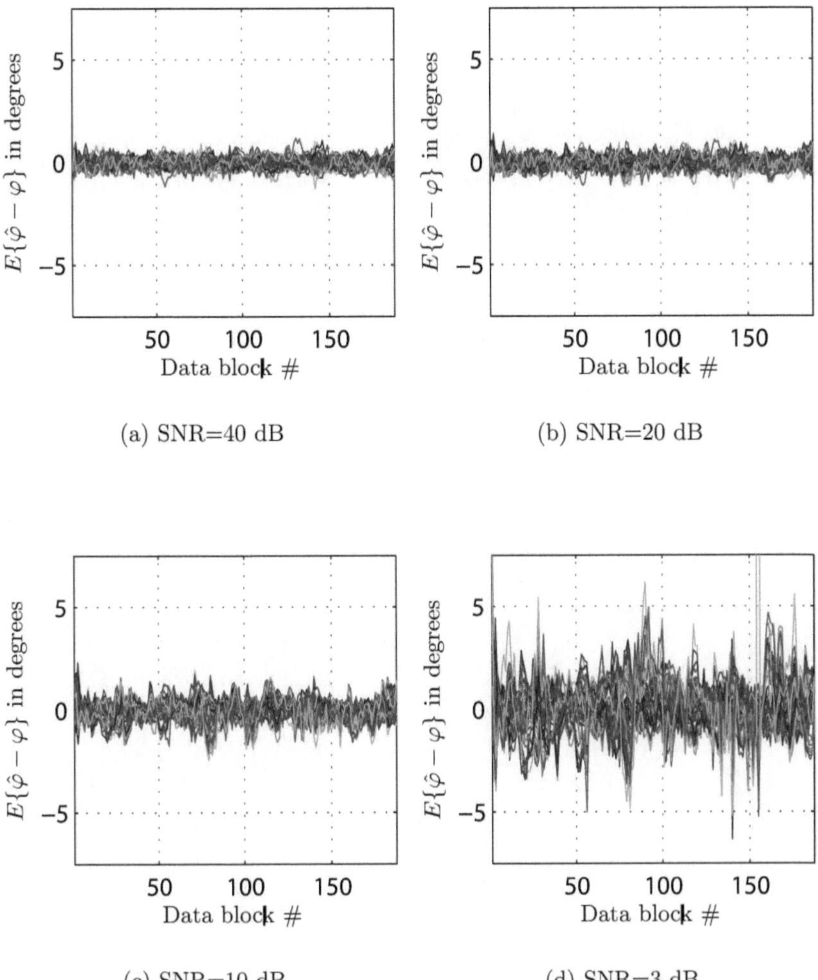

Fig. F.3. Measured DOA estimation performance of circular EB-ESPRIT using *noise signal* played back separately on all 48 loudspeakers with *all curtains closed*, where $f = 300 \ldots 3000$ Hz, $M = 10$, $\mathcal{N} = 3$, data block length is 1024 samples

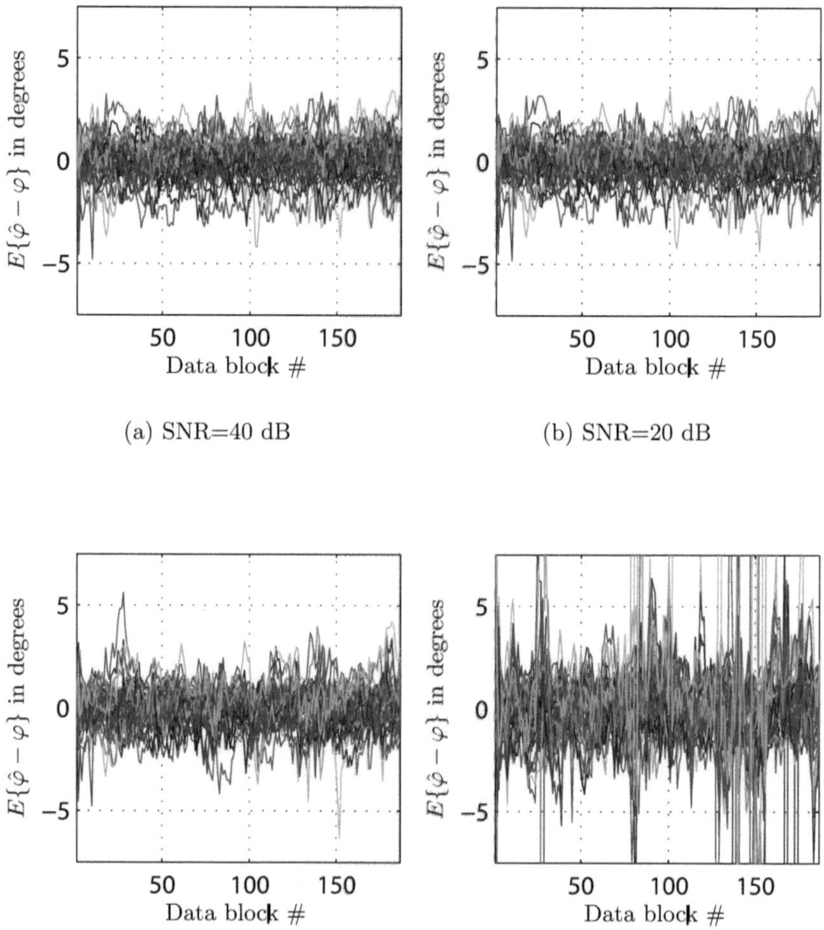

(a) SNR=40 dB

(b) SNR=20 dB

(c) SNR=10 dB

(d) SNR=3 dB

Fig. F.4. Measured DOA estimation performance of circular EB-ESPRIT using *noise signal* played back separately on all 48 loudspeakers with *all curtains opened*, where $f = 300 \ldots 3000$ Hz, $M = 10$, $\mathcal{N} = 3$, data block length is 1024 samples

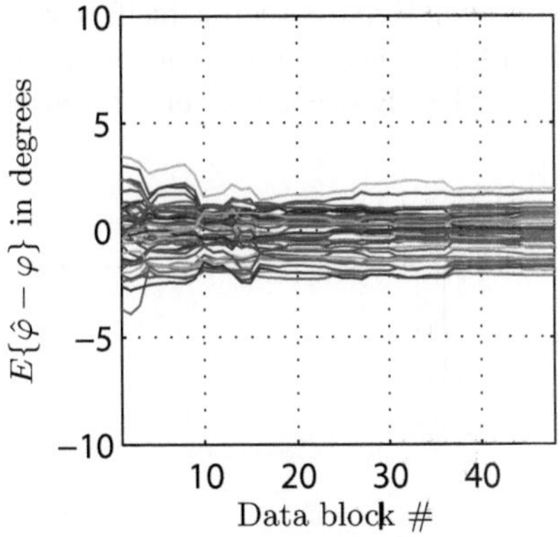

(a) SNR=40 dB

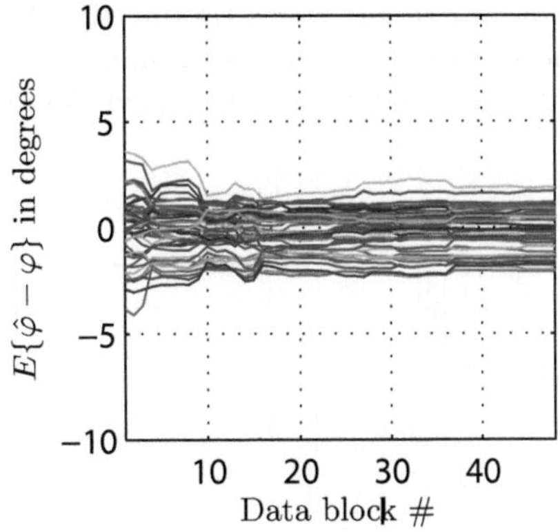

(b) SNR=20 dB

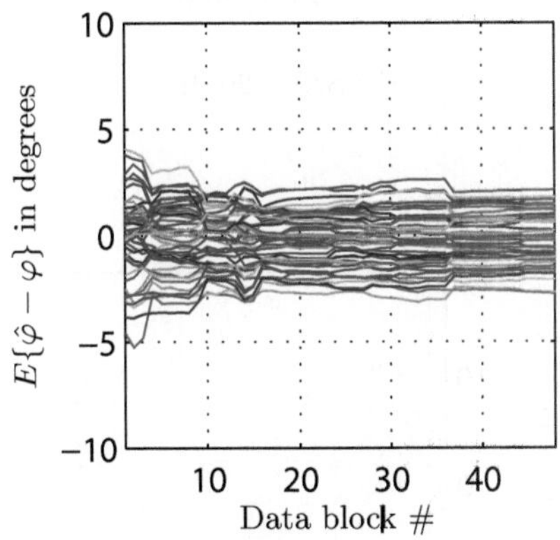

(c) SNR=10 dB

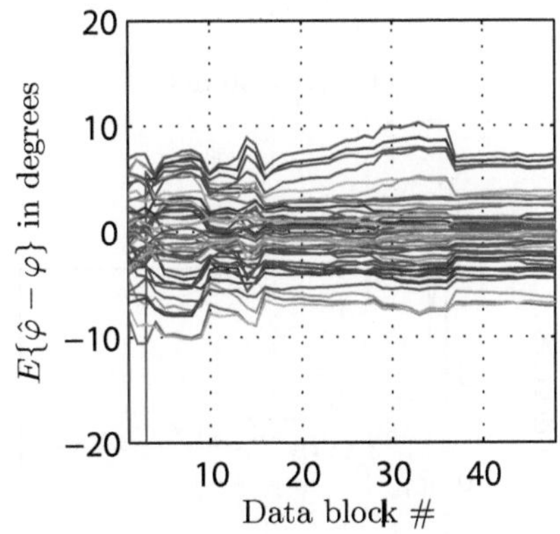

(d) SNR=3 dB

Fig. F.5. Measured DOA estimation performance of circular EB-ESPRIT using *speech signal* played back separately on all 48 loudspeakers with *all curtains closed*, where $f = 300 \dots 3000$ Hz, $M = 10$, $\mathcal{N} = 3$, data block length is 4800 samples with the ACO scenario. Note that no speech activity detector has been used in order to avoid errors produced by the speech activity detector itself. Instead, the speech signal has been manually edited for speech pause removal and the data block-length was increased. Again, several SNR scenarios were considered

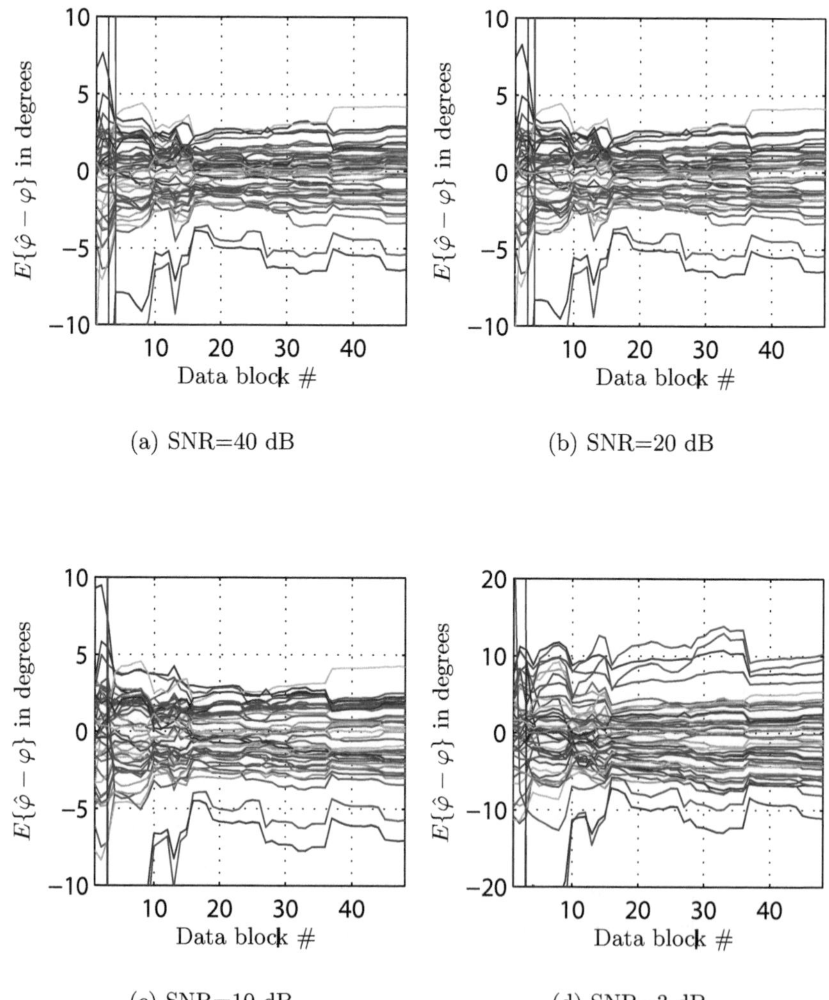

Fig. F.6. Measured DOA estimation performance of circular EB-ESPRIT using *speech signal* played back separately on all 48 loudspeakers with *all curtains opened*, where $f = 300 \ldots 3000$ Hz, $M = 10$, $\mathcal{N} = 3$, data block length is 4800 samples

by applying cylindrically diffuse noise to the sensor signals. Note that for SNR=3 dB a threshold effect already obvious from Fig. 6.7 becomes apparent, where DOA estimation suddenly becomes unreliable.

The second set of evaluations is concerned with the performance of multiple plane-waves, where one source is held fixed at one loudspeaker position, LS 1, while the other source is successively played by all 48 loudspeakers. Figures F.7 and F.8 present the results for both the ACC and ACO acoustic environment, respectively. The excitation signals were uncorrelated bandpass-filtered white noise signals. As can be seen, the estimation performance depends to a large extent on the relative distance between the two loudspeaker positions. For large values of the SNR, estimation becomes reliable as soon as the angular distance between the two loudspeakers exceeds about $\pi/6$, equivalent to a spacing of at least five loudspeakers. Note the distinct difference in estimation performance of the two sources as their spacing gets closer and closer. Remember that EB-ESPRIT applied to two sources returns two estimates at all times. As the two sources get too close, one estimate returns a "virtual" source which sits right in between the two sources. This estimate is actually quite accurate, as can be seen in Figs. F.7 and F.8. The other estimate returned by EB-ESPRIT, however, is rather arbitrary, as is evident from the figures.

The following set of evaluations deals with the estimation of the numbers of active signals in a real acoustic wavefield (ACC, ACO) containing the speech signals as shown on the right-hand side of Fig. F.1. The speech signals were played back using LS 1 and LS 25, respectively. Compared to the same scenario using noise signals shown in Fig. 6.9, the performance is deteriorated. However, provided moderate levels of SNR can be assumed, detection becomes possible after appropriate post-processing that distinguishes the relevant from the spurious peaks in the estimated impulse response.

As a last, and more qualitative, example of the localization capability, Fig. F.10 depicts snapshots of the real-time DOA estimation using EB-ESPRIT for $I > 1$. Figure F.10 basically displays a computer generated birds-eye view of Fig. 6.2.

The background shows the loudspeaker array, operated in wavefield synthesis (WFS) mode. Note that the difference here is that the sources are now not simulated using individual loudspeakers but using several loudspeakers simultaneously as dictated by the theory of WFS [BdV93, Vog93, STKR04]. For example, the loudspeakers used for synthesizing source 2 in Fig F.10(a) are indicated by filled loudspeaker symbols. This mode of operation has been chosen because by utilizing the notion of WFS, the sources can be continuously placed anywhere around the loudspeaker array. The graphical user interface that facilitates a simple source placement by appropriately calculating the loudspeaker driving signals in real-time has been designed at the

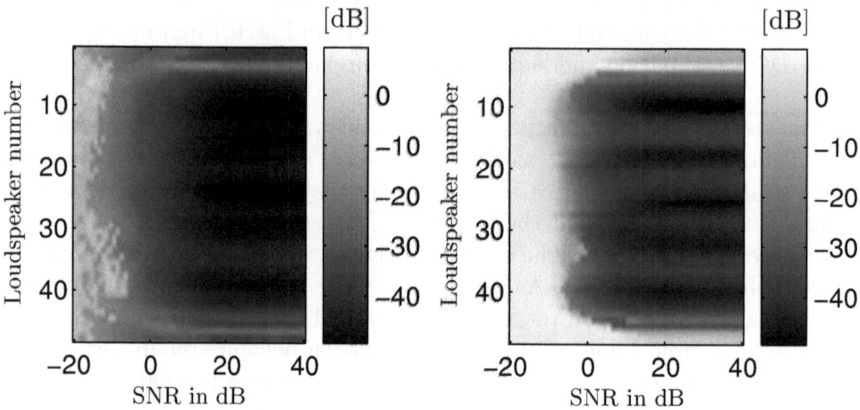

(a) Estimator variance for source 1

(b) Estimator variance for source 2

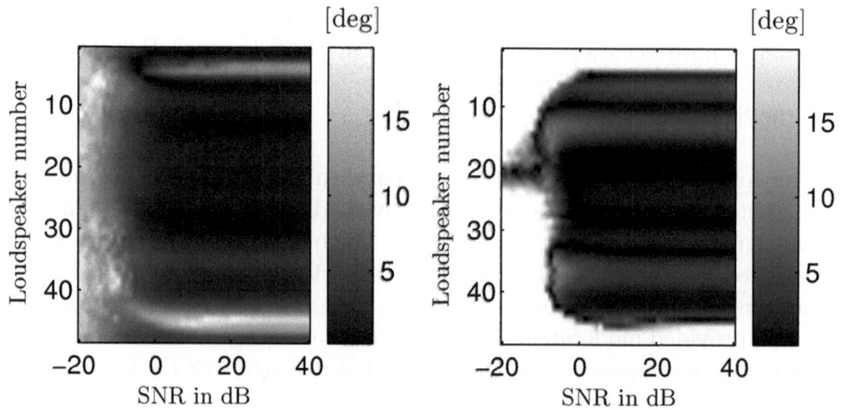

(c) Mean estimation error for source 1

(d) Mean estimation error for source 2

Fig. F.7. Measured DOA estimation performance of circular EB-ESPRIT using two noise signals with *all curtains closed*, where $f = 1000 \ldots 3000$ Hz, $M = 10$, $\mathcal{N} = 3$, data block length is 1024 samples

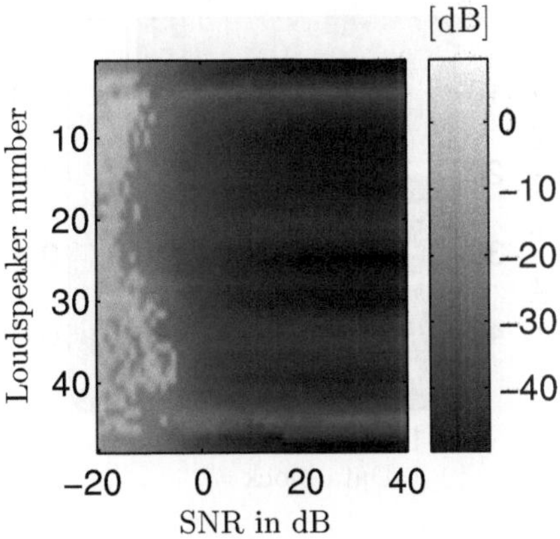

(a) Estimator variance for source 1

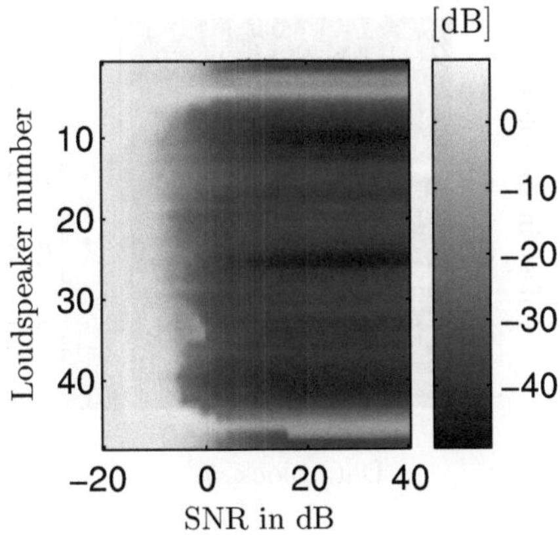

(b) Estimator variance for source 2

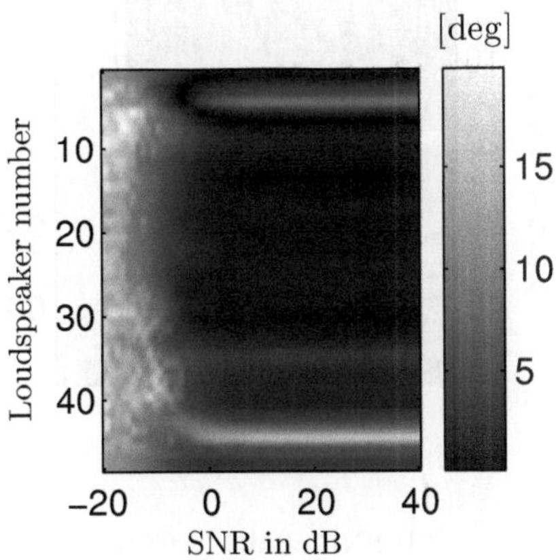

(c) Mean estimation error for source 1

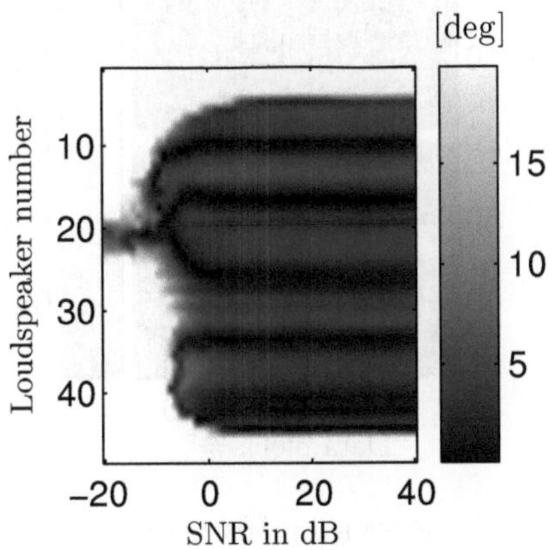

(d) Mean estimation error for source 2

Fig. F.8. Measured DOA estimation performance of circular EB-ESPRIT using two noise signals with *all curtains open*, where $f = 1000 \dots 3000$ Hz, $M = 10$, $\mathcal{N} = 3$, data block length is 1024 samples

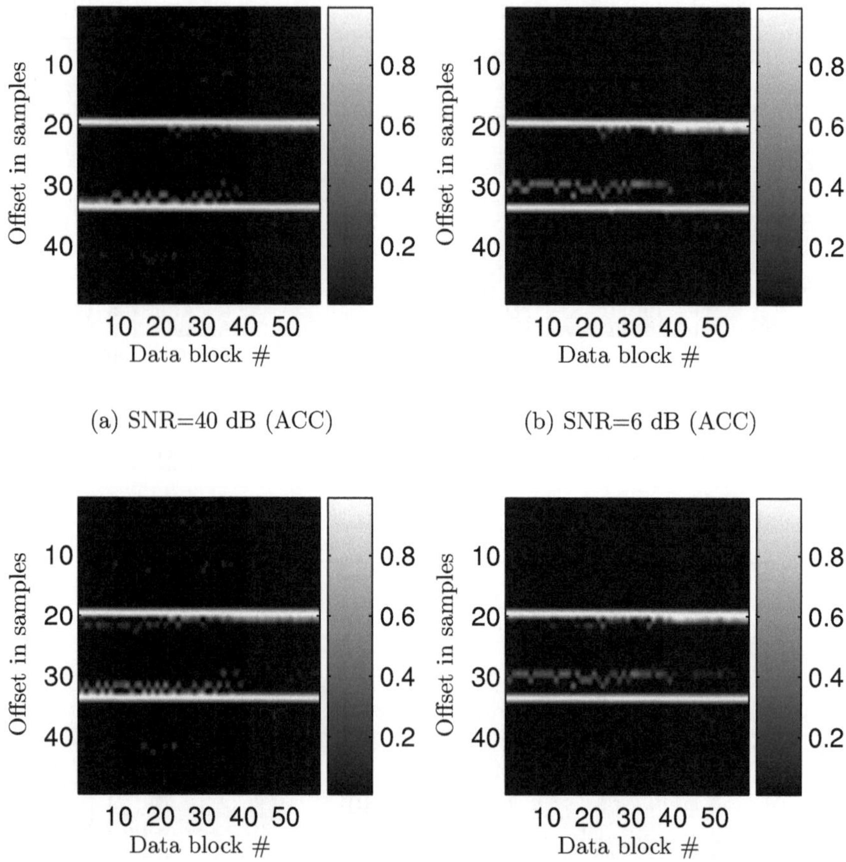

(a) SNR=40 dB (ACC)

(b) SNR=6 dB (ACC)

(c) SNR=40 dB (ACO)

(d) SNR=6 dB (ACO)

Fig. F.9. Number of active speech signals in real acoustic environments using loudspeakers LS 1 and LS 25, where $f_{\max} = 7$ kHz, $M = 10$, $\mathcal{N} = 5$, data blocklength is 8192 samples

Chair of Multimedia Communications and Signal Processing, University of Erlangen-Nuremberg, Germany [STR02, LMS].

The polar-plot in the foreground shows the result of the real-time implementation of the EB-ESPRIT algorithm. The microphone array is symbolized here by a filled circle. Note that the proportions are not to scale. Although not further investigated in previous sections for the sake of brevity, cases are considered here where the number of sources exceeds two. As mentioned previously, by decomposing a wavefield into a total of $2\mathcal{N} + 1$ harmonics, the

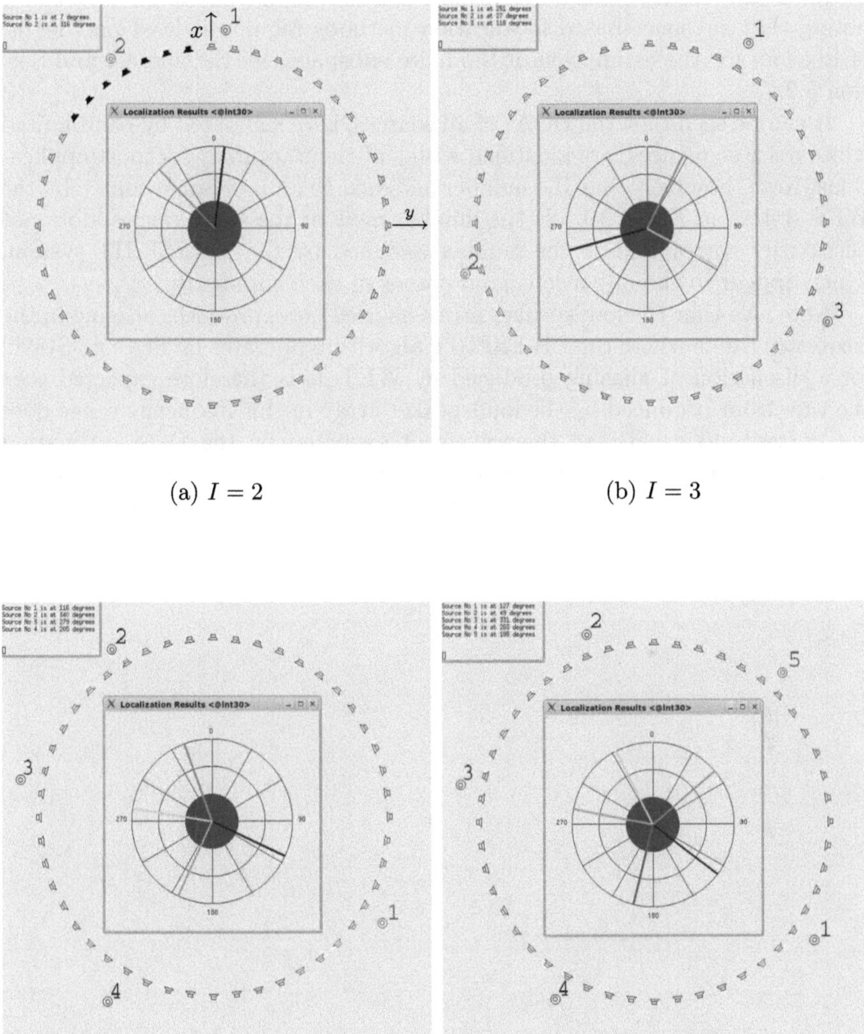

(a) $I = 2$

(b) $I = 3$

(c) $I = 4$

(d) $I = 5$

Fig. F.10. Snapshots of the real-time DOA estimation using EB-ESPRIT for $I > 1$, where $f = 1000 \ldots 3000$ Hz, $M = 10$, $\mathcal{N} = 3$, data block length is 1024 samples

DOAs of $2\mathcal{N}$ sources can be identified. This relationship can be verified by noting that subspace-based localization methods require at least one degree of freedom for the estimation of the noise subspace, see Section 4.4 and Section 5.2.

It can be seen that the DOAs of all sources, here simulated by bandlimited white noise sequences, are identified with sufficient accuracy for most applications. Note, however, that the number assigned to an individual source by the WFS system in Fig. F.10, i.e. the number next to the small circles, does not necessarily correspond to the number assigned by the EB-ESPRIT system, which appear in the upper left-hand corner of each sub-figure.

Note also that the loudspeaker array, as used here, produces aliasing in the frequency range where the EB-ESPRIT algorithm operates in. See e.g. [Sta96] for a discussion of aliasing produced by WFS. It is therefore expected that the wavefront produced by the loudspeaker array in this frequency range does not correspond exactly to the wavefront expected by the DOA estimation algorithm. The fact that all sources can be identified by the EB-ESPRIT algorithm anyway stresses the system's robust estimation capabilities.

References

AB79. J. B. Allen and D.A. Berkeley. Image method for efficiently simulating small-room acoustics. *J. Acoust. Soc. Am*, 65(4):943–950, April 1979.

Aka74. H. Akaike. A new look at the statistical model identification. *IEEE Trans. Automat. Contr.*, 19(6):716–723, December 1974.

AKG. AKG Acoustics. http://www.akg.com.

ALS. Advanced Linux Sound Architecture. http://www.alsa-project.org.

And50. V.C. Anderson. Sound scattering from a fluid sphere. *J. Acoust. Soc. Am.*, 22(4):426–431, July 1950.

And63. T.W. Anderson. Asymptotic theory for principal component analysis. *Ann. J. Math. Stat.*, 34:122–148, January 1963.

Arf85. G. Arfken. *Mathematical Methods for Physicists.* Academic Press, 3rd edition, 1985.

AS72. M. Abramowitz and I.A. Stegun, editors. *Handbook of Mathematical Functions.* Dover, 1972.

BAK04. H. Buchner, R. Aichner, and W. Kellermann. Blind source separation for convolutive mixtures: A unified treatment. In Y. Huang and J. Benesty, editors, *Audio Signal Processing for Next-Generation Multimedia Communication Systems*, chapter 10, pages 255–293. Kluwer Academic Publishers, 2004.

BAK05a. H. Buchner, R. Aichner, and W. Kellermann. A generalization of blind source separation algorithms for convolutive mixtures based on second order statistics. *IEEE Trans. Speech and Audio Processing*, 13(1):120–134, January 2005.

BAK05b. H. Buchner, R. Aichner, and W. Kellermann. Relation between blind system identification and convolutive blind source separation. In *Proc. Joint Workshop on Hands-Free Communication and Microphone Arrays*, pages d–3–d–4, Piscataway, NJ, USA, March 2005.

Bal91. G. Ballou, editor. *Handbook for Sound Engineers – The New Audio Cyclopedia.* Focal Press, 2nd edition, 1991.

Bar83. A.J. Barabell. Improving the resolution performance of eigenstructure-based direction-finding algorithms. In *IEEE Proc. Int. Conf. Acoustics, Speech, and Signal Processing (ICASSP)*, pages 336–339, Boston, MA, USA, April 1983.

BAS+05. H. Buchner, R. Aichner, J. Stenglein, H. Teutsch, and W. Kellermann. Simultaneous localization of multiple sources using blind adaptive mimo filtering. In *IEEE Proc. Int. Conf. Acoust., Speech and Signal Processing*, pages III–97 – III–100, Philadelphia, PA, USA, March 2005.

BBK03. H. Buchner, J. Benesty, and W. Kellermann. Multichannel frequency-domain adaptive filtering with application to acoustic echo cancellation. In J. Benesty and Y. Huang, editors, *Adaptive Signal Processing: Application to Real-world Problems*, pages 95–128. Springer, 2003.

BdV93. A.J. Berkhout, D. de Vries, and P. Vogel. Acoustic control by wave field synthesis. *J. Acoust. Soc. Am.*, 93(5):2764–2778, January 1993.

Ben00. J. Benesty. Adaptive eigenvalue decomposition algorithm for passive acoustic source localization. *J. Acoust. Soc. Am.*, 107(1):384–391, January 2000.

Bla00. D.T. Blackstock. *Fundamentals of Physical Acoustics.* John Wiley & Sons, 2000.

Bra95. M.S. Brandstein. *A Framework for Speech Source Localization Using Sensor Arrays.* PhD thesis, Brown University, 1995.

BS01. J. Bitzer and K.U. Simmer. Superdirective microphone arrays. In M.S. Brandstein and D.B. Ward, editors, *Microphone Arrays – Signal Processing Techniques and Applications*, chapter 2, pages 19–38. Springer, 2001.

BW01. M. Brandstein and D. Ward, editors. *Microphone Arrays – Signal Processing Techniques and Applications.* Springer, 2001.

CA02. A. Cichocki and S. Amari. *Adaptive Blind Signal and Image Processing.* Wiley & Sons, 2002.

CBS96. B. Champagne, S. Bédard, and A. Stéphenne. Performance of time-delay estimation in the presence of room reverberation. *IEEE Trans. Speech Audio Processing*, 4(2):148–152, March 1996.

Cho95. T. Chou. Frequency-independent beamformer with low response error. In *Proc. IEEE Int. Conf. on Acoustics, Speech, and Signal Processing (ICASSP)*, volume 5, pages 2995–2998, May 1995.

CK98. D. Colton and R. Kress. *Inverse Acoustic and Electromagnetic Scattering Theory.* Springer, 2nd edition, 1998.

CZK86. H. Cox, R.M. Zeskind, and T. Kooij. Practical supergain. *IEEE Trans. Acoust., Speech , and Signal Processing*, 34(3):393–398, June 1086.

CZO87. H. Cox, R.M. Zeskind, and M.M. Owen. Robust adaptive beamforming. *IEEE Trans. Acoust. Speech Signal Processing*, 35(10):1365–1375, October 1987.

Dav70. W.B. Davenport. *Probability and Random Processes.* McGraw-Hill, 1970.

DB88. J.H. Doles and F.D. Benedict. Broad-band array design using the asymptotic theory of unequally spaced arrays. *IEEE Trans. Ant. Prop.*, 36(1):27–33, January 1988.

DD94. M.A. Doron and E. Doron. Wavefield modeling and array processing, part I – spatial sampling. *IEEE Trans. Signal Processing*, 42(10):2549–2559, October 1994.

DHS00. R.O. Duda, P.E. Hart, and D.G. Stork. *Pattern Classification.* Wiley Interscience, 2nd edition, 2000.

DJ94. J.R. Driscoll and D.M. Healy Jr. Computing Fourier transforms and convolutions on the 2-sphere. *Advances in Applied Mathematics*, 15:202–250, June 1994.

Doc03. S. Doclo. *Multi-microphone Noise Reduction and Dereverberation Techniques for Speech Applications*. PhD thesis, Katholieke Universiteit Leuven, 2003.

Dol46. C.L. Dolph. A current distribution for broadside arrays which optimize the relationship between beam width and side-lobe level. *Proc. I.R.E.*, 34(6):335–348, June 1946.

DR84. P.J. Davis and P. Rabinowitz. *Methods of Numerical Integration*. Academic Press, 2nd edition, 1984.

DSB01. J.H. DiBiase, H.F. Silverman, and M.S. Brandstein. Robust localization in reverberant rooms. In M.S. Brandstein and D.B. Ward, editors, *Microphone Arrays – Signal Processing Techniques and Applications*, chapter 8, pages 158–180. Springer, 2001.

DW92. M.A. Doron and A.J. Weiss. On focusing matrices for wide-band array processing. *IEEE Trans. Signal Processing*, 40(6):1295–1302, June 1992.

EKM03. G.W. Elko, R.A. Kubli, and J.M. Meyer. Audio system based on at least second-order eigenbeams. US Patent Application US2003/0147539A1, 2003.

ELA. ELAC Electroacoustic. http://www.elac.com/en/index.html.

Elk01. G.W. Elko. Spatial coherence functions for differential microphones in isotropic noise. In M.S. Brandstein and D.B. Ward, editors, *Microphone Arrays – Signal Processing Techniques and Applications*, chapter 4, pages 61–85. Springer, 2001.

Elk04. G.W. Elko. Differential microphone arrays. In Y. Huang and J. Benesty, editors, *Audio Signal Processing for Next-Generation Multimedia Communication Systems*, chapter 2, pages 11–65. Kluwer, 2004.

EM84. Y. Ephraim and D. Malah. Speech enhancement using a soft-decision noise suppression filter. *IEEE Trans. Acoust., Speech, Signal Processing*, 32(6):1109–1121, December 1984.

FD52. W. Franz and W. Deppermann. Theorie der Beugung am Zylinder unter Berücksichtigung der Kriechwelle. *Annalen der Physik*, 10:361–373, 1952.

Feu04. C. Feuillade. Animations for visualizing and teaching acoustic impulse scattering from spheres. *J. Acoust. Soc. Am.*, 115(5):1893–1904, May 2004.

FJZE85. J.L. Flanagan, J.D. Johnston, R. Zahn, and G.W. Elko. Computer-steered microphone arrays for sound transduction in large rooms. *J. Acoust. Soc. Am.*, 78(5):1508–1518, November 1985.

FK97. K. Furuya and Y. Kaneda. Two-channel blind deconvolution of nonminimum phase FIR systems. *IEICE Trans. Fundamental*, E80-A(5):804–808, May 1997.

Fla85. J.L. Flanagan. Beamwidth and usable bandwidth of delay-steered microphone arrays. *AT&T Technical Journal*, 64(4):983–995, April 1985.

FM96. J. Fliege and U. Maier. A two-stage approach for computing cubature formulae for the sphere. Ergebnisberichte Angewandte Mathematik, No. 139T. Fachbereich Mathematik, Universität Dortmund, Dortmund, Germany, 1996.

Fro72. O.L. Frost. An algorithm for linearly constrained adaptive array processing. *Proc. IEEE*, 60(10):926–935, August 1972.

Gau14. C.F. Gauss. Methodus nova integralium valores per approximationem inveniendi. *Commentationes Societatis Regiae Scientarium Gottingensis Recentiorcs 2 [Werke III]*, pages 163–196, 1814.

GE93. M.M. Goodwin and G.W. Elko. Constant beamwidth beamforming. In *Proc. IEEE Int. Conf. on Acoustics, Speech, and Signal Processing (ICASSP)*, volume 1, pages 169–172, Minneapolis, Minnesota, May 1993.

GJ82. L.J. Griffiths and C.W. Jim. An alternative approach to linearly constrained adaptive beamforming. *IEEE Trans. Antennas Propag.*, AP-30(1):27–34, January 1982.

GL89. G. H. Golub and C. F. Van Loan. *Matrix Computations.* Johns Hopkins University Press, 2nd edition, 1989.

GM55. E.N. Gilbert and S.P. Morgan. Optimum design of directive antenna arrays subject to random variations. *Bell Syst. Tech. J.*, pages 637–663, May 1955.

GR65. I.S. Gradshteyn and I.M. Ryzhik. *Table of Integrals, Series, and Products.* Academic Press, 4th edition, 1965.

Hay02. S. Haykin. *Adaptive Filter Theory.* Prentice Hall, 4th edition, 2002.

Her05. W. Herbordt. *Sound Capture for Human / Machine Interfaces – Practical Aspects of Microphone Array Signal Processing.* Springer, 2005.

HK88. H. Hung and M. Kaveh. Focusing matrices for coherent signal subspace processing. *IEEE Trans. Acoust., Speech, Signal Processing*, 36(8):1272–1281, August 1988.

HN76. W.S. Hodgkiss and L.W. Nolte. Covariance between Fourier coefficients representing the time waveforms observed from an array of sensors. *J. Acoust. Soc. Am*, 59(3):582–590, March 1976.

HN95. M. Haardt and J.A. Nossek. Unitary ESPRIT: How to obtain increased estimation accuracy with a reduced computational burden. *IEEE Trans. Signal Processing*, 43(5):1232–1242, May 1995.

HRM96. D. Healy, D. Rockmore, and S. Moore. An FFT for the 2-sphere and applications. In *Proc. IEEE Int. Conf. on Acoustics, Speech, and Signal Processing (ICASSP)*, volume 3, pages 1323–1326, May 1996.

HS69. M.L. Harbold and B.N. Steinberg. Direct experimental verification of creeping waves. *J. Acoust. Soc. Am.*, 45(3):592–603, March 1969.

HS96. R.H. Hardin and N.J.A. Sloan. McLaren's improved snub cube and other new spherical designs in three dimensions. *Discrete Computational Geometry*, 15:429–441, 1996.

Hua01. Y. Huang. *Real-time Acoustic Source Localization with Passive Microphone Arrays.* PhD thesis, Georgia Institute of Technology, 2001.

IPP. Intel Integrated Performance Primitives. http://www.intel.com/software/products/ipp.

JAC. Jack Audio Connection Kit. http://www.jackaudio.org.

JD82. D.H. Johnson and S.R. DeGraaf. Improving the resolution of bearing in passive sonar arrays by eigenvalue analysis. *IEEE Trans. Acoust, Speech, and Signal Processing*, 30(4):638–647, August 1982.

JD93. D.H. Johnson and D.E. Dudgeon. *Array Signal Processing – Concepts and Techniques.* Prentice Hall, New Jersey, 1993.

JF93. M.C. Junger and D. Feit. *Sound, Structures, and Their Interaction.* Acoustical Society of America, 1993.

KC76. C.H. Knapp and G.C. Carter. The generalized correlation method for estimation of time delay. *IEEE Trans. Acoust., Speech, Signal Processing*, 24(4):320–327, August 1976.

Kel05. W. Kellermann. Personal communication, 2005.

KV96. H. Krim and M. Viberg. Two decades of array signal processing research – the parametric approach. *IEEE Signal Processing Mag.*, 13(4):67–94, July 1996.

Lam79. H. Y-F. Lam. *Analog and Digital Filters – Design and Realization.* Prentice Hall, 1979.

LMS. Chair of Multimedia Communications and Signal Processing, University of Erlangen-Nuremberg, Germany. http://www.LNT.de/lms.

MAT. MATLAB – A Matrix Algebra Program. http://www.mathworks.com.

ME02. J. Meyer and G.W. Elko. A highly scalable spherical microphone array based on a orthonormal decomposition of the soundfield. In *Proc. Int. Conf. Acoustics, Speech, and Signal Processing (ICASSP)*, pages II–1781–II–1784, May 2002.

ME04. J. Meyer and G.W. Elko. Spherical microphone arrays for 3D sound recording. In Y. Huang and J. Benesty, editors, *Audio Signal Processing for Next-Generation Multimedia Communication Systems*, chapter 3, pages 67–89. Kluwer, 2004.

MF53. P.M. Morse and H. Feshbach. *Methods of Theoretical Physics.* McGraw-Hill, 1953.

MI68. P.M. Morse and K.U. Ingard. *Theoretical Acoustics.* McGraw-Hill, 1968.

Moh99. M.J. Mohlenkamp. A fast transform for spherical harmonics. *The Journal of Fourier Analysis and Applications*, 5(2/3):159–184, 1999.

MPE. MPEG – Moving Pictures Experts Group. http://www.mpeg.org.

MS69. V.H. MacDonald and P.M. Schultheiss. Optimum passive bearing estimation in a spatially incoherent noise environment. *J. Acoust. Soc. Am.*, 46(1):37–45, July 1969.

MZ94. C.P. Mathews and M.D. Zoltowski. Eigenstructure techniques for 2-D angle estimation with uniform circular arrays. *IEEE Trans. Signal Processing*, 42(9):2395–2407, September 1994.

OS75. A.V. Oppenheim and R.W. Schafer. *Digital Signal Processing.* Prentice Hall, New Jersey, 1975.

OS98. A.V. Oppenheim and R.W. Schafer. *Discrete-Time Signal Processing.* Prentice Hall, 2nd edition, 1998.

OVK91. B. Ottersten, M. Viberg, and T. Kailath. Performance analysis of the total least squares ESPRIT algorithm. *IEEE Trans. Signal Processing*, 39(5):1122–1135, May 1991.

Pap65. A. Papoulis. *Probability, Random Variables, and Stochastic Processes.* Mc-Graw-Hill, 1965.

PGH00. M. Pesavento, A.B. Gershman, and M. Haardt. Unitary root-MUSIC with a real-valued eigendecomposition: A theoretical and experimental performance study. *IEEE Trans. Signal Processing*, 48(5):1306–1314, May 2000.

Pie89. A. Pierce. *Acoustics – An Introduction to Its Physical Principles and Applications.* Acoustical Society of America, 1989.

PK89. S.U. Pillai and B.H. Kwon. Forward/backward spatial smoothing techniques for coherent signal identification. *IEEE Trans. Acoust., Speech, and Signal Processing*, 37(1):8–15, January 1989.

PS00. L. Parra and C. Spence. Convolutive blind separation of non-stationary sources. *IEEE Trans. Speech Audio Processing*, 8(3):320–327, May 2000.

PTVF02. W.H. Press, S.A. Teukolsky, W.T. Vetterling, and B.P. Flannery. *Numerical Recipes in C++ – The Art of Scientific Computing*. Cambridge University Press, second edition, 2002.

Raf05. B. Rafaely. Analysis and design of spherical microphone arrays. *IEEE Trans. Speech and Audio Processing*, 13(1):135–143, January 2005.

RD82. T. Rahim and D.E.N. Davis. Effect of directional elements on the directional response of circular antenna arrays. *IEE Proc., Pt. H*, 129(1):18–22, January 1982.

RH89. B.D. Rao and K.V.S. Hari. Performance analysis of root-MUSIC. *IEEE Trans. Acoust., Speech, and Signal Processing*, 37(12):1939–1949, December 1989.

RK89. R. Roy and T. Kailath. ESPRIT – estimation of signal parameters via rotational invariance techniques. *IEEE Trans. Acoust., Speech, and Signal Processing*, 37(7):984–995, July 1989.

RME. RME – Intelligent Audio Solutions. http://www.rme-audio.de.

Ros57. M.E. Rose. *Elementary Theory of Angular Momentum*. John Wiley, 1957.

SBR04. S. Spors, H. Buchner, and R. Rabenstein. Efficient active listening room compensation for wave field synthesis. In *Proc. 116th Audio Engineering Society (AES) Convention*, Berlin, Germany, 2004.

Sch78. G. Schwartz. Estimating the dimension of a model. *Ann. Stat.*, 6(2):461–464, 1978.

Sch81. R.O. Schmidt. *A Signal Subspace Approach to Multiple Emitter Location and Spectral Estimation*. PhD thesis, Stanford University, November 1981.

Sch86. R.O. Schmidt. Multiple emitter location and signal parameter estimation. *IEEE Trans. Antennas Propagat.*, AP-34(3):276–280, March 1986.

SHS. N.J.A. Sloane, R.H. Hardin, and W.D. Smith. Tables of spherical codes. http://www.research.att.com/~njas/packings.

SK92. M.M. Sondhi and W. Kellermann. Adaptive echo cancellation for speech signals. In S. Furui and M.M. Sondhi, editors, *Advances in Speech Signal Processing*, pages 327–356. Marcel Dekker Inc., 1992.

Sko70. M.I. Skolnik, editor. *Radar Handbook*. McGraw-Hill, 1970.

SLG01. P. Stoica, E.G. Larsson, and A.B. Gershman. The stochastic CRB for array processing: a textbook derivation. *IEEE Signal Processing Letters*, 8(5):148–150, May 2001.

SM93. P.M. Schultheiss and H. Messer. Optimal and suboptimal broad-band source location estimation. *IEEE Trans. Signal Processing*, 41(9):2752–2763, September 1993.

SN89. P. Stoica and A. Nehorai. MUSIC, maximum likelihood, and Cramer-Rao bound. *IEEE Trans. Signal Processing*, 37:720–741, May 1989.

Son67. M.M. Sondhi. An adaptive echo canceller. *Bell Syst. Tech. J.*, 46(3):497–511, 1967.

SS90. P. Stoica and K.C. Sharman. Maximum likelihood methods for direction-of-arrival estimation. *IEEE Trans. Acoust., Speech, and Signal Processing*, 38(7):1132–1143, July 1990.

Sta96. E.W. Start. Application of curved arrays in wave field synthesis. In *Proc. 100th AES Convention.* Audio Engineering Society, May 1996.

STKR04. S. Spors, H. Teutsch, A. Kuntz, and R. Rabenstein. Sound field synthesis. In Y. Huang and J. Benesty, editors, *Audio Signal Processing for Next-Generation Multimedia Communication Systems*, chapter 12, pages 323–344. Kluwer, 2004.

STR02. S. Spors, H. Teutsch, and R. Rabenstein. High-quality acoustic rendering with wave field synthesis. In *Proc. Vision, Modeling and Visualization (VMV),*, pages 101–108, Erlangen, Germany, 2002.

SWK85. T-J. Shan, M. Wax, and T. Kailath. On spatial smoothing fo direction-of-arrival estimation of coherent signals. *IEEE Trans. Acoust., Speech, and Signal Processing*, ASSP-33(4):806–811, August 1985.

TE01a. H. Teutsch and G.W. Elko. An adaptive close-talking microphone array. In *Proc. 2001 IEEE Workshop on Applications of Signal Processing to Audio and Acoustics*, pages 163–166, October 2001.

TE01b. H. Teutsch and G.W. Elko. First- and second-order adaptive differential microphone arrays. In *Proc. Int. Workshop on Acoustic Echo and Noise Control*, pages 35–38, September 2001.

TK05a. H. Teutsch and W. Kellermann. EB-ESPRIT: 2D localization of multiple wideband acoustic sources using eigen-beams. In *IEEE Proc. Int. Conf. Acoust., Speech and Signal Processing*, pages III–89 – III–92, March 2005.

TK05b. H. Teutsch and W. Kellermann. Eigen-beam processing for direction-of-arrival estimation using spherical apertures. In *Proc. Joint Workshop on Hands-Free Communication and Microphone Arrays*, pages c–13–c–14, March 2005.

TK05c. H. Teutsch and W. Kellermann. Estimation of the number of acoustic wideband sources using eigen-beam processing for circular apertures. In *Proc. IEEE Workshop on Applications of Signal Processing to Audio and Acoustics*, pages 110–113, October 2005.

Tre02. H. L. Van Trees. *Optimum Array Processing – Part IV of Detection, Estimation, and Modulation Theory.* Wiley-Interscience, 2002.

TSH+03. H. Teutsch, S. Spors, W. Herbordt, W. Kellermann, and R. Rabenstein. An integrated real-time system for immersive audio applications. In *Proc. IEEE Workshop on Applications of Signal Processing to Audio and Acoustics (WASPAA)*, pages 67–70, October 2003.

ÜDM66. H. Überall, R.D. Doolittle, and J.V. McNicholas. Use of sound pulses for a study of circumferential waves. *J. Acoust. Soc. Am.*, 39:564–578, March 1966.

Vog93. P. Vogel. *Application of Wave Field Synthesis in Room Acoustics.* PhD thesis, Delft University of Technology, 1993.

VSd96. M. Van der Wal, E.W. Start, and D. de Vries. Design of logarithmically spaced constant-directivity transducer arrays. *Journal of the Audio Engineering Society (AES)*, 44(6):497–507, June 1996.

vVB88. B.D. van Veen and K.M. Buckley. Beamforming: A versatile approach to spatial filtering. *IEEE ASSP Magazine*, 5(2):4–24, April 1988.

WA01. D.B. Ward and T.D. Abhayapala. Reproduction of a plane-wave sound field using an array of loudspeakers. *IEEE Trans. Speech and Audio Processing*, 9(6):697–7007, September 2001.

Wei03. E.W. Weisstein. *CRC Concise Encyclopedia of Mathematics*. Chapman & Hall/CRC, 2nd edition, 2003.

Wie47. F.M. Wiener. Sound diffraction by rigid spheres and circular cylinders. *J. Acoust. Soc. Am.*, 19(3):444–451, May 1947.

Wil99. E.G. Williams. *Fourier Acoustics: Sound Radiation and Nearfield Acoustic Holography*. Academic Press, London, 1999.

WK85a. H. Wang and M. Kaveh. Coherent signal-subspace processing for the detection and estimation of angles of arrival of multiple wide-band sources. *IEEE Trans. Acoust., Speech, and Signal Processing*, ASSP-33:823–831, August 1985.

WK85b. M. Wax and T. Kailath. Detection of signals by information theoretic criteria. *IEEE Trans. Acoust., Speech, and Signal Processing*, ASSP-33(2):387–392, April 1985.

WKW95. D.B. Ward, R.A. Kennedy, and R.C. Williamson. Theory and design of broadband sensor arrays with frequency-invariant far-field patterns. *J. Acoust. Soc. Am.*, 97(2):1023–1034, Febuary 1995.

WKW01. D.B. Ward, R.A. Kennedy, and R.C. Williamson. Constant directivity beamforming. In M.S. Brandstein and D.B. Ward, editors, *Microphone Arrays: Signal Processing Techniques and Applications*, chapter 1, pages 3–17. Springer, Berlin, 2001.

WW83. A.J. Weiss and E. Weinstein. Fundamental limitations in passive time delay esitmation – part I: Narrow-band systems. *IEEE Trans. Acoust., Speech, and Signal Processing*, 31(2):472–486, April 1983.

WWW84. E. Weinstein, A.J. Weiss, and E. Weinstein. Fundamental limitations in passive time delay esitmation – part II: Wide-band systems. *IEEE Trans. Acoust., Speech, and Signal Processing*, 32(5):1064–1078, October 1984.

WZ89. M. Wax and I. Ziskind. On unique localization of multiple sources by passive sensor arrays. *IEEE Trans. Acoust, Speech, and Signal Processing*, 37(7):996–1000, July 1989.

XRK94. G. Xu, R.H. Roy, and T. Kailath. Detection of number of signals via exploitation of centro-symmetry property. *IEEE Trans. Signal Processing*, 42(1):102–112, January 1994.

You59. R.W. Young. Sabine reverberation equation and sound power calculations. *J. Acoust. Soc. Am.*, 31(7):912–921, July 1959.

Index

Lecture Notes in Control and Information Sciences

Edited by M. Thoma and M. Morari

Further volumes of this series can be found on our homepage:
springer.com

Vol. 347: Assawinchaichote, W.; Nguang, S.K.; Shi, P.
Fuzzy Control and Filter Design for Uncertain Fuzzy Systems
178 p. 2006 [3-540-37011-0]

Vol. 346: Tarbouriech, S.; García, G.; Glattfelder, A.H. (Eds.)
Advanced Strategies in Control Systems with Input and Output Constraints
430 p. 2006 [3-540-37009-9]

Vol. 345: Huang, D.-S.; Li, K.; Irwin, G.W. (Eds.)
Intelligent Computing in Signal Processing and Pattern Recognition (ICIC 2006)
1179 p. 2006 [3-540-37257-1]

Vol. 344: Huang, D.-S.; Li, K.; Irwin, G.W. (Eds.)
Intelligent Control and Automation (ICIC 2006)
1121 p. 2006 [3-540-37255-5]

Vol. 343: Jalali, A.A.; Sims, C.S.; Famouri, P.
Reduced Order Systems
170 p. 2006 [3-540-34358-X]

Vol. 342: Hasegawa, Y.; Suzuki, T.
Realization Theory and Design of Digital Images
275 p. 2006 [3-540-36115-4]

Vol. 341: Commault, C.; Marchand, N. (Eds.)
Positive Systems
448 p. 2006 [3-540-34771-2]

Vol. 340: Diehl, M.; Mombaur, K. (Eds.)
Fast Motions in Biomechanics and Robotics
500 p. 2006 [3-540-36118-9]

Vol. 339: Alamir, M.
Stabilization of Nonlinear Systems Using Receding-horizon Control Schemes
325 p. 2006 [1-84628-470-8]

Vol. 338: Tokarzewski, J.
Finite Zeros in Discrete Time Control Systems
325 p. 2006 [3-540-33464-5]

Vol. 337: Blom, H.; Lygeros, J. (Eds.)
Stochastic Hybrid Systems
395 p. 2006 [3-540-33466-1]

Vol. 336: Pettersen, K.Y.; Gravdahl, J.T.; Nijmeijer, H. (Eds.)
Group Coordination and Cooperative Control
310 p. 2006 [3-540-33468-8]

Vol. 335: Kozłowski, K. (Ed.)
Robot Motion and Control
424 p. 2006 [1-84628-404-X]

Vol. 334: Edwards, C.; Fossas Colet, E.; Fridman, L. (Eds.)
Advances in Variable Structure and Sliding Mode Control
504 p. 2006 [3-540-32800-9]

Vol. 333: Banavar, R.N.; Sankaranarayanan, V.
Switched Finite Time Control of a Class of Underactuated Systems
99 p. 2006 [3-540-32799-1]

Vol. 332: Xu, S.; Lam, J.
Robust Control and Filtering of Singular Systems
234 p. 2006 [3-540-32797-5]

Vol. 331: Antsaklis, P.J.; Tabuada, P. (Eds.)
Networked Embedded Sensing and Control
367 p. 2006 [3-540-32794-0]

Vol. 330: Koumoutsakos, P.; Mezic, I. (Eds.)
Control of Fluid Flow
200 p. 2006 [3-540-25140-5]

Vol. 329: Francis, B.A.; Smith, M.C.; Willems, J.C. (Eds.)
Control of Uncertain Systems: Modelling, Approximation, and Design
429 p. 2006 [3-540-31754-6]

Vol. 328: Loría, A.; Lamnabhi-Lagarrigue, F.; Panteley, E. (Eds.)
Advanced Topics in Control Systems Theory
305 p. 2006 [1-84628-313-2]

Vol. 327: Fournier, J.-D.; Grimm, J.; Leblond, J.; Partington, J.R. (Eds.)
Harmonic Analysis and Rational Approximation
301 p. 2006 [3-540-30922-5]

Vol. 326: Wang, H.-S.; Yung, C.-F.; Chang, F.-R.
H_∞ Control for Nonlinear Descriptor Systems
164 p. 2006 [1-84628-289-6]

Vol. 325: Amato, F.
Robust Control of Linear Systems Subject to Uncertain Time-Varying Parameters
180 p. 2006 [3-540-23950-2]

Vol. 324: Christofides, P.; El-Farra, N.
Control of Nonlinear and Hybrid Process Systems
446 p. 2005 [3-540-28456-7]

Vol. 323: Bandyopadhyay, B.; Janardhanan, S.
Discrete-time Sliding Mode Control
147 p. 2005 [3-540-28140-1]

Vol. 322: Meurer, T.; Graichen, K.; Gilles, E.D. (Eds.)
Control and Observer Design for Nonlinear Finite and Infinite Dimensional Systems
422 p. 2005 [3-540-27938-5]

Vol. 321: Dayawansa, W.P.; Lindquist, A.; Zhou, Y. (Eds.)
New Directions and Applications in Control Theory
400 p. 2005 [3-540-23953-7]

Vol. 320: Steffen, T.
Control Reconfiguration of Dynamical Systems
290 p. 2005 [3-540-25730-6]

Vol. 319: Hofbaur, M.W.
Hybrid Estimation of Complex Systems
148 p. 2005 [3-540-25727-6]

Vol. 318: Gershon, E.; Shaked, U.; Yaesh, I.
H_∞ Control and Estimation of State-muliplicative Linear Systems
256 p. 2005 [1-85233-997-7]

Vol. 317: Ma, C.; Wonham, M.
Nonblocking Supervisory Control of State Tree Structures
208 p. 2005 [3-540-25069-7]

Vol. 316: Patel, R.V.; Shadpey, F.
Control of Redundant Robot Manipulators
224 p. 2005 [3-540-25071-9]

Vol. 315: Herbordt, W.
Sound Capture for Human/Machine Interfaces: Practical Aspects of Microphone Array Signal Processing
286 p. 2005 [3-540-23954-5]

Vol. 314: Gil', M.I.
Explicit Stability Conditions for Continuous Systems
193 p. 2005 [3-540-23984-7]

Vol. 313: Li, Z.; Soh, Y.; Wen, C.
Switched and Impulsive Systems
277 p. 2005 [3-540-23952-9]

Vol. 312: Henrion, D.; Garulli, A. (Eds.)
Positive Polynomials in Control
313 p. 2005 [3-540-23948-0]

Vol. 311: Lamnabhi-Lagarrigue, F.; Loría, A.; Panteley, E. (Eds.)
Advanced Topics in Control Systems Theory
294 p. 2005 [1-85233-923-3]

Vol. 310: Janczak, A.
Identification of Nonlinear Systems Using Neural Networks and Polynomial Models
197 p. 2005 [3-540-23185-4]

Vol. 309: Kumar, V.; Leonard, N.; Morse, A.S. (Eds.)
Cooperative Control
301 p. 2005 [3-540-22861-6]

Vol. 308: Tarbouriech, S.; Abdallah, C.T.; Chiasson, J. (Eds.)
Advances in Communication Control Networks
358 p. 2005 [3-540-22819-5]

Vol. 307: Kwon, S.J.; Chung, W.K.
Perturbation Compensator based Robust Tracking Control and State Estimation of Mechanical Systems
158 p. 2004 [3-540-22077-1]

Vol. 306: Bien, Z.Z.; Stefanov, D. (Eds.)
Advances in Rehabilitation
472 p. 2004 [3-540-21986-2]

Vol. 305: Nebylov, A.
Ensuring Control Accuracy
256 p. 2004 [3-540-21876-9]

Vol. 304: Margaris, N.I.
Theory of the Non-linear Analog Phase Locked Loop
303 p. 2004 [3-540-21339-2]

Vol. 303: Mahmoud, M.S.
Resilient Control of Uncertain Dynamical Systems
278 p. 2004 [3-540-21351-1]

Vol. 302: Filatov, N.M.; Unbehauen, H.
Adaptive Dual Control: Theory and Applications
237 p. 2004 [3-540-21373-2]

Vol. 301: de Queiroz, M.; Malisoff, M.; Wolenski, P. (Eds.)
Optimal Control, Stabilization and Nonsmooth Analysis
373 p. 2004 [3-540-21330-9]

Vol. 300: Nakamura, M.; Goto, S.; Kyura, N.; Zhang, T.
Mechatronic Servo System Control
Problems in Industries and their Theoretical Solutions
212 p. 2004 [3-540-21096-2]

Vol. 299: Tarn, T.-J.; Chen, S.-B.; Zhou, C. (Eds.)
Robotic Welding, Intelligence and Automation
214 p. 2004 [3-540-20804-6]

Vol. 298: Choi, Y.; Chung, W.K.
PID Trajectory Tracking Control for Mechanical Systems
127 p. 2004 [3-540-20567-5]

Vol. 297: Damm, T.
Rational Matrix Equations in Stochastic Control
219 p. 2004 [3-540-20516-0]

Vol. 296: Matsuo, T.; Hasegawa, Y.
Realization Theory of Discrete-Time Dynamical Systems
235 p. 2003 [3-540-40675-1]

Vol. 295: Kang, W.; Xiao, M.; Borges, C. (Eds)
New Trends in Nonlinear Dynamics and Control, and their Applications
365 p. 2003 [3-540-10474-0]

Vol. 294: Benvenuti, L.; De Santis, A.; Farina, L. (Eds)
Positive Systems: Theory and Applications (POSTA 2003)
414 p. 2003 [3-540-40342-6]

Vol. 293: Chen, G. and Hill, D.J.
Bifurcation Control
320 p. 2003 [3-540-40341-8]

Vol. 292: Chen, G. and Yu, X.
Chaos Control
380 p. 2003 [3-540-40405-8]

Printing: Krips bv, Meppel
Binding: Stürtz, Würzburg

MIX
Papier aus verantwortungsvollen Quellen
Paper from responsible sources
FSC® C105338

If you have any concerns about our products,
you can contact us on
ProductSafety@springernature.com

In case Publisher is established outside the EU,
the EU authorized representative is:
Springer Nature Customer Service Center GmbH
Europaplatz 3, 69115 Heidelberg, Germany

Printed by Libri Plureos GmbH
in Hamburg, Germany